Recent Advances in Security, Privacy, and Trust for Internet of Things (IoT) and Cyber-Physical Systems (CPS)

Recent Advances in Security, Privacy, and Trust for Internet of Things (IoT) and Cyber-Physical Systems (CPS)

Edited by
Kuan-Ching Li
Brij B. Gupta
Dharma P. Agrawal

CRC Press is an imprint of the
Taylor & Francis Group, an **informa** business
A CHAPMAN & HALL BOOK

First Edition published 2021
by CRC Press
6000 Broken Sound Parkway NW, Suite 300, Boca Raton, FL 33487-2742

and by CRC Press
2 Park Square, Milton Park, Abingdon, Oxon, OX14 4RN

Library of Congress Cataloging-in-Publication Data

Names: Li, Kuan-Ching, editor. | Gupta, Brij, 1982- editor. | Agrawal, Dharma P. (Dharma Prakash), 1945- editor.
Title: Recent advances in security, privacy and trust for Internet-of-Things (IoT) and cyber-physical systems (CPS) / edited by Kuan-Ching Li, Brij B. Gupta, Dharma P. Agrawal.
Description: First edition. | Boca Raton : CRC Press, 2021. | Includes bibliographical references and index.
Identifiers: LCCN 2020038707 | ISBN 9780367220655 (hardback) | ISBN 9780429270567 (ebook)
Subjects: LCSH: Internet of things--Security measures. | Cooperating objects (Computer systems)--Security measures.
Classification: LCC TK5105.8857 .R43 2021 | DDC 005.8--dc23
LC record available at https://lccn.loc.gov/2020038707

ISBN: 978-0-367-22065-5 (hbk)
ISBN: 978-0-429-27056-7 (ebk)

Typeset in Minion
by KnowledgeWorks Global Ltd.

Dedicated to my family and friends for their constant support during the course on the preparation of this book.

— Kuan-Ching Li
— Brij B. Gupta
— Dharma P. Agrawal

Contents

Preface

The Internet of Things (IoT) and Cyber-Physical Systems (CPS) provide mechanisms that are monitored and controlled by computing algorithms that are tightly coupled among users and the Internet. That is, the hardware and the software entities are intertwined, and they typically function on different time- and location-based scales. Security, privacy, and trust in IoT and CPS are different from conventional security. Different security concerns revolve around the collection, aggregation, or transmission of data over the network. Analysis of cyber-attack vectors and the provision of appropriate mitigation techniques are important research areas for these systems. Adoption of best practices and maintaining a balance between ease of use and security are again crucial for the effective performance of these systems.

This book will discuss present techniques and methodologies, as well as a wide range of examples and illustrations, to effectively show the principles, algorithms, challenges, and applications of security, privacy, and trust for IoT and CPS. It will be full of valuable insights into security, privacy, and trust in IoT and CPS and will cover most of the essential security aspects and current trends that are missed in other books. The material will prepare readers for exercising better protection/defense in terms of understanding the motivation of attackers and how to deal with it, as well to mitigate the situation in a better manner. The proposed book will guide readers to follow their paths of learning yet structured in distinctive modules that permit flexible reading. It will be a well-informed, revised, and comprehensible educational and informational book that addresses not only professionals but also students or any public interested in security, privacy, and trust of IoT and CPS. Specifically, this book contains discussions on the following topics:

- An Overview of the Integration between Cloud Computing and Internet of Things (IoT) Technologies
- Cyber-Physical Systems in Healthcare: Review of Architecture, Security Issues, Intrusion Detection, and Defenses
- The Future of Privacy and Trust on the Internet of Things (IoT) for Healthcare: Concepts, Challenges, and Security Threat Mitigations
- Toward the Detection and Mitigation of Ransomware Attacks in Medical Cyber-Physical Systems (MCPSs)

- Security Challenges and Requirements for Industrial IoT Systems
- Network Intrusion Detection with XGBoost
- Anomaly Detection on Encrypted and High-Performance Data Networks by Means of Machine Learning Techniques
- Deep Learning for Network Intrusion Detection: An Empirical Assessment
- SPATIO: end-uSer Protection Against ioT IntrusiOns
- Low Power Physical Layer Security Solutions for IoT Devices
- Some Research Issues of Harmful and Violent Content Filtering for Social Networks in the Context of Large-Scale and Streaming Data with Apache Spark

Acknowledgment

Many people have contributed significantly to this book on *Recent Advances in Security, Privacy, and Trust for Internet of Things (IoT) and Cyber-Physical Systems (CPS)*. As editors of this reference work, we would like to acknowledge all of them for their valuable help and liberal ideas in improving the quality of this book. With our feelings of gratitude, we would like to introduce them in turn. The first mention is the authors and reviewers of each chapter of this book. Without their outstanding expertise, constructive reviews, and devoted effort, this comprehensive book would become something without content. The second mention is the CRC Press/Taylor and Francis Group staff, especially Randi Cohen and her team, for constant encouragement, continuous assistance, and untiring support. Without their technical and administrative support, this book would not be completed. The third mention is the Editors' families for being the source of endless love, unconditional support, and prayers for this work and throughout our lives. Last but far from least, we express our heartfelt thanks to the Almighty for bestowing over us the courage to face the complexities of life and complete this work.

Kuan-Ching Li
Brij B. Gupta
Dharma P. Agrawal

Contributors

Rangel Arthur
University of Campinas (UNICAMP)
São Paulo, Brazil

Anastasios N. Bikos
Computer Engineering & Informatics
University of Patras
Hellas, Greece

João Bota
Instituto Superior Tecnico
Universidade de Lisboa
Lisboa, Portugal

Alberto Huertas Celdrán
Telecommunications Software & Systems Group
Waterford Institute of Technology
Waterford, Ireland

Félix J. García Clemente
Departamento de Ingeniería y Tecnología de Computadores
University of Murcia
Murcia, Spain

Miguel Correia
INESC-ID, Instituto Superior Técnico
Universidade de Lisboa
Lisboa, Portugal

Phuc Do
University of Information Technology (UIT)
Vietnam National University (VNU-HCM)
Ho Chi Minh City, Vietnam

Reinaldo Padilha França
University of Campinas (UNICAMP)
São Paulo, Brazil

Arnaldo Gouveia
IBM Belgium
Brussels, Belgium
and
INESC-ID, Instituto Superior Técnico
Universidade de Lisboa
Lisboa, Portugal

Fouad Amine Guenane
Senior Consultant
France

Yuzo Iano
University of Campinas (UNICAMP)
São Paulo, Brazil

Lorenzo Fernandez Maimo
Department of Computer Engineering
University of Murcia
Murcia, Spain

Ahmed Mehaoua
Beamap, Sopra Steria Group LIPADE
Université Paris Descartes Paris
Paris, France

Melody Moh
San José State University
San Jose, California

Ana Carolina Borges Monteiro
University of Campinas (UNICAMP)
São Paulo, Brazil

Gil Mouta
Instituto Superior Tecnico
Universidade de Lisboa
Lisboa, Portugal

Miguel L. Pardal
Instituto Superior Tecnico
Universidade de Lisboa
Lisboa, Portugal

Jyoti Patel
Design & Manufacturing
Pandit Dwarka Prasad Mishra Indian Institute of Information Technology
Jabalpur, Madhya Pradesh, India

Gregorio Martinez Perez
Departamento de Ingeniería y Tecnología de Computadores
University of Murcia
Murcia, Spain

Phu Pham
University of Information Technology (UIT)
Vietnam National University (VNU-HCM)
Ho Chi Minh City, Vietnam

Trung Phan
University of Information Technology (UIT)
Vietnam National University (VNU-HCM)
Ho Chi Minh City, Vietnam

Chithraja Rajan
Design & Manufacturing
Pandit Dwarka Prasad Mishra Indian Institute of Information Technology
Jabalpur, Madhya Pradesh, India

Robinson Raju
San José State University
San Jose, California

Dip Prakash Samajdar
Design & Manufacturing
Pandit Dwarka Prasad Mishra Indian Institute of Information Technology
Jabalpur, Madhya Pradesh, India

Dheeraj Sharma
Design & Manufacturing
Pandit Dwarka Prasad Mishra Indian Institute of Information Technology
Jabalpur, Madhya Pradesh, India

Nicolas Sklavos
Computer Engineering & Informatics
University of Patras
Hellas, Greece

Valentin Vallois
Beamap, Sopra Steria Group LIPADE
Paris Descartes University
Paris, France

CHAPTER 1

An Overview of the Integration between Cloud Computing and Internet of Things (IoT) Technologies

Reinaldo Padilha França, Ana Carolina Borges Monteiro, Rangel Arthur and Yuzo Iano

CONTENTS

1.1 INTRODUCTION

The high level of connectivity between the most varied devices in society has caused an explosion in data volume. The ability to analyze and manage this large flow of information currently generated has allowed companies to make more assertive decisions and improve their relationship with their customers. Coupled with the massive increase in the number of devices and the amount of data transmitted requires changes in the architecture and implementation of technologies, making data management and balancing the biggest challenge for companies. In order to control these demands of infrastructure and data traffic and storage, corporations have been betting on the cloud, which will, therefore, have more and more direct influence on the increase of quality and safety level (Hashem et al. 2015).

When it comes to cloud computing, it refers to the sharing of information, data, images, and services over the Internet anytime and anywhere, where cloud storage and data processing services are provided over the cloud. Cloud technology has gained importance in every possible scenario. Whereas in the past, almost every investment focused on infrastructure to be able to store virtual documents and execute different software options, today information technology has a more strategic role. In addition, cloud computing allows storage terms and bandwidth to be tailored to seasonal business needs, where planning becomes more flexible and easier to execute (Rittinghouse and Ransome 2017).

Thus, Cloud computing is the way to use programs and documents that, instead of being located on a physical computer, are in the cloud, that is, on an external server, accessed over the Internet. One reason for its rapid expansion was the ability of this technology to lower infrastructure costs. Since unlike maintaining a physical server, where the user pays for the entire space, even if it is empty, in cloud computing the user pays only for what is used. The more space he needs, the more he would pay, bringing simplicity, having a greater economic character for the company or certain application projects (Rittinghouse and Ransome 2017).

The concept of the Internet of Things (IoT) refers to common electronic devices in our daily lives. Its highlight is the way it connects and interacts with other smart electronics using the Internet, being a technological revolution that has been a big bet on big companies. It seeks to connect electronic devices, which we use in our daily lives, to the Internet. Examples of such electronics are appliances, industrial machines, small appliances, and means of transportation. In addition, other objects are part of this technology, such as clothes, shoes, lamps, and pens. Its main focus is to establish the communication of the virtual world with the physical world, giving objects the ability to see and hear through sensors. That is, it tends to meet the communication networks with the real world of things, besides being able to detect problems in advance without the dependence on the human factor, proposing to revolutionize our daily lives and make our lives simpler, quickly and efficiently (Wortmann and Flüchter 2015).

In the age of artificial intelligence (AI) or IoT where the Internet is not only present in computers, phones, and tablets, but also in vehicles, shoes, clothes, pens, among many other objects that we could not even imagine a while ago, the cloud computing has become a solution with an increasingly crucial role with all this virtualization. Today, innovation can be found in a variety of ways, from industrial machinery, appliances, small appliances, and transportation to mixing with accessories, since that what sets these technological products apart from normal objects lies in their ability to exchange information through sensors; in this sense, the user relies on smart devices to automate their tasks (Russell and Norvig 2016).

The concept of cloud computing aims to facilitate data access and program execution using the Internet, gives the ability to use services and applications without the need for an installation as everything, or almost everything, will run on servers, where data access is possible from any devices as long as they are connected to the Internet and have the permission of the person responsible. The devices need to be connected to the network and exchange information with each other, especially the most complex devices. Taking as an example the case of a surveillance camera that requires a strong database to warn the owner of something suspicious, something quite complicated for a data center of its own, as the cost of maintaining this base becomes high. In this sense, cloud computing has the function of replacing the infrastructure of large companies, where there is no concern about high energy costs, since depending on the object, such as security cameras, should be kept on all day, or maintenance (Rittinghouse and Ransome 2017).

The smart device is responsible for capturing and processing the information, and cloud technology, on the other hand, allows collected information to be stored and transferred to another device—complementing each other's actions. However, the operation of smart objects requires full performance, such as a camera that broadcasts to smartphones and can detect when something unusual is happening, this object is capable of receiving and transmitting information in real time, so it must have 100% availability for its good performance. This technology has as its basic principle, connect devices to the World Wide Web to make their use easier and even expand the possibilities of resource delivery to users. To better understand the concept, other examples such as sensors that can be installed on trucks for speed control and fleet tracking, smart locks that only open with a smartphone, or wristbands that record users' activities as distance traveled, speed, pulse rate, among others. In addition, doing everything "indoors" requires more time and high investment value, because maintaining an internal data center is not cheap since there is constant maintenance, cooling needs, upgrades, among other conditions (Lom et al. 2016).

Cloud computing has a robust, high-performance infrastructure in qualified data centers that are prepared to handle a large amount of data and applications, with high-speed processing, being the best option for the IoT data that will be processed in the cloud. Therefore, the greater the number of interconnected objects communicating with each other, the number of data increases simultaneously and the need for storage will also be greater (Lom et al. 2016).

Unfortunately, smart devices are still not immune when it comes to their own security. For both loss and intrusion, devices often have little or no protection for compromised data, so for this reason, the cloud serves as a great addition to storing the information collected, having your infrastructure more robust for both security and backup itself. Thus, in case of loss, cloud technology has the ability to retrieve what has been lost and encrypt it if data is stolen, meaning any malicious person is unlikely to have access to data (Hossain et al. 2015).

In the IoT, there are devices that do data collection, and in the cloud, this data is stored, that is, it completes each other. With all this data traffic, security is paramount, it is totally out of the question to put this volume of information at risks, such as virus attacks or malicious attacks, for example, since these innovations need the guarantee of security and privacy, and the cloud computing service ensures security with its continuity, backup, disaster recovery capability, and privacy by encrypting data. Another important point is that these devices IoT must have flexibility, as there are some objects that are practical and can be moved from one place to another, is another condition that the cloud can handle, its mobility allows the user to have the continuity of their device anytime, anywhere, and from any device (Bokefode et al. 2016).

Considering the highlights, it can be seen that the age of the IoT is fully associated with Cloud services, as the cloud facilitates the entire process with its automated tools, keeping devices up-to-date and service fast, affordable, and flexible. On the other hand, the cloud allows this data to be synchronized in real time to generate smart reports and make them available later. One of the factors that catch the eye of companies in this combination is its cost-effectiveness, both devices and the Internet, if properly configured, can be a low cost to the enterprise (Hsu and Lin 2016).

A practical example of where both technologies work together is enterprise data collection using web services from companies, such as AWS (Amazon Web Services), which relies on IoT technology on a cloud platform that enables smart devices to securely interact with other applications and devices connected to the enterprise to cloud. This practice allows to measure, for example, how much is spent on electricity in equipment, this way, the company can use tools, that are also available on the platform as a SaaS (Software-as–a-Service), IaaS (Infrastructure as a Service), or even PaaS (Platform as a Service) to analyze all the results obtained, improve its decision-making and thus reduce company costs (Hsu and Lin 2016)..

If cloud computing is a consolidated trend today, IoT's greatest influence on IT is to reinforce this return less path toward a virtualized IT architecture. It is no longer possible to imagine the world without technology. This is why IT is indispensable in the management of the IoT, because that cannot put information and applications in the cloud without thinking about how to keep it safe. As the IoT trend grows, IoT security challenges increase. Increasingly, the world becomes mobile and automated, since, with each passing day, more devices become connected by sensors, communicating with each other. But while this technology has been very favorable for people and businesses, on the other hand, it is amazing how much data needs to be stored for everything to work as planned. In this way, the use of cloud resources becomes necessary as the IoT becomes more and more present in people's reality, being the Cloud proves to be the ideal resource for storing and making available a large amount of data that needs to be handled in real time (Hossain et al. 2015).

Another important point regarding Cloud computing's partnership with IoT is data security, which means that from the moment that having a universal smart device connectivity environment, it is necessary to create a secure environment through which information must be trafficked. So, there is no denying that both Cloud computing and the IoT have become prevalent technologies today and represent trends for the future. Therefore, this chapter aims to provide an updated review and overview of IoT and Cloud computing, showing its relationship and integrations approaching its success relation, with a concise bibliographic background, categorizing and synthesizing the potential of both technologies (Singh et al. 2016).

1.2 CLOUD COMPUTING BACKGROUND

It is used to store files and data of all kinds as well as to use applications on-premise, that is, installed on our own computers or devices. However, in modern corporate environments, this scenario is different as long as applications available on servers that can be accessed by any authorized terminal are used. The main advantage of on-premise is that it is possible, at least most of the time, to use applications even without access to the Internet or the local network. In other words, there is a possibility of using these resources offline. On the other hand, in the on-premise model, all data generated is restricted to a single device, except when there is network sharing, which is not very common in the home environment. Even in the corporate environment, this practice can have some limitations, such as the need to have a certain software license for each computer (Bassi and Chaudhary 2015).

The constant evolution of computer technology and telecommunications is making Internet access ever wider and faster. This scenario creates the perfect condition for the popularization of cloud computing, as it spreads the concept around the world. The concept for what cloud computing is the possibility of offering content or service through online pages, being tied exactly to the need to be tied to something, needing only the Internet to perform the operations. Since it is possible to access files, edit and view photos and videos from anywhere in the world as if a person was at home or in the office, or even just with a mobile device, having a browser like Mozilla Firefox, Google Chrome, or Opera and an Internet connection for access to all cloud computing. Thus, with cloud computing, many applications, as well as files and other related data, no longer need to be installed or stored on a user's computer or a nearby server. This content becomes available in the clouds, that is, on the Internet facilities (Changchit 2015).

Cloud computing brings a huge paradigm shift, since no longer is it necessary or has to install proprietary offline file editing software. Instead, it is possible to connect to large business services and use their applications like Google Docs and do it online without even writing a file to your hard drive. Most of the time the hiring is done through rent and in the same way the user can have access to the service through a monthly payment, that is, he pays for what he uses. This also characterizes as a constant circulation of capital since the user not only makes a payment for the purchase of the service, but a smaller and constant payment according to the use (Rafaels 2015).

The functioning of the cloud uses concepts of distributed computing, and it is because of this feature that this technology has been named. It is unknown where your files and services are stored. It can be either on a server next to your home or in a data center in China. But there is no need to worry about security or privacy, as access to files and services will only be allowed to those who have their login and password. A good example of cloud computing is websites and applications for photo and video editing, because nowadays there is a vast diversity of applications and software of this type and for this purpose and in online versions, that is, they do not need to be downloaded for use (Sen 2015).

Another example is that applications that exist in desktop, mobile (Android and/or IOS) versions, can also be accessed directly by the browser, and in all three options meet basic tasks of photo and video editing without change from one platform to the other. So, summarizing the concept of what cloud computing is, it is access and management of data through a network connection without the need for a download. Within the idea of what cloud computing is, there are still some divisions to separate different types that can be offered through cloud storage. They are primarily divided between deployment and services, followed by subdivision into these first two (Sen 2015).

The deployment of cloud computing needs special attention because as it is online access, various companies and customer information will be saved in clouds, that is, on the Internet. In addition, charges may vary for each storage cloud service deployment license. Private cloud is suitable for larger companies that need a large amount of storage in the cloud, that is, constantly generating a lot of data, it is the best option for only the company to have access to that network and server as a share could disrupt the connection of company and/or customer. From the user's point of view, the private cloud offers virtually the same benefits as the public cloud, but the difference is essentially behind the scenes, since the equipment and systems used to build the cloud are within the corporation's own infrastructure (Chang 2015).

1.2.1 Cloud Computing Models

The need for security and privacy is one of the reasons why an organization adopts a private cloud. In third party services, contractual clauses and protection systems are the features offered to prevent unauthorized access or improper sharing of data. A private cloud can also offer the advantage of being precisely shaped to company needs, especially for large companies. In other words, the company makes use of a private cloud built and maintained within its domains. But the concept goes further: the private cloud also considers corporate culture, so that policies, goals, and other aspects inherent in the company's activities are respected (Rittinghouse and Ransome 2017, Chang 2015).

The public cloud is suitable for smaller, less data-intensive businesses that need to invest less in the storage cloud service, where while it is smaller and requires a lower investment when infrastructure is installed, it is still a good fit for businesses that need to make apps and some functions of your business off-site (Almorsy et al. 2016).

Hybrid cloud is when a single company makes use of the other two storage cloud options at the same time, that is, when the company hosts, produces, and manages the heaviest and most sensitive materials through a private cloud, and uses the public cloud for everyday data and transfers. Being the best choice for the business that needs more advanced data security without losing the quality of service for lighter access to the simplest business tasks. In them, certain applications are directed to public clouds, while others, usually more critical, remain under the responsibility of your private cloud. There may also be features that work on local systems (on-premise), complementing what is in the clouds. Hybrid cloud deployment can be done to meet continuous demand as well as to meet a temporary need (Diaby and Rad 2017).

1.2.2 Cloud Computing Infrastructure

IaaS is the infrastructure of the Internet-managed computing network, more simply, it is like your online server, it is only avoided to hire a more expensive service and have access to identical functions. It will be your storage cloud data center. The focus is on the structure of hardware or virtual machines, with the user even having access to operating system resources (Malawski 2015)

PaaS serves to serve the services the company intends to offer, being a platform that receives, manages, and spreads content through this online server that supports the application and access to it. This is a broader type of solution for certain applications, including all (or nearly all) of the features required for the operation, such as storage, database, scalability (automatic storage or processing capacity increase), language support programming, security, and so on. Who hires this mode avoids server costs for the distribution of your app and also avoids the maintenance headache and access licenses. A modern example can be seen on Netflix, which has the platform to receive all its content and allow users to access it online without downloading the app (Van Eyk et al. 2018).

SaaS is aimed at companies that need to make software hosted on an online server allowing users access. In essence, it is a way of working in which software is offered as a service, so, there is no need to purchase installation licenses or even purchase computers or servers to run it. As modern examples of SaaS can be seen as Google Apps like Google Docs, which provides access to a tool made entirely online and allows the user to create and store their documents. In this mode, at most, one pays a periodic amount—as if it were a subscription—only for the resources used and/or the time of use (Safari et al. 2015).

In today's market, there are also concepts derived from SaaS that are used by some companies to differentiate their services. Database as a Service (DaaS), this modality is aimed at providing services for storage and access of data volumes. The advantage here is that the application holder has more flexibility to expand the database, share information with other systems, and facilitate remote access by authorized users, among others. Testing as a Service (TaaS) provides an appropriate environment for the user to remotely test applications and systems, simulating their behavior at run level (Sousa et al. 2016, Villanes et al. 2015).

1.2.3 Benefits of Cloud Computing

One of the advantages of cloud computing is access to applications from the Internet, without having them installed on specific computers or devices. But there are other significant benefits such as the user being able to access applications regardless of their operating system or equipment used; not having to worry about the structure to run the application—hardware, backup procedures, security control, maintenance, and more. Information sharing and collaborative work become easier as all users access applications and data from the same place, the cloud. In a cloud structure, the user can count on high availability, that is, if one server stops working, the others that are a part of the structure continue to offer the service (Hashem 2015, Rittinghouse and Ransome 2017).

Many cloud applications are free, and when a person needs to pay, it's just paid for the features to be used or for the precise time required. It is not necessary to have to pay for a full usage license as it is done in the traditional model of software; depending on the application that a person is using, all or most of the processing (and even data storage) is up to the clouds. It is worth mentioning that regardless of the application, with cloud computing the user does not need to know the entire structure behind it, that is, he/she does not need to know how many servers run a particular tool, which hardware configurations are used, how the scheduling is done, and where is the physical location of the data center, anyway. What matters is knowing that the app is available in the clouds. There is still a lot of work to be done, since the very idea of certain information being stored on third-party computers (in this case, service providers), even with documents ensuring privacy and confidentiality, worries people and, especially, companies, why this aspect needs to be better studied (Tarhini et al. 2018).

But storage cloud computing is not always going to be the best option, since as it is an Internet-dependent service it can be compromised depending on the business and user connection. In addition, of course, the ease of access and encounter with your company through browsers that are such a constant use in most people's lives today. The constant expansion of Internet access services and the advent of mobile devices (smartphones, tablets, smartwatches, and the like) increasingly open space for cloud applications (Jain and Mahajan 2017).

1.3 IoT BACKGROUND

IoT is the way physical objects are connected and communicating with each other and the user, through intelligent sensors and software that transmit data to a network, an interpretation would be like a large nervous system enabling the exchange of information between two or more points, resulting in a smarter and more responsive planet. The IoT cannot be seen as a single, massive technology, but rather a set of factors that determine how the concept is constituted, since essentially three components that need to be combined to get an IoT application are devices, networks communication, and control systems (Olson 2016).

From a clock or a refrigerator to cars, machines, computers, and smartphones, which are devices ranging from large items to small objects such as light bulbs and watches, the important thing being that these devices are equipped with the items certain to provide communication such as chips, sensors, antennas, among others. Allowing a person to talk to each other via technology over communication networks such as Wi-Fi, Bluetooth, and Near Field Communication (NFC), which is a new wireless communication technology compared to Wi-Fi, both of which can being and having use for IoT; however, since these networks have a limited range, certain applications depend on mobile networks like 3G and 4G/LTE (Olson 2016).

Also considering that today's mobile networks (with regard to 2G, 3G, and 4G) are targeted at devices such as smartphones, tablets, and laptops, with a focus on the text, voice, image, and video applications, but this aspect is not preventing these networks from being used for IoT. As a consequence, it generates more comfort, productivity, information, and practicality, and its uses may include health monitoring, providing real-time information about city traffic or the number of parking spaces available and in which direction they are, even recommending activities, reminders, or content on your connected devices, that is, any tool that can theoretically and in practice be connected and transmitting information, thus entering the world of the IoT (Pan and McElhannon 2017, Olson 2016).

1.3.1 IoT Devices and Connectivity

Connectivity serves to make objects more efficient or receive complementary attributes. In this sense, the Internet refrigerator could warn a person when food is about to run out and at the same time search the web which markets offer the best prices for that item. The fridge could also search and display recipes for that person. As it turns out, creativity can bring really interesting applications. A simple thermostat can check on the Internet what the weather conditions are in your neighborhood to make the air conditioner at the ideal temperature for when a person gets home, the device can still send information to the user's smartphone through a specific application so that have reports showing how air conditioners are being used or apply custom settings for energy savings (Lee and Lee 2015).

Still considering smartphone usage, users can know the location of a particular bus, since sensors can also help the company discover that a vehicle has mechanical defects, as well as know-how, is meeting schedules, which indicates the need or not to strengthen the fleet. And when it comes to the fleet, sensor data installed on trucks, containers, and even individual boxes combined with traffic information gives the information for a company to define the best routes and logic, choosing the most suitable trucks for a given area, which orders to distribute between the active fleet (Greengard 2015).

In this sense, everyday things become intelligent and have their functions expanded by data crossing, in this context enter the virtual assistants crossing data from connected devices in order to inform, even without request or request, a simple example in this regard is about the time it will take a person to get to work when he sits in his car to leave home. In this scenario, wearable devices such as smartwatches, sensor accessories, and exercise monitoring headsets are only recently being widely adopted and used by people, and classic objects are examples of connected devices that integrate the IoT (Greengard 2015).

Security cameras that, online, allow a person to monitor your home remotely or watch your store when the property is closed. Practical applications of the IoT in the organization of traffic, the speeding up of medical treatments, and also the preservation of the environment, always conditioned on the human capacity to analyze the data that the connected devices generate. However, there are a number of other possibilities that are often overlooked, such as aircraft parts or oil and gas rig platform structures that can be connected to the Internet for accident prevention and real-time problem detection with regard to in general, if an object is electronic, it has the potential to be integrated into the IoT (Greengard 2015, Olson 2016).

Cities that suffer a lot from pollution have been directing efforts to improve air and water qualities. Modern cities are distributing small air pollution measuring devices to citizens, which can be plugged into cars and bicycles, circulating with vehicles around the city, working on actions to reduce air pollution. The sensors transmit the information to the company application. The app, in turn, consolidates information on a single server, allowing Londoners to check a digital air quality map at every point in the city (Weber et al. 2017).

The IoT can help a person to measure machine productivity in real time or indicate which sectors of the plant need the most equipment or supplies. And in the sales market, using smart shelves can tell a person in real time when a particular item is starting to run out, which product is having the least output where the manager can make promotions or at what times certain items sell the most, aiding in crafting sales strategies. From agricultural technology to air cleaning, smart devices act as important allies in solving humanity's problems (Caputo et al. 2016).

1.3.2 IoT Benefits and Applicability

The field also benefits from the IoT, as in environments plagued by drought that harms local farmers, using aerial imagery drones and soil quality sensors have helped growers identify the best places to plant new crops. Also taking into account that scattered sensors can give very accurate information on temperature, soil moisture, rainfall probability, wind speed, and other information essential for good crop yield. Similarly, sensors connected to animals can help control cattle through a chip placed in the bull's ear can track the animal, report its vaccine history, and so on (Sharma et al. 2018).

Still considering that millions of people still suffer from hunger and malnutrition in the poorest countries, and one-third of food produced annually for human consumption is lost or spoiled somewhere in the supply chain, using the IoT can be reduced. The extent of the problem, acting directly in the rural environment, since it is possible to monitor processes such as irrigation, pollination, and soil fertilization, and to provide reports to farmers, also enabling management for producers to manage their sales and prevent losses in the transportation of goods (Greengard 2015).

Connected sensors are also already being used in medicine, since wearable devices that measure patients' heart rate, pulse, and blood pressure are already used in many countries, leaving physicians informed all the time. This is not only in hospitals but also in patients' own homes for those who are at constant risk. Still considering the possibility of controlling epidemics of contagious diseases, where there are devices that measure risk indicators in people with the virus. With data transmitted via Bluetooth, physicians' need for physical interaction with infected patients has been reduced, helping to control disease transmission (Greengard 2015, Olson 2016).

It is not enough for the device to connect to the Internet or exchange information with other objects. This data needs to be processed, that is, it must be sent to a system that handles it. Data from cameras, fire alarms, air conditioners, lamps, and other items are sent to a system that controls every aspect. This system can be a cloud service, ensuring access from anywhere, as well as relieving the homeowner of the task of upgrading it. However, a problem faced by many companies is the amount of information that all these devices produce, where they will have to find ways to store, track, analyze, and make use of this large amount of data. And to make sense of all this data, big data analytics play a critical role, regardless of the size and size of the business, the IoT has the ability to accelerate business processes (Hassanien et al. 2015).

If the IoT describes a scenario where almost everything is connected, however, there are risks associated with it. IoT has a greater concern about the security and privacy of the sensors used and the data they store, yet the integration of devices to transfer all critical data also presents problems, since there are billions of devices connected to each other, and with there must be guarantees that the information will remain secure. The risks are not just an individual, there may be problems of collective order wherein a city that has all the traffic lights connected. The traffic management system intelligently controls each to reduce traffic jams, provide detours on accident-blocked roads, and create alternate routes when major events occur. If this system gets attacked or crashes, city traffic will become chaos in minutes (Wortmann and Flüchter 2015).

Industry must, therefore, define and follow criteria to ensure service availability (including rapid recovery from failure or attacks here), communications protection (which, in enterprise applications, must include strict protocols and audit processes), definition privacy, data confidentiality (no one can access data without proper authorization), integrity (ensuring that data will not be improperly modified), among others. Many industry segments are already dealing with such issues, but this is a work in constant development. Objects, people, and even nature emitted a lot of data, we just couldn't see, hear, or make sense of them. It is common for us to think about how, throughout the history of humanity, our technology has advanced far enough that we could perceive smaller and smaller things: atoms, protons, electrons, quarks, among others (Gilchrist 2016).

However, the IoT and the data we generate is one of the examples of the giant things we come to see, understand, and use to our advantage as technology advances. This is what IoT has come to change in our reality, because now everything around us has intelligence, and is interconnected, so that we now have access to data, or rather a piece of information. Deep down, we had a sea of data, which we are now able to put intelligence into and transform it into information, knowledge, and ultimately wisdom (Soldatos et al. 2015).

Once we can understand the patterns of all this data, society will become more efficient, increasing productivity, improving the quality of life of people and our planet itself. With that, we can generate new insights, new activities, and of course further fuel innovation. The bridge between data collection and appropriate sharing of data, with security and protection for all parties, remains a key challenge in the evolution of this industry. Nonetheless, it is an exciting segment and one that we must follow closely (Soldatos et al. 2015).

1.4 CLOUD COMPUTING AND IoT INTEGRATION

IoT is the future of devices since this technology has as its core principle, connecting devices to the worldwide computer network to make their use easier and even expanding the possibilities of delivering resources to users, ranging from smart locks that they only open with a smartphone, sensors installed on trucks to control the speed and location of a fleet, or even wristbands that record users activities such as distance traveled, speed, heart rate, among many others. Like IoT, Cloud computing technology is also expanding worldwide due to its many advantages that have taken companies to another technological level, being currently a cost reduction engine, the ease and agility in scalability, and security of data are hallmarks that have made cloud computing expand worldwide (Botta et al. 2016).

Thus, the Cloud and the IoT are inseparable due to the rapid technological development in which IoT is responsible, since every day more devices become connected by sensors, communicating with each other, generating an astronomical volume of data that must be captured, stored, and turned into knowledge for mankind. So, cloud computing comes into play as a facilitator of this process related to this amazing amount of data that needs to be stored for everything to work as scheduled. In this way, the use of cloud resources becomes necessary as the IoT becomes more and more present in people's reality. Thus, Cloud is the ideal resource for storing and making available a large amount of data that needs to be handled in real time (Díaz et al. 2016).

Cloud is the fuel of the IoT because it is through the cloud that it is possible to realize all the potential benefits of IoT, since the cloud offers not only storage to maintain this data volume but also efficient to manage it. In addition, scaling analysis is possible, which helps to decongest local infrastructure. Another important point regarding Cloud computing's partnership with IoT is data security. This means that from the moment we have a universal smart device connectivity environment, it is necessary to create a secure environment through which information must be transported. If there is an environment where a person has invested in high-end security, this environment is cloud computing. With cloud computing, organizations do not have to deploy extensive hardware, let alone configure and manage networks and infrastructure in IoT deployments. The cloud enables companies to extend their storage and processing capabilities to their needs, all without setting up additional hardware and infrastructure. These factors help accelerate the process of deploying IoT solutions, as well as reduce costs, as companies do not have to invest in purchasing or provisioning servers or infrastructures as they pay only for resources consumed (Aazam et al. 2016).

A business example of the interaction between technologies is in the area of people management, with electronic time registration systems simplifying the routine of the human resources and personnel departments, but with some unique IoT advantages, such as time communication between the devices on which time recording is made, through marking software such as a mobile phone, notebook, or even a clock, among other devices. Registering this point in the company or outside, and he/she receives this information at the same time, a possibility that facilitates the control of scales and overtime. With IoT it is possible to easily access time card information from anywhere in the world, a valuable resource for companies that have branches in various regions of the country or abroad. All done online (Sajid et al. 2016).

Since this information is stored in the cloud, since it is possible to do analysis anywhere, there is no need to be on a specific computer that has a certain software installed, providing more flexibility and independence to the personal department, just an Internet connection. Another advantage in the business environment is that of advertising and marketing, exploring the union of IoT and Cloud, in terms of knowledge extracted from information in the physical world, especially regarding people's behavior and their interactions with things. And are essential in determining advertising campaigns, product definition, customer segmentation, and affirmative and brand value actions. Just as several business modalities receive real-time insights from devices around the world through cloud technology, coupled with IoT services, allowing specific analysis of this easily collected device data to be performed using advanced analytics technologies and applying technologies like Machine Learning (Cai et al. 2016).

The combination of both technologies enables the construction of smart cities and buildings, creating a comprehensive solution that encompasses billions of sensors and devices that bring a new level of intelligence and automation to homes, buildings, or entire cities. Since there may be ground sensors that use an advanced combination of infrared and magnetic technologies to safely record a vehicle's arrival and communicate with a gate or gate barrier, allowing operators to obtain real-time information about the use of the environment and managing their capacity by allowing parking operators to identify whether vehicles parked for limited periods of time have exceeded the limits, photographing license plates to identify such vehicles, notifying and taking actions such as issuing infringement notices, or even provide automated guidance and signage to customers and users about the number of parking spaces available at each level of a parking structure or even the street area of a city. Everything is managed by a cloud—primarily due to support for sophisticated big data management and machine learning in serverless architecture, all processing at an incredible rate, providing real-time messaging, enabling a serverless environment to enable data delivery streaming, and connecting managing parking device data, yet with the ability to identify trends at all customer parking locations and make recommendations to operators (Alioto 2017).

Also taking into account that IoT systems are based on resource-constrained networks and devices, that is, there is a lack of computational resources to leverage and leverage the information generated, which leads to applications with high capacity reduction and not are enabled to perform large processing such as the use of data analytics and/or big data. Since IoT-associated devices have various storage, network, and processing limitations that combine the need for complex analysis, scalability, and data access, IoT needs Cloud computing technology as the most widely used alternative, where IoT system sends data for cloud processing, storage, and analysis. Noting that this combination of both technologies follows a trend related to the evolution of processing power with the ability to transmit information (Amadeo et al. 2016).

However, one of the main issues related to IoT platforms is the interoperability issue, as it relates to a single platform connection centralizing all types of protocols, devices, and applications that can use these services. A major challenge in technology synergy is related to the wide heterogeneity of devices, operating systems, platforms, and services available and possibly used for new or improved applications. Cloud services typically come with proprietary interfaces, so feature integration tends to be properly customized based on specific providers. So, the difficulty is that for each possible application/service, they have to examine target scenarios, analyze requirements, select hardware and software environments, integrate heterogeneous subsystems, develop, provide computing infrastructure, and provide service maintenance (Lucero. 2016).

Cloud IoT applications often ensure specific performance and Quality of Service (QoS) requirements at various levels for communication, computing, and storage aspects. When these applications are deployed in resource-limited environments, there are a number of challenges related to device failure or inability to reach this device; however, cloud capabilities outweigh some of these challenges as it increases device reliability because of lets to the device discharge heavy processing tasks by extending battery life. These applications allow being created innovative applications that target the integration and embedded analysis of real-world information. However, deploying IoT devices sometimes makes monitoring tasks more difficult as they have to deal with dynamic latency and connectivity issues (Khodkari et al. 2016).

Coupled with the ubiquity of mobile devices and sensor diffusion, it directly reaches scalable computing platforms, with huge volumes of data bytes being created, their handling naturally being a key challenge, highly dependent on the properties of data management technologies such as big data. Relating that it is in developing countries that there are the highest growth rates in IoT investment, and of course this is where there is the most room for creating and implementing these new systems. Since most IoT and Cloud applications are in developed countries, and since traditional systems are fully deployed, the cost of moving it to a new IoT paradigm is exorbitant. In short, IoT and the growing number of devices, technologies, and platforms in this area have made it a global and extended technology in many other areas; however, it has limitations and the need for complex resources to meet the many existing demands. Cloud computing is appropriate as an integrative add-on (Formisano et al. 2015).

Since the adoption of this aggregator, the paradigm has made possible several new applications, which derives from the success in the main challenges faced by each one of them. The main conclusion is that when considering the concept of IoT, not as island systems or as large silos, but as something that connects to the Internet and talks to each other, as in a large network, it is impossible to imagine them without the use of resources a cloud (Olson 2016).

1.5 DISCUSSION

At first, computers began to think later, talking to each other, nowadays with IoT being the connection between several devices, it is possible to collect information, share different databases, and generate significant advances in the production system, logistics, and customer experience to name just a few fields.

Numerous technological resources have been employed to provide devices and artifacts that go beyond Bluetooth, proximity field communication (CPP) is also a feature used in IoT. However, the increasingly innovative features of Radio Frequency IDentification (RFID) are the technical and functional basis for IoT.

However, the most interesting thing is not only to connect objects to the Internet, but also to realize the main potential of IoT that is to realize the communication between objects and people. This concept, of a practical nature, can also be understood as transmission of data and information. Thus, via the Internet, things (artifacts, devices, among others) exchange signals with each other, that is, mobile and fixed objects gain autonomy to interact with each other.

In agriculture, sensors scattered across crops can feed a system with data on temperature, humidity, wind speed, and direction, as well as soil conditions. Parameters can automatically trigger the irrigation system, pest control, and other aspects of farm management. Already in the herd, chips in the animals' ear send their location, history of vaccines, and treatment protocols.

In factories, the reasoning is the same. Combining monitoring and automation increases efficiency, reduces waste, reduces costs, and minimizes errors. But we can already see the IoT on the daily commute to work using public transportation. Sensors on buses inform the location if there is any mechanical breakdown or any other unforeseen event. Thus, through an application, we can know if the collective will reach the point on time and what is the best combination of modes to reach a certain destination.

In the corporate universe, it is possible to identify the same movement, as understanding about IoT is focused on generating insights through the sea of data that is collected. Attached to cloud computing that enables the storage and processing of so much of data. And if data is being generated exponentially, then it takes devices to process it. And it is not just state-of-the-art AI advanced computers that can do it, but also devices with intelligent software that have the function of connecting to the Internet, accessing specific databases, learning, and improving their performance or user experience.

In healthcare, IoT applications with AI are innumerable from the information continuously generated in the modules of a hospital management system; administrators and physicians can make much more assertive decisions, thus contributing to diagnostic processes, identifying predispositions to diseases and treatment prescriptions. Using AI to assist physicians in the diagnostic process, based on data obtained from standard medical devices when connected to the Internet, broadens their ability to collect data and broadens insight into symptoms and trends such as data capture multiparameter monitors such as heart rate, respiratory rate, blood pressure, and temperature of patients, wherefrom this information, intelligence tools will, for example, classify risk to patients, aiming to optimize care and shorten the long waits that can be today very common in the public health network, integrating IoT using information collected by hospital equipment for AI learning, promoting greater integration between systems and equipment, and feeding the AI database. As a result, better services will be provided to patients by recording and tracking data by regularly monitoring health status.

Two other technologies are driving IoT growth, AI is ensuring more autonomy and learning for objects connected to the Internet, and the blockchain promotes more security, so that objects connected to networks are not hacked. Like the big booster of IoT, it is Big Data, since the more objects connected, the more data produced and extracted for use. As a result, the accumulation, analysis, and use of Big Data will be more significant, especially for companies, where they have the most expressive data production with IoT, because they have a large number of objects that can be connected or already connected.

In short, digital transformation is shaping new behavioral patterns, along with IoT, coupled with other technologies such as Cloud, and AI derivations such as machine learning and Big Data, is an important part of this process, as through it other technologies in devices that connect to each other, modifying and simplifying the way they perform routine tasks.

1.5.1 Challenges, Issues, and Implications

Cloud computing and IoT appear to be two completely distant technologies. However, they are complementary, even more than you might think. The spread of cloud computing and the IoT has created ample space for innovative applications, since one needs the other to deliver more and better resources to users.

They are technologies that work to increase the efficiency of everyday tasks and both have a complementary relationship, the IoT is largely responsible for capturing a huge amount of extremely useful data for business strategies. With the great versatility of devices, this technology is able to meet the needs of any business. On the other hand, Cloud computing is responsible for the data, enabling the transport, processing, storage, and security of this information, making the IoT a financially viable solution.

Just as cloud computing is built on the principles of speed and scalability, IoT applications are built on the principle of mobility and the broad network of contacts. The cloud makes the IoT financially viable, since IoT devices are responsible for capturing a multitude of data, which needs to be processed and stored.

Since only with IoT, a specific storage and processing capacity would be necessary, and when transferring processing and storage to Cloud computing, the possibilities of using IoT devices are much greater. This is because the cloud is highly flexible, since it is possible to increase or decrease the bandwidth and the processing and storage capacity as needed, extracting the maximum from the IoT devices in different scenarios.

The implications of collaborating these technologies are based on four premises related to *flexibility* where smart devices are used in different locations, since it is not convenient to depend on a local network for storage and processing, given the demand for a lot of work at each space change physical, and so these IoT devices need nothing more than an Internet connection to take advantage of the cloud, and in this way, the cloud acts as a bridge in the form of a mediator or facilitator of communication, allowing the devices to "talk" both with the processing plants and with each other.

Collaboration with the view that cloud computing allows for better collaboration between developers, facilitating remote data access, allowing changes and optimizations to be implemented quickly and easily. *Scalability* with respect to IoT devices being able to generate a large flow of data and, therefore, require a large storage capacity, which is provided by the cloud.

And finally, *Security* and *Privacy*, since one of the great challenges of IoT is dealing with the security of the data collected, and concentrating the storage of data on the IoT device itself can lead companies to face serious security problems, as long as these connected "things" are small and easily transportable, making them potential targets for theft and robbery. Still considering that IoT devices do not have improved data protection systems, also reflecting that the technology would become very expensive.

In that sense, Cloud computing providers are able to offer a high level of data storage security through firewalls, encryption, antivirus, infrastructure, and appropriate settings, in addition to being a great solution against cyber-attacks, in addition to simplifying and optimizing disaster recovery activities and backup.

As previously shown, the possibilities for using IoT devices are vast, with implementations in the field, in the means of transport, in houses, on the shop floor, in the transit system, among many others, making all the data collected and even devices, can be easily accessed from anywhere, thanks to the cloud.

This flexibility is essential for companies and their technological solutions to be able to react quickly to real-world events and, thus, improve processes, productivity, safety, and many more.

Still considering that the implications of collaborating these technologies can be extended with the sum of other technologies, such as AI and machine learning, making it possible to obtain a high degree of automation, making companies much more effective and responsive in their processes, meaning moving toward a reality in which the IoT transforms and automates daily tasks and processes in companies, regardless of their size.

With the increasing development of IoT, there is also the challenge of developing new forms of efficient communication that adapt to the concept, using low energy consumption and computational power, being found in Cloud technology.

The TCP/IP model is used for protocols on the Internet, framing with IoT because it was initially based on the Internet, and so this model is able to describe naturally the layers used in it. In the network layer, it is also defined as physical, that is, it corresponds to the technologies used for the exchange of packets, for the IoT, the communication in the network layer is based on wireless communication.

In the Internet layer, it is to deliver packages where they are needed, with the purpose of the protocols of this layer is to address, deliver, and avoid congestion. In the transport layer, its purpose is to allow entities to depart from the source and destination hosts, maintaining communication. And in the application layer, there are specific service protocols for data communication on a process-by-process level, it is important to note that if IoT is used alone, it should use the Hypertext Transfer Protocol (HTTP) to provide web services, but it has a lot of computational complexity and has high energy consumption for IoT devices. Therefore, these two technologies (IoT + Cloud) must work together to exchange their services in order to work properly exploring its potentials.

The data generated at the network layer must be forwarded via the Internet layer and transported to a cloud application, present in the application layer, according to the TCP/IP model, since in turn, in the application, the data must be aggregated with others, processed and analyzed, so that a specific application can make a decision and perform actions in relation to this data.

And in the same sense that despite efforts to standardize and unify communication protocols, with the development of IoT objects carried out by several manufacturers that use different protocols in the higher layers, such as application, interoperability proved to be one of the biggest challenges faced.

1.6 TRENDS

It will be even more the age of smart products, from appliances that interact with the user to wearables that make life easier in homes and businesses. New sensors will allow an even wider range of situations and events to be detected as current sensors will be packaged in new ways to support new applications, and new algorithms will emerge to extract and deduce more information from sensor technologies. AI will be the fundamental ingredient needed to understand the vast amount of data collected today and will increase its business value. It will be applied to a wide range of IoT information, including video, still images, speech, traffic activity, network, and sensor data (Schmidt et al. 2015).

As IoT matures and becomes widely adopted, a wide range of social, legal, and ethical issues will grow in importance, where these points will include ownership of data and deductions made from it, adding algorithmic bias, privacy, and compliance with new laws that will emerge based on General Data Protection Regulation (GDPR) (Young 2017).

The cloud, however, is and will find its limits even more related to fast and constant connectivity, especially with regard to connected vehicles or installations in remote areas, latency between data sending, processing, and response is not always compatible with certain applications, and higher storage costs even for data that is not necessarily indispensable. With this, edge computing will become increasingly important and increasingly intelligent, thanks to machine-optimized chips and solutions capable of bringing AI algorithms locally. It will be increasingly natural to produce preventive digital twins, that is, simulations of objects or plants that do not yet exist, we are using physical modeling algorithms and, as paradoxical as it may seem, AI systems that replace sensors in feedback simulations leading companies to take their data analysis and processing processes to the edge of their networks (Erl et al. 2015).

New IoT wireless networking technologies, since an IoT network must involve balancing a competing set of requirements such as power consumption, bandwidth, endpoint cost, latency, connection density, cost quality of service, and frequency range of the connection. Where one of the advantages this migration will allow is cost savings and network latency. This will make it necessary to invest in better-consolidated IoT platforms so that efficient data processing support can be provided (Hossain et al. 2017).

1.7 CONCLUSION

IoT has been closely linked mainly to infrastructure and control, affecting many areas of life, such as cars, buildings, and medicine, in a way that cannot yet be fully described, similar to what it was like to think about how electricity would be used before its existence, because it will be less dependent on the human factor, and can detect several problems even before it can be perceived by someone, its malfunction, making various practical and rapid activities in various fields, and can be a great market for the emergence of various services.

Cloud computing sees this as the future of the Internet and services, where the data that customers use would be stored in locations that could be distributed across multiple countries and servers, resulting in a huge reduction in hardware cost without the need to store data.

Where IoT has required companies to handle high data demands, driving migration to the cloud, allowing them to relinquish their servers and only use a service that does this storage, but that means security, because no company wants to have your data disclosed or stored, for example, in the same location as your direct rival.

Even as cloud computing is gaining more and more space in many industries, many companies still do not understand the different solutions that can be used to leverage their business. Today, cloud computing is divided into three main categories of IaaS, PaaS, and SaaS services, each with specificities that can impact business results. IaaS typically provides an automated and scalable IT infrastructure—storage, hosting, and networking—from its own global servers, charging only for what the user consumes. PaaS, on the other hand, provides all the basics of IaaS, as well as the tools and resources needed to safely develop and manage applications without having to worry about infrastructure. Finally, SaaS can be defined as the location where software is hosted by third parties and can be accessed over the web.

These two technological fronts will be used together, each with their importance, will be a significant technological advance and once implemented will bring several services that will emerge around its use, services that are not yet designed, since this union has been shown the best choice for companies to deliver many benefits such as automation, security, scalability, mobility, and cost savings.

REFERENCES

Aazam, M., Huh, E. N., St-Hilaire, M., Lung, C. H., & Lambadaris, I. (2016). Cloud of things: integration of IoT with cloud computing. In Robots and Sensor Clouds (pp. 77–94). Springer, Cham.

Alioto, M. (Ed.). (2017). Enabling the Internet of Things: from integrated circuits to integrated systems. Springer, Cham.

Almorsy, M., Grundy, J., & Müller, I. (2016). An analysis of the cloud computing security problem. arXiv preprint arXiv:1609.01107.

Amadeo, M., Campolo, C., Quevedo, J., Corujo, D., Molinaro, A., Iera, A., & Vasilakos, A. V. (2016). Information-centric networking for the Internet of Things: challenges and opportunities. IEEE Network, 30(2), 92–100.

Bassi, S., & Chaudhary, A. (2015). Cloud computing data security—background and benefits. International Journal of Computer Science & Communication, 6(1), 34–40.

Bokefode, J. D., Bhise, A. S., Satarkar, P. A., & Modani, D. G. (2016). Developing a secure cloud storage system for storing IoT data by applying role based encryption. Procedia Computer Science, 89, 43–50.

Botta, A., De Donato, W., Persico, V., & Pescapé, A. (2016). Integration of cloud computing and Internet of Things: a survey. Future Generation Computer Systems, 56, 684–700.

Cai, H., Xu, B., Jiang, L., & Vasilakos, A. V. (2016). IoT-based big data storage systems in cloud computing: perspectives and challenges. IEEE Internet of Things Journal, 4(1), 75–87.

Caputo, A., Marzi, G., & Pellegrini, M. M. (2016). The Internet of Things in manufacturing innovation processes: development and application of a conceptual framework. Business Process Management Journal, 22(2), 383–402.

Chang, V. (2015). Towards a Big Data system disaster recovery in a Private Cloud. Ad Hoc Networks, 35, 65–82.

Changchit, C. (2015). Cloud computing: should it be integrated into the curriculum? International Journal of Information and Communication Technology Education (IJICTE), 11(2), 105–117.

Diaby, T., & Rad, B. B. (2017). Cloud computing: a review of the concepts and deployment models. International Journal of Information Technology and Computer Science, 9(6), 50–58.

Díaz, M., Martín, C., & Rubio, B. (2016). State-of-the-art, challenges, and open issues in the integration of Internet of Things and cloud computing. Journal of Network and Computer Applications, 67, 99–117.

Erl, T., Cope, R., & Naserpour, A. (2015). Cloud computing design patterns. Prentice-Hall Press.

Formisano, C., Pavia, D., Gurgen, L., Yonezawa, T., Galache, J. A., Doguchi, K., & Matranga, I. (2015). The advantages of IoT and cloud applied to smart cities. In 2015 3rd International Conference on Future Internet of Things and Cloud (pp. 325–332). IEEE.

Gilchrist, A. (2016). Industry 4.0: the industrial Internet of Things. Apress.

Greengard, S. (2015). The Internet of Things: MIT Press, Cambridge, MA.

Jain, A., & Mahajan, N. (2017). Introduction to cloud computing. In The Cloud DBA-Oracle (pp. 3–10). Apress, Berkeley, CA.

Hashem, I. A. T., Yaqoob, I., Anuar, N. B., Mokhtar, S., Gani, A., & Khan, S. U. (2015). The rise of "big data" on cloud computing: review and open research issues. Information Systems, 47, 98–115.

Hassanien, A. E., Azar, A. T., Snasael, V., Kacprzyk, J., & Abawajy, J. H. (2015). Big data in complex systems. Springer International Publishing, Switzerland.

Hossain, M. M., Fotouhi, M., & Hasan, R. (2015). Towards an analysis of security issues, challenges, and open problems in the Internet of Things. In 2015 IEEE World Congress on Services (pp. 21–28). IEEE.

Hossain, M. S., Xu, C., Li, Y., Pathan, A. S. K., Bilbao, J., Zeng, W., & El-Saddik, A. (2017). Impact of next-generation mobile technologies on IoT-Cloud convergence. IEEE Communications Magazine, 55(1), 18–19.

Hsu, C. L., & Lin, J. C. C. (2016). Factors affecting the adoption of cloud services in enterprises. Information Systems and e-Business Management, 14(4), 791–822.

Khodkari, H., Maghrebi, S. G., & Branch, R. (2016). Necessity of the integration Internet of Things and cloud services with quality of service assurance approach. Bulletin de la Société Royale des Sciences de Liège, 85(1), 434–445.

Lee, I., & Lee, K. (2015). The Internet of Things (IoT): applications, investments, and challenges for enterprises. Business Horizons, 58(4), 431–440.

Lom, M., Pribyl, O., & Svitek, M. (2016). Industry 4.0 as a part of smart cities. In 2016 Smart Cities Symposium Prague (SCSP) (pp. 1–6). IEEE.

Lucero, S. (2016). IoT platforms: enabling the Internet of Things. White paper.

Malawski, M., Juve, G., Deelman, E., & Nabrzyski, J. (2015). Algorithms for cost- and deadline-constrained provisioning for scientific workflow ensembles in IaaS clouds. Future Generation Computer Systems, 48, 1–18.

Olson, N. (2016). The Internet of Things. New Media & Society, 18(4), 680–682.

Pan, J., & McElhannon, J. (2017). Future edge cloud and edge computing for Internet of Things applications. IEEE Internet of Things Journal, 5(1), 439–449.

Rafaels, R. J. (2015). Cloud computing: from beginning to end. CreateSpace Independent Publishing Platform, Scotts Valley, California.

Rittinghouse, J. W., & Ransome, J. F. (2017). Cloud computing: implementation, management, and security. CRC Press, Boca Raton, FL

Russell, S. J., & Norvig, P. (2016). Artificial intelligence: a modern approach. Pearson Education Limited, Malaysia.

Safari, F., Safari, N., & Hasanzadeh, A. (2015). The adoption of software-as-a-service (SaaS): ranking the determinants. Journal of Enterprise Information Management, 28(3), 400–422.

Sajid, A., Abbas, H., & Saleem, K. (2016). Cloud-assisted IoT-based SCADA systems security: a review of the state of the art and future challenges. IEEE Access, 4, 1375–1384.

Schmidt, R., Möhring, M., Härting, R. C., Reichstein, C., Neumaier, P., & Jozinović, P. (2015). Industry 4.0—potentials for creating smart products: empirical research results. In International Conference on Business Information Systems (pp. 16–27). Springer, Cham.

Sen, J. (2015). Security and privacy issues in cloud computing. In Cloud technology: concepts, methodologies, tools, and applications (pp. 1585–1630). IGI Global, Hershey, PA.

Sharma, V., Choudhary, G., Ko, Y., & You, I. (2018). Behavior and vulnerability assessment of drones-enabled Industrial Internet of Things (IIoT). IEEE Access, 6, 43368–43383.

Singh, S., Jeong, Y. S., & Park, J. H. (2016). A survey on cloud computing security: issues, threats, and solutions. Journal of Network and Computer Applications, 75, 200–222.

Soldatos, J., Kefalakis, N., Hauswirth, M., Serrano, M., Calbimonte, J. P., Riahi, M. & Skorin-Kapov, L. (2015). OpenIoT: Open source Internet of Things in the cloud. In Interoperability and open-source solutions for the Internet of Things (pp. 13–25). Springer, Cham.

Sousa, J. A. R. D. (2016). Database as a Service (DaaS) or RDBMS? (Doctoral dissertation).

Tarhini, A., Al-Gharbi, K., Al-Badi, A., & AlHinai, Y. S. (2018). An analysis of the factors affecting the adoption of Cloud computing in higher educational institutions: a developing country perspective. International Journal of Cloud Applications and Computing (IJCAC), 8(4), 49–71.

Van Eyk, E., Toader, L., Talluri, S., Versluis, L., Uţă, A., & Iosup, A. (2018). Serverless is more: from PaaS to present cloud computing. IEEE Internet Computing, 22(5), 8–17.

Villanes, I. K., Costa, E. A. B., & Dias-Neto, A. C. (2015). Automated mobile testing as a service (AM-TaaS). In 2015 IEEE World Congress on Services (pp. 79–86). IEEE.

Weber, M., Lučić, D., & Lovrek, I. (2017). Internet of Things context of the smart city. In 2017 International Conference on Smart Systems and Technologies (SST) (pp. 187–193). IEEE.

Wortmann, F., & Flüchter, K. (2015). Internet of Things. Business & Information Systems Engineering, 57(3), 221–224.

Young, T. (2017). PRIMER: General Data Protection Regulation. International Financial Law Review.

CHAPTER 2

Cyber-Physical Systems in Healthcare

Review of Architecture, Security Issues, Intrusion Detection, and Defenses

Robinson Raju and Melody Moh

CONTENTS

2.1 INTRODUCTION

2.1.1 CPS Definition

Cyber-Physical Systems (CPSs), in simple terms, are physical systems that are integrated deeply with cyber systems. CPS is "intelligent" as it is able to modify the properties in the physical world by gathering inputs from the physical world. The physical system has sensors to take inputs from the physical world, connections to connect to the network of communication and computation nodes, and also actuators to act on the physical world based on instructions from the computation nodes. A formal definition by Rajkumar et al. in their paper titled "Cyber-physical systems: the next computing revolution", defines CPS as "*physical and engineered systems whose operations are monitored, coordinated, controlled and integrated by a computing and communication core*"[1].

2.1.2 Rapid Evolution of Technology

One just needs to look at how humanity got to this point of having intelligent systems that can interact with the physical world, to understand the impact of CPS currently and for the future. In his book, "*The Singularity Is Near: When Humans Transcend Biology*"[2], the futurist Ray Kurzweil talks about the *Law of Accelerating Returns*, that the rate of progress accelerates with time. If one looks at the tremendous developments in the past decades in comparison to the past centuries or the advancements in the past century to the past millennium, one can see the acceleration of progress in action. The progress in the last couple of centuries, since the Industrial Revolution, was at a scale much more massive than the millennia prior. Industrial Revolution made it possible to support a world population of over 7 billion people currently, since there are now machinery and automation to grow more food, raise more livestock, travel long distances effortlessly, and communicate instantly across the planet like never before. Information Revolution that succeeded the Industrial Revolution has, in two decades, produced more growth than what the Industrial Revolution did in 200 years. It has led to 4.5 billion people being connected to the Internet[3], over 2 billion people on Facebook and related social media[4], and trillions of dollars of business transactions occurring online across the globe. The next big revolution is the Internet of Things (IoT), where physical things connected to the Internet are able to sense their surroundings and relay data to remote computing devices. "Things" that can think and act are a result of computing devices becoming more miniaturized, Machine Learning algorithms getting more sophisticated, and other advances in embedded systems, sensor technology, and communication protocols. By 2020, the Internet would have over 50 billion connected things[5], and as these gain more capabilities, it is like the population of working "beings" across the world increased by tenfold.

2.1.3 Ubiquity of CPS

In addition to things becoming smart and recording data from their surroundings, and terms like "smart homes", "smart medical appliances", "smart bridges", "smart power grids", "smart machinery", and so forth, becoming commonplace, connected devices gave rise to CPSs, which are able to modify properties of their surroundings. Today, there are examples of CPSs in action almost in every domain. Be it in power grids that provide efficient energy management to a grid with millions of connected power equipment, or in homes that provide automated lighting, heating, air conditioning, fire prevention, and security, or in hospitals that enable continuous patient monitoring through sensors in assisted living centers, operation theatres, and devices hospital rooms, or in unmanned aircraft systems that help in rescue efforts and goods delivery; CPS is everywhere. CPSs have enhanced the capability of the physical systems in Manufacturing, Construction, Traveling, Healthcare, and Agriculture multifold. They have improved safety, productivity, and efficiency in physical systems that almost seems magical. As physical systems become smart things, the potential for growth and development is unprecedented.

2.1.4 CPS in Healthcare

One area where CPSs could make a considerable impact is in healthcare. To give some statistics, more than 90 million Americans live with chronic illnesses. Chronic diseases account for 70% of all US deaths and around $1.4 trillion in medical costs[1]. In the United

States, there were 35 million inpatient hospital stays in 2013. Also, the number of people over the age of 65 is expected to exceed 70 million by 2030[6]. The numbers are much larger across the world. Hence the need for automated health monitoring and assistance is on the increase since there are not enough nurses and doctors to take care of all these people. Smart devices placed on or implanted inside people have enabled less expensive and more effective care. These devices, called Wearable Body Sensor Networks (BSNs) and Implantable Medical Devices (IMDs), have enhanced the mortality and functioning of many impaired people through continuous health monitoring via a real-time medical data feed. In many cases, one could argue, CPSs are more efficient and react to situations much faster; for instance, automated injection of insulin or automated electric shocks require detection and reaction to be very fast. CPSs are enabling chronically ill and older adults to live independently in their homes while being monitored remotely[7]. There have been many examples in recent times where Apple Watch saved people's lives when it alerted them about a stroke or a rise in blood pressure[8]. In "*Sapiens: A Brief History of Humankind*"[9], Yuval Harari talks about the biotechnological revolution being the next significant milestone in the history of humanity, leading to bio-engineered humans. If the direction of evolution is pointing toward a cyber-human where micro-devices within the body are connected to the Internet, it is essential to understand the security and privacy threats of the current CPSs in use in the healthcare domain.

2.1.5 Overview of Security Issues in CPS

Many CPSs evolved from Industrial Control Systems (ICSs) that had efficient mechanisms to perform industrial operations through a closed-loop system. Though the transition of these systems from something that was working efficiently in an industrial plant to devices connected to the Internet seems natural, it should be noted that these systems predated the Internet and were designed without considering access from a public network. The ICS requirements focused on performance, reliability, and safety and not on network security[10]. The security for these systems was on air-gapping, where physical access was controlled through strict authentication. But CPSs have both physical and networking aspects and hence are more complex to secure. They depend on information networking infrastructures, which makes them a target to cyber-attacks and network vulnerabilities[10]. They are also subject to physical vulnerabilities like bad weather, intentional destruction of sensors, and so forth. In the case of BSNs, the sensors generally have low power, limited memory, and computational capabilities, which sometimes result in data communication via cleartext or insecure channels. Since the data in question is private patient data that is highly sensitive, the risk is not only for the well-being of the patient but also for privacy. In some cases, it can even harm people if the device that is compromised is an implantable device that administers medicines to a person.

Moreover, the focus of newly introduced IoT or CPS devices is almost always convenience and utility, on how the latest developments are ushering in another wave of comfort, ease, and luxury at our fingertips. However, the remote connectivity that enables the convenience also opens the door to adversaries to gain wrongful access and cause harm. Security and privacy vulnerabilities are sometimes found after they are exploited or when the hacking becomes public. In many cases, the average user is willing to trade off security for convenience, sometimes unconsciously because the systems are complex to understand.

AlTawy and Youssef[10] share various examples of the tradeoff between utility and security. People are okay to store sensitive data on the cloud since it is convenient, and they are fine with governments accessing their emails or social network activity if that could help to prevent terrorism. Another example was of a European luxury car that had a safety feature to unlock automatically if enough pressure was applied on the roof, for instance, if the car topples which was exploited by a car thief who just jumped on top of the car. CPS systems had many safety mechanisms built in to prevent attacks, and most of them focused on physical access and modifications. A safety mechanism to lock out a room in case of tampering could hamper legitimate recovery efforts. Since mechanisms that are meant to prevent cyber-attacks could conflict with a critical requirement of a CPS, the National Institute of Standards and Technology (NIST) and other such regulatory bodies have specific guidelines integrating cybersecurity features in CPS[10]. In "*A survey on concepts, applications, and challenges in cyber-physical systems*", Gunes et al.[11] list Interoperability, Predictability, Reliability, Sustainability, Dependability, and Security as challenges in designing and deploying CPSs with Security being the most critical challenge of all[11].

2.1.6 Chapter Outline

The principal goal of this chapter is to serve as an introductory guide to comprehend CPS for anyone new to it, and to understand the security and privacy issues in CPS. The chapter is broadly divided into two parts. The first part introduces CPS in simple terms and then reviews the design and architecture of prevalent CPS systems. After that, it reviews communication protocols in ICS and CPS devices, including wired and wireless technologies. In order to have a more personal understanding of the effects of CPS on society, the healthcare domain is taken as a focus area. Existing CPS health monitoring systems are reviewed to get an understanding of types of medical sensors, implantable devices, and commercial products in use. The second part focuses on security threats on CPS systems and the techniques that are being used to overcome the threats. It introduces security requirements for CPS and categorizes the security challenges. After that, a brief history of security issues in medical systems is reviewed for over four periods from the 1980s to the present. Then, CPS architectures proposed to secure medical monitoring devices are reviewed. After that, the focus is on looking at existing solutions for security in CPS, including biometric and cryptographic solutions. Of all the prevention techniques for control system devices, Intrusion Detection is one of the most commonly used. Hence, the next section in the chapter is about Intrusion Detection. Current research and the challenges in implementing it in CPS are discussed. Finally, the chapter goes over machine learning techniques to prevent intrusions on IoT/CPS devices.

2.2 CPS DESIGN AND ARCHITECTURE FOR MEDICAL MONITORING

2.2.1 Architectural Styles for CPS in Healthcare

A typical architecture of a CPS system includes sensors, actuators, a CPS network, and Control Units. The models are mostly described by using variations of Industrial Control Systems (ICS). A comprehensive architecture for CPS Healthcare is difficult to define due to the variety of applications and complexity of the design. In "*A security analysis of cyber-physical systems architecture for healthcare*"[6], the authors list the history of CPS architectures for healthcare. They mention that though different architectural styles exist, the architectural standards are yet to be defined.

1. **Process Control (Closed-Loop) Architecture**—A simplistic architecture that shows the data flow from sensors to the control unit and the action on the actuators by the control unit. This was used in industrial control systems and environmental management systems.

2. **Publish and Subscribe Architecture**—Uses a discovery protocol to discover which sensors are in the current environment and either publishes or/and subscribes messages. CodeBlue software developed at Harvard University used this architecture.

3. **Layered Architecture**—Describes the system in layers, each layer taking care of particular functionality. Seifert and Reza[6] cite proposals of layered architecture from researches Sasi and Min, and also Lu and Fu. Companies like MobiHealth and AlarmNet are examples where layered architecture was used[6].

4. **Blackboard Architecture**—In this, sensors submitted readings to a blackboard data system. The initial proposal was from Winograd.

Each of these patterns is reviewed in detail in the section "Review of CPS Architectures", and they help understand the CPS system from different viewpoints.

2.2.2 Nonfunctional Requirements of CPS

AlTawy and Youssef[10] list Safety, Security, and Availability as the key requirements for a functional CPS. Seifert and Reza[6] list Heterogeneity, Reactivity, Adaptability, Scalability, Uncertainty, and Robustness as the key qualities against which the effectiveness of a CPS architecture could be measured.

Safety is a crucial feature of CPS since it needs to ensure the well-being of the users who are interacting with the system directly or indirectly. The design of CPS should account for scenarios that could cause harm and offer protection against it. For example, if a driverless car is involved in a crash, there must be mechanisms and policies to ensure the least danger to life[10]. **Security** of CPS should take into account physical security as well as cybersecurity. It needs to ensure authorized access, traceability of access and actions, user's privacy, and information integrity. There are several instances of security breaches of sensors and networks. Failing to secure the infrastructure may lead to occurrences like Water treatment plant being hacked and chemical mix changed for tap supplies, Nuclear power plant breach in Ukraine, Video Baby Monitors hacked, Data from wearable devices used to plan robberies, and so forth[5]. A 2015 report of the "Internet of Things" research study done by Hewlett Packard reported that 80% of devices raised privacy concerns and found that on an average, 25 vulnerabilities were found per device, totaling 250 vulnerabilities[5]. **Availability** is another key functionality, especially in CPS for healthcare, since CPSs are mostly used to provide critical functionality. Also, they are increasingly used for eldercare.

Hence downtimes are not acceptable, and the architecture must support usage or device whose resource consumption is minimal. For example, the 2017 DDOS attack on Dyn was made possible by a large number of unsecured IoT devices, such as home routers and surveillance cameras. The attackers employed thousands of such devices that had been infected with malicious code to form a botnet. The devices themselves were not powerful, but collectively they generated a massive amount of traffic to overwhelm targeted servers[5].

Heterogeneity is a unique challenge for CPS. Many integration issues are not known until the system is installed and integrated into an environment. CPS healthcare devices could be deployed in different types of facilities with different physical characteristics and access patterns. So, the ability to handle **uncertainty** is essential. The system must be **adaptable** to new configurations and environments. Another key metric to evaluate is **reactivity,** which measures how quickly the system can react to changing conditions that are being monitored. Pacemakers or insulin pumps would need to react in a near real-time manner to be effective. Another critical quality is **robustness**. The system should continue to function effectively or not crash even if some minor values are missing or corrupted. Robustness can be built by having redundancies and sophisticated algorithms[6].

2.2.3 Review of CPS Architectures

2.2.3.1 Closed Loop/Control Loop Architecture

Closed Loop Architecture is the most commonly used architecture to describe industrial systems, as shown in Figure 2.1[6]. The sensors are continually measuring some aspects of

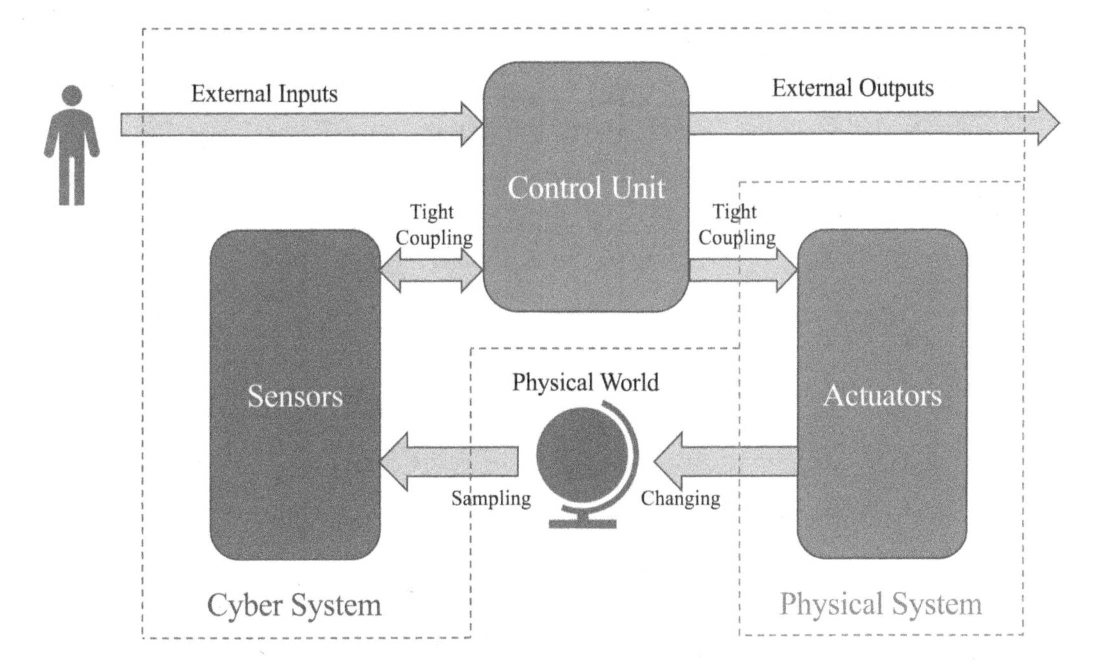

FIGURE 2.1 Control loop architecture.

Source: Seifert, Darren, and Hassan Reza. "A security analysis of cyber-physical systems architecture for healthcare"

the physical world and relaying them to the Control unit. The control unit determines if the values have deviated from the range of expected values. If it has, the control system sends signals to the actuators to act on the physical system to modify the reading that is being measured. In a CPS, the Control unit is connected to the Internet and gets external inputs.

2.2.3.2 Layered Architecture

Layered architecture is generally used to describe IoT systems. It is easy to abstract, common functionalities together and to explain the features and responsibilities of devices at every level. Han et al. describe a common model of the layered architecture that includes a Physical layer, Sensor layer, Network layer, Control layer, and Information layer[13]; an example is shown in Figure 2.2. Seifert and Reza[6] describe a layered architecture that consists of a Physical-event layer, Physical observation layer, Sensor event layer, Cyber-physical event layer, and Cyber event layer. Moh and Raju[5] describe a layered architecture for IoT consisting of layers Sensing, Network, Service, and Interface while describing the security threats at every layer. In all these architectures, the physical objects being monitored are in the physical layer; the sensors connect to the physical layer and are responsible for measuring the physical system; the network layer uses the appropriate communication protocols to convey the measured data to the control layer; the control layer uses the information to take the necessary action and conveys the message back to the sensor/actuator layer to act upon the physical world.

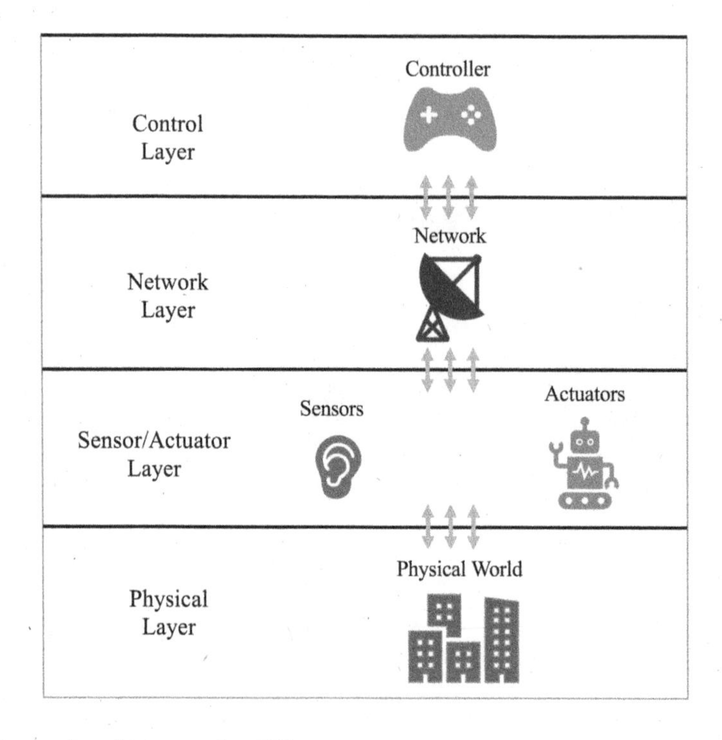

FIGURE 2.2 Layered architecture for CPS.

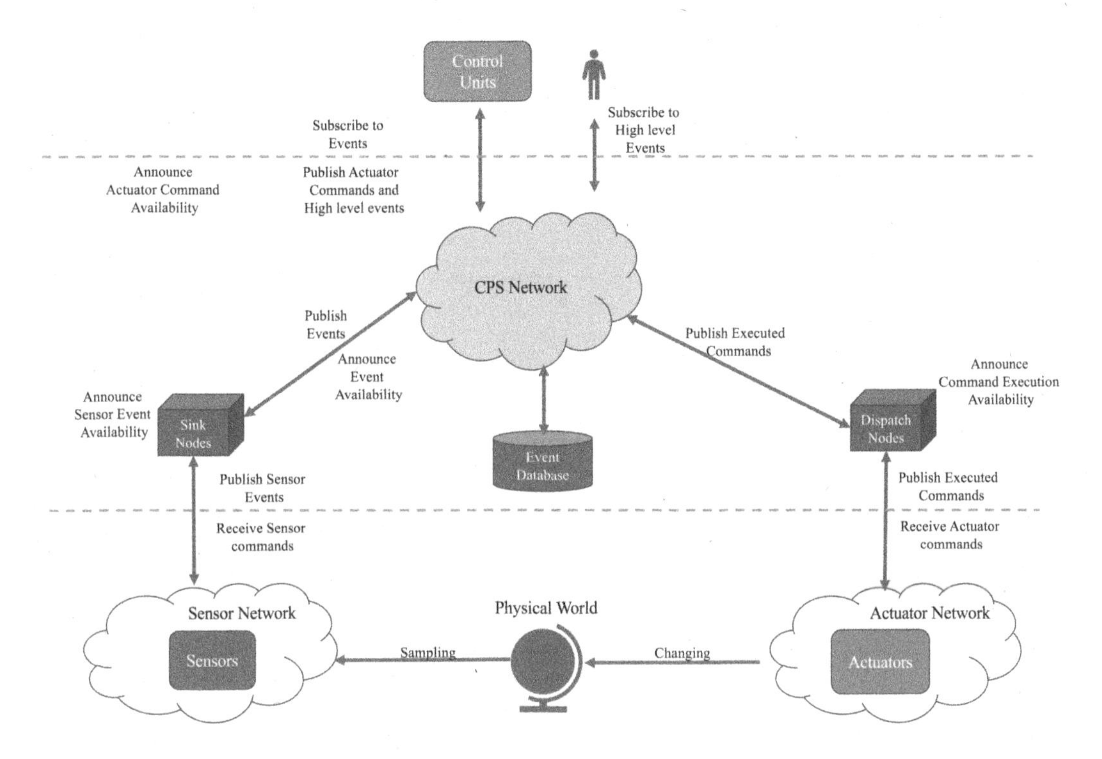

FIGURE 2.3 Publish-and-subscribe architecture for CPS.

Source: Seifert, Darren, and Hassan Reza. "A security analysis of cyber-physical systems architecture for healthcare"

2.2.3.3 Publish-and-Subscribe Architecture

In this architecture, each sensor or actuator is a node independent of each other, and all of them connect to the CPS network, as shown in Figure 2.3[6]. As nodes do not need prior knowledge of each other, additional nodes could be added to the network as needed[6]. A node could subscribe to listen to messages or broadcast messages.

2.2.3.4 Blackboard Architecture

The blackboard architecture, as shown in Figure 2.4[6], is similar to the publish-and-subscribe architecture; note that it also contains sensor and actuator nodes that are independent of each other. The key difference is that they do not communicate with each other but communicate to local "knowledge sources" called blackboard nodes. Blackboards serve as the knowledge sources, and a controller manages the data movement to the CPS network and eventually to the nodes.

2.2.4 CPS Architectures Compared against the Nonfunctional Requirements

If the architectures are ranked on the qualities of heterogeneity, reactivity, adaptability, scalability, uncertainty, and robustness, one could see that all the architectures do not inherently support all the qualities.

1. **Control Loop Architecture**—The control loop architecture has drawbacks when it comes to heterogeneity, adaptability, uncertainty, and robustness. It does not account

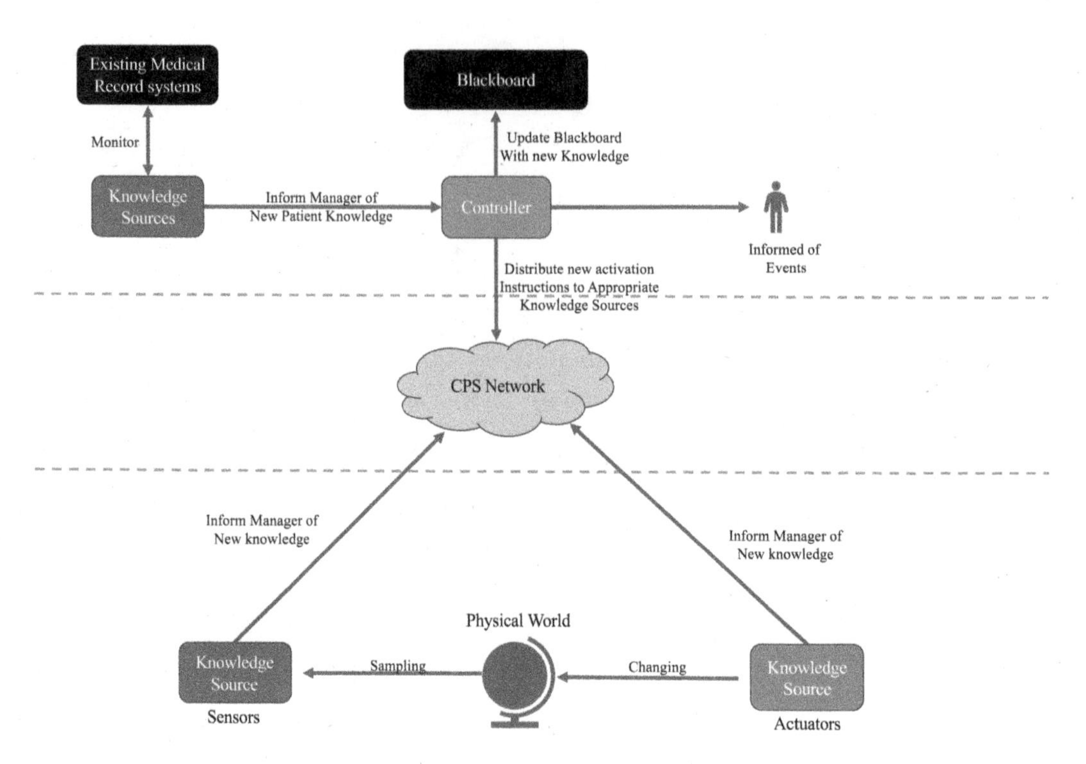

FIGURE 2.4 Blackboard architecture for CPS.

Source: Seifert, Darren, and Hassan Reza. "A security analysis of cyber-physical systems architecture for healthcare"

for the complex nature of diversity of sensors, thus affecting heterogeneity. It also does not account for how nodes could adapt to changes in the environment. Uncertainty is another issue since it is only able to deal with incremental corrections[6].

2. **Layered Architecture**—While evaluating the effectiveness of the architecture, one of the key concerns with the layered architecture is reactivity because the data must be sent across different layers before the decision to act is made[6]. Another concern is regarding adaptability across different environments since the reading may be different for the same sensors in different situations. The architecture scores high on other characteristics like robustness and scalability.
3. **Publish-and-Subscribe Architecture**—The fact that nodes are independent and the system adapts to the addition of new nodes gives this architecture robustness and adaptability. Also, since events in a node as broadcasted, the reactivity could be much faster than a system that has to consolidate data and then transmit to another layer or node for consumption. One issue with the architecture is around scalability. Though the system may not feel the impact of adding a few nodes, if many nodes get added, there could be a possibility to have network traffic congestion, and this could affect availability.
4. **Blackboard Architecture**—The architecture is designed to work with different data sources and is able to react fast and handle uncertainty well[6]. It scores well on all the

quality metrics of robustness since it is built to handle a variety of different redundant sensors. It is scalable since nodes can be added at will, and the traffic to and from the nodes is modulated by the controller.

Seifert and Reza[6] used a 3-point scale to compare the architectures and found that publish-and-subscribe Blackboard architectures are most suited for healthcare CPS. They proposed a modified architecture of both models to explain how they could be adapted to being more secure. The secured architectures are discussed in a later section in this chapter.

2.3 CPS COMMUNICATION AND CONTROL PROTOCOLS

2.3.1 Communication Protocols in Industrial Control Systems

ICSs preceded CPSs, and a lot of the communication protocols in CPSs are ICS protocols that have been adapted to work with internet connectivity. ICSs used many specialized protocols designed for industrial automation and control, with the key goal of efficiency and reliability. The operational requirements for ICSs were real-time synchronization to support precision operation and deterministic communication of both monitoring and control data. The ICS devices initially only worked over serial channels like RS232 at low speeds, but many have evolved to operate over Ethernet networks using routable protocols like TCP/IP and UDP/IP. In general, the industrial protocols can be divided into two common categories, Fieldbus protocols and backend protocols.

Fieldbus is a broad category of protocols that are commonly found in process and control. They are commonly deployed to connect process connected devices, like sensors or actuators. Examples include Modicon Communication Bus or Modbus and the Distributed Network Protocol or DNP3. Backend protocols are protocols that are deployed on or above supervisor networks and are used to provide efficient system-to-system communication. Backend protocols connect an ICS from one supplier to another supplier's systems. Examples include the Open Process Communication, OPC, and the Inter-Control Center Protocol, ICCP.

2.3.1.1 Fieldbus Protocols

1. **The Modbus (Modicon communication bus) protocol,** designed in 1979 by Modicon, is one of the most popular protocols used in ICS architectures, and it enables process controllers to communicate with real-time computers. Modbus is an open standard, is widely adopted, and has also been enhanced over the years into several distinct variants. Modbus communicates raw messages without the restriction of authentication or obsessive overhead. Modbus is an application layer protocol and allows for communication between interconnected assets based on the request-reply methodology. This enables extremely simple devices, such as sensors and motors, to use Modbus to communicate with more complex computers. During communication, it uses three distinct protocol data units, or PDUs, which are Modbus request,

Modbus response, and Modbus exception response[14]. A transaction begins with the transmission of an initial function code and a data request within a request PDU. The receiving device responds in one of two ways. If there are no errors, it will respond with a function code and data response within a response PDU. If there are errors, the device will respond with an exception function code. An exception code within a Modbus exception response. Modbus can be implemented on either an RS-232C, which is point to point or RS-45, which is a multidrop physical layer. Up to 32 devices can be implemented on a single RS45 serial link, requiring each device communication via Modbus to be assigned a unique address. A command is addressed to a specific Modbus address, and while other devices may receive a message, only the addressed device will respond. Data are represented in Modbus using four primary tables. They include the discrete input data table, the coil table, the input register table, and the holding register table. The method of handling each of these tables is device-specific, as some may offer a single data table for all types, while others offer unique tables.

Modbus has several variants. These include Modbus RTU, and Modbus ASCII, which support binary and ASCII transmissions over serial buses, respectively. Additionally, Modbus TCP is a variant of Modbus developed to operate on modern networks. And Modbus Plus is a variant designed to extend the reach of Modbus via interconnected buses using token passing techniques. Here we have the Modbus RTU and Modbus ASCII frames. The similar variance of Modbus is used in asynchronous serial communication, and they are the simplest of the variants based on the original spec. Modbus RTU uses a binary representation, whereas Modbus ASCII uses ASCII characters to represent data when transmitting over the serial to your link.

2. Like Modbus, the **distributed network protocol or DNP3** began as a serial protocol for use between master stations or control stations, and slave devices called outstations. It is also commonly used to connect RTUs configured as master stations to IED outstations in electric substations. DNP3 was initially introduced in 1990 by Westronic and was based on early drafts of the IEC 60780-5 standard. The primary motivation for this protocol was to provide reliable communications in environments common within the electric utility industry that include a high level of electromagnetic interference. DNP3 was extended to work over IP via encapsulation in TCP or UDP packets in 1998. And is now widely used in not only electric utility, but also oil and gas, water, and wastewater industries. DNP3 was based on the IEC One of the leading reasons for some industry migration from Modbus to DNP3, includes features that apply to these other industries.

2.3.2 CPS Protocols and Corresponding Ports

In their paper on Network security and privacy for CPS, Henze et al.[15] presented the evolution of CPS and note that dating back to the year 2000 and before, radio frequency identification (RFID) technology bridged the gap between the physical and the cyber world, ushering in an era of connected devices[15]. The year 2005 onward, WSNs (Wireless Sensor

TABLE 2.1 List of CPS Protocols and Corresponding Ports

CPS Communication Protocols	Port Number	Type
BACnet	47,808	Registered
DNP3	19,999/20,000	Registered
ICCP	102	Well-known
IEC-104	2404	Registered
Johnson Controls Metasys N1	11,001	Registered
Modbus	502	Well-known
MQ Telemetry Transport	1,883	Registered
Niagara Fox	1,911/4,911	Registered
PROFINET	34,962/24,963/34,964	Registered
Red Lion	789	Well-known
ROC Plus	4,000	Registered
Siemens Spectrum Power TG	50,001/50,018/50,020, etc.	Dynamic/Private

Networks) formed the next step of the evolution. Industrial Control Systems (ICSs) relied mostly on one or more specialized protocols, including vendor-specific proprietary protocols or nonproprietary protocols, like Modbus/TCP (Transmission Control Protocol), or DNP3. CPSs have grown out of ICSs, and many of the protocols been adapted to operate over the standard Ethernet link layer, using the Internet protocol with both User Datagram Protocol (UDP) and TCP as transports.

In "*Passive inference of attacks on CPS communication protocols*", Bou-Harb et al.[16] discuss CPS security issues focusing on protocol vulnerabilities. They list the following as CPS communication protocols and their corresponding port numbers[16], summarized in Table 2.1.

2.3.3 Radio Technologies for Wireless Sensors

The main LPWN (Low Power Wireless Network) standards for WSNs (Wireless Sensor Networks) are IEEE 802.15.4, IEEE 802.15.6, and Bluetooth[7].

- IEEE 802.15.4 and Zigbee—This standard has been widely used in CPSs in industrial systems, home automation, and health monitoring[7]. The most commonly used frequency is the ISM band, which is 2.4 GHz.
- Zigbee—This is an open specification that complements IEEE 802.15.4 standard with network and security layers[7].
- Bluetooth—Bluetooth, also known as IEEE 802.15.1, is used in a variety of electronic devices for short-range wireless communication. The technology works by emitting radio waves in the 2.4–2.485 GHz spectrum and can adapt to 70+ frequencies with a 1 MHz space between each other. This enables communication without interference. One drawback is that Bluetooth devices need to go through a "pairing" process. To expand the reach of Bluetooth, especially for devices with resource constraints like in CPS, BLE (Bluetooth low energy) was introduced.

- UWB and IEEE 802.15.6—Ultra-wideband (UWB) standard was developed by the IEEE 802.15.4a committee and included a WSN scenario. IEEE 802.15.6 defines an access control layer that supports several types of physical layers, such as narrowband (NB), UWB, and human body communication[7].

2.3.4 Telemetry

Implantable Medical Devices (IMDs) communicate via Telemetry, which commonly refers to wireless data transfer, usually between 10 and 20 m. The communication is usually done using Radio Frequency (RF) signals through a specific frequency spectrum[10]. The following key standards regulate Telemetry for IMDs—Wireless Medical Telemetry Services (WMTS), Medical Implant Communication Systems (MICS), and Medical Device Radiocommunications Service (MedRadio)[10].

2.4 EXISTING CPS HEALTH MONITORING SYSTEMS

2.4.1 Medical Sensors and Implantable Medical Devices (IMDs)

Medical devices that contain physical sensors to measure health parameters can be connected to a network to monitor patient health digitally. The devices include body temperature monitor, blood pressure monitor, electrocardiogram (ECG/EKG) monitor, heart rate monitor, glucose monitor, pulse oximeter, hemoglobin monitor, smart garments, sleep monitor, medication management, food contamination detection device, early warning system, and so forth. Beyond the wearable and out of the body sensors, there are implantable medical devices inserted inside the body that are highly sophisticated, miniaturized, and reliable. Some examples are pacemakers that stimulate the heart muscles to regulate its rhythm, deep brain stimulation (DBS) systems which act as neurostimulators that provide highly controlled electrical impulses into the brain to help people with Parkinson's disease, cochlea implants that stimulate electrodes placed inside the inner-ear to help with hearing[17], infusion pumps such as insulin pumps, body area networks which are composed of biosensors to trace various biological functions, and gastric simulators on gastroparesis patients to send electrical pulses to the nerves and stomach to decrease nausea[10]. Table 2.2[12]

TABLE 2.2 Commonly Used Sensors in BSNs

Sensors	Signal Type	Frequency	Position
Accelerometer	Continuous	High	Wearable
Artificial Cochlea	Continuous	High	Implantable
Artificial Retina	Continuous	High	Implantable
Blood-Pressure Sensor	Discrete	Low	Wearable
Camera Pill	Continuous	High	Implantable
Carbon Dioxide Sensor	Discrete	Low/Very low	Wearable
ECG/EEG/EMG Sensor	Continuous	High	Wearable
Gyroscope	Continuous	High	Wearable
Humidity Sensor	Discrete	Very low	Wearable
Blood Oxygen Saturation Sensor	Discrete	Low	Wearable
Pressure Sensor	Continuous	High	Wearable/Surrounding
Respiration Sensor	Continuous	High	Wearable
Temperature Sensor	Discrete	Very low	Wearable

Source: Raju, Robinson, Melody Moh, and Teng-Sheng Moh. "Compression of wearable body sensor network data using improved two-threshold-two-divisor data chunking algorithms"

lists commonly used sensors in BSNs (Body Sensor Networks) that gives an idea of the range of medical devices in use.

2.4.2 Commercial Products

Čaušević et al.[7] list many CPS applications for continuous health monitoring in their discussion on data security and privacy in CPS. Haque et al.[18] also list many CPS applications for healthcare in their review of CPS in healthcare", which are as follows:

1. CodeBlue—As discussed earlier in the section on architecture, this system was developed at Harvard University and consisted of many sensors, placed on the human body, that transmit medical data using wireless communication[18].
2. UbiMon—Provides a monitoring environment for wearable and implantable medical devices (IMDs) and captures outlier measurements that are life-threatening[7].
3. LifeGuard—Developed as a joint effort between NASA Ames Research Centre and Stanford University[7], is a personal physiological monitor for extreme environments.
4. Alarm-Net—Developed by researchers at the University of Virginia[7], helps in patient-health monitoring in the assisted-living and home environment.
5. MobiCare—A wide-area mobile patient monitoring providing continuous monitoring of a person's health attributes[7].
6. SATIRE—A wearable personal monitoring service embedded in user garments[7].
7. MEDiSN—Developed as a join-effort between researchers at John Hopkins University and those in the University of Latvia, used for automating the process of patient monitoring[7].
8. LAURA—A lightweight wireless sensor-based system for monitoring of patients within nursing institutions[7].
9. Apple Watch, Fitbit, and other wearable watches have the capability to measure heart rate, monitor activity, and track other metrics to help people to be aware of their health.
10. EMR (Electronic Medical Records) is the design of a cyber-physical interface for automated vital sign readings[18].

2.5 SECURITY THREATS IN CPS

2.5.1 Introduction

A security threat is a potential possibility of a deliberate unauthorized attempt to access data, manipulate the system, or render it nonfunctional[13]. CPSs support continuous monitoring and operations at every scale, and a successful compromise of the network or any of the devices in the system can be used directly to impact the services. The consequences can be anything like an interruption of operations, taking sensors offline, modifying the sensor readings, delaying data transmission, blocking, or even sabotaging the actuators to cause harm. Large-scale sabotage has far-reaching consequences and can result in environmental damage like oil spills or fires, or injury or loss of life, and the loss of critical services, blackouts, and other disruptions. Moreover, when it affects critical services like power or healthcare, companies can incur penalties for regulatory noncompliance. Another thing to note is that safety is not the same as security. A system could be built with a lot of safety and fail-safe mechanisms in place, but if it is not secure, even if the device does not fail, there could be other consequences like data modification, delay or blocking of communications, etc. Also, compliance with rules does not guarantee security. More often than not, the law and regulations are slower to catch up to the latest cyber threats. If companies follow just the guidelines set up by organizations, for example, NIST Special Publication (SP) 800-82, Guide to Industrial Control Systems (ICS) Security, they may fall short of protecting their devices and end-users.

2.5.2 Fundamental Security Requirements of a CPS Health Monitoring System

The fundament requirements of a health monitoring system from a security standpoint are—data integrity, data confidentiality, authentication, authorization, and data freshness[7].

- Data Integrity—Ensures the integrity of patient data. Makes sure that the data is not corrupted or lost in transmission or manipulator by unauthorized entities.
- Data Confidentiality—Ensures that data is private and is accessible only to individuals who are authorized to view/use the data.
- Authentication—Ensures that the data is sent by a trusted sender.
- Authorization—Ensures that the data is viewable only by persons with the right security clearance and authorizations.
- Data Freshness—Ensures that there is no delay in data transmission, and the data that is sent/received is in order.

2.5.3 Common Cyber Attacks on CPS Devices

Some examples of common cyber-attacks are man-in-the-middle attacks, information harvesting, denial-of-service attacks, and replay attacks. The primary reason for this is a combination of insecure communications protocols, little device-to-device authentication, and also less computing power in embedded devices.

- **Man-in-the-Middle Attack**: In a man-in-the-middle attack, the attacker seeks to get in the middle of the communication between devices, and if the connection lacks encryption and authentication, the attacker can read the data. The adversary then is able to impersonate the device to the hub and vice versa. This could lead to the wrong data being sent to the physician or the monitoring system.
- **Information Harvesting**: Information harvesting is a significant threat to CPS healthcare devices than CPSs in other domains. Of all the personal data that is available online, personal health information is deemed as a big gold mine. In the black market, the value

of a person's health record is estimated to be $50, compared to $3 for a social security number and $1.50 for a credit card[10] and a lot more focus is on SSN and credit card fraud while the big danger posed by IMDs and other medical sensors is increasing by the day. Disclosure threat, Identity theft, and patient's prescription leakage could be the outcomes of information harvesting from the devices or network. Many times, the information gathered may be used for other purposes. For example, data from a wearable device like Fitbit could be used to commit burglaries since that could give information about when the person is not available at their homes. The same goes for video footage by unmanned aircraft or driverless cars. The data that is used to enable Machine Learning can also be exploited for malintent in these cases. Moreover, if the data is from influential people or high net worth individuals, then it is more valuable, and people could be subject to extortion or other threats if their private data is compromised.

- **Denial-of-Service (DoS) Attack**: Denial-of-service attack is a broad category of attacks that include crashing critical devices or flooding the network with a deluge of data, resulting in the loss of availability of the system thereby preventing the system from doing its job. The attacker could try to request power-consuming tasks that might drain CPS devices and networks that are generally LPWN (Low power wireless network) devices. An attacker might use a signal jamming device to scramble responses from a device to the hub, rendering the sensor useless from the hub's perspective. In the case of IMDs with magnetic switches, an attacker could exert a magnetic field near the patient to trigger automatic shut off[10]. DOS attacks could escalate into catastrophic damages in CPSs, unlike online services or websites. A CPS system could be controlling the flow of crude oil in a pipeline, converting steam into electricity, or controlling ignition timing in an automobile engine, and if that gets disrupted, the results can be disastrous. It could lead to not only the hospital or the industry where it is deployed but also cause environmental damages. When a system fails as a result of an attack, one of the following failure modes could be activated: (i) Fail-stop where the system operation stops, (ii) Fail-safe where the system enters a safe mode to avoid any hazardous effects, (iii) Fail-loud where the system sounds an alarm, or (iv) Fail-quiet where the system allows unauthorized access so that the pattern can be studied[10].
- **Replay Attack**: This is similar to man-in-the-middle attack but more dangerous. Here the attacker eavesdrops on the channel and replays the traffic many times by modifying the data. A replay attack could be made even if the message is encrypted. This results in the behavior of the system being altered by manipulating the actions of the controllers.
- **Tracking**: When CPS devices that are wireless, especially IMDs are discoverable, an attacker who has different types of hubs at hand can track the movement of people who have the devices on them. This poses a violation of privacy and similar concerns as Information Harvesting.

The OWASP (Open Web Application Security Project) summarization of the attack surface areas for IoT[5] is very relevant for CPS also. Table 2.3 has the mapping of attack surfaces and the corresponding vulnerabilities.

TABLE 2.3 OWASP IoT Attack Surface Areas

Attack Surface	Vulnerability
Ecosystem Access Control	Implicit trust between components Enrollment security Lost access procedures
Device Memory	Cleartext usernames and passwords Third-party credentials Encryption keys
Device Web Interface	SQL injection Cross-site scripting Cross-site Request Forgery Username enumeration Weak passwords Account lockout Known default credentials
Device Firmware	Hardcoded credentials Sensitive information disclosure Sensitive URL disclosure Encryption keys Firmware version display and/or last update date
Device Network Services	Information disclosure User CLI Administrative CLI Injection Denial of Service Unencrypted Services Poorly implemented encryption Vulnerable UDP Services DoS
Administrative Interface	SQL injection Cross-site scripting Cross-site Request Forgery Username enumeration Weak passwords Account lockout Known default credentials Logging options Two-factor authentication Inability to wipe the device

2.5.4 Categorization of Security Challenges in CPS

Due to the entangled nature of cyber and physical systems in CPS, the system is vulnerable to multiple types of threats. As shown in Figure 2.5[10], the threats can be categorized into two groups—Cyber and Physical. Cyber threats can be active or passive. Active threats like Impersonation, Relaying, and DoS (Denial of Service) violate confidentiality, integrity, authentication, availability, and sustainability. Passive threats like Tracking or Information harvesting violate confidentiality. A few of the cyber threats have been discussed in the prior section. Physical threats could be from Environmental factors, Social crises, or Intentional threats, and these violate availability and safety.

The same security threats could also be viewed from different layers of the CPS system. Han et al. list security threats against CPS as shown in Table 2.4[13].

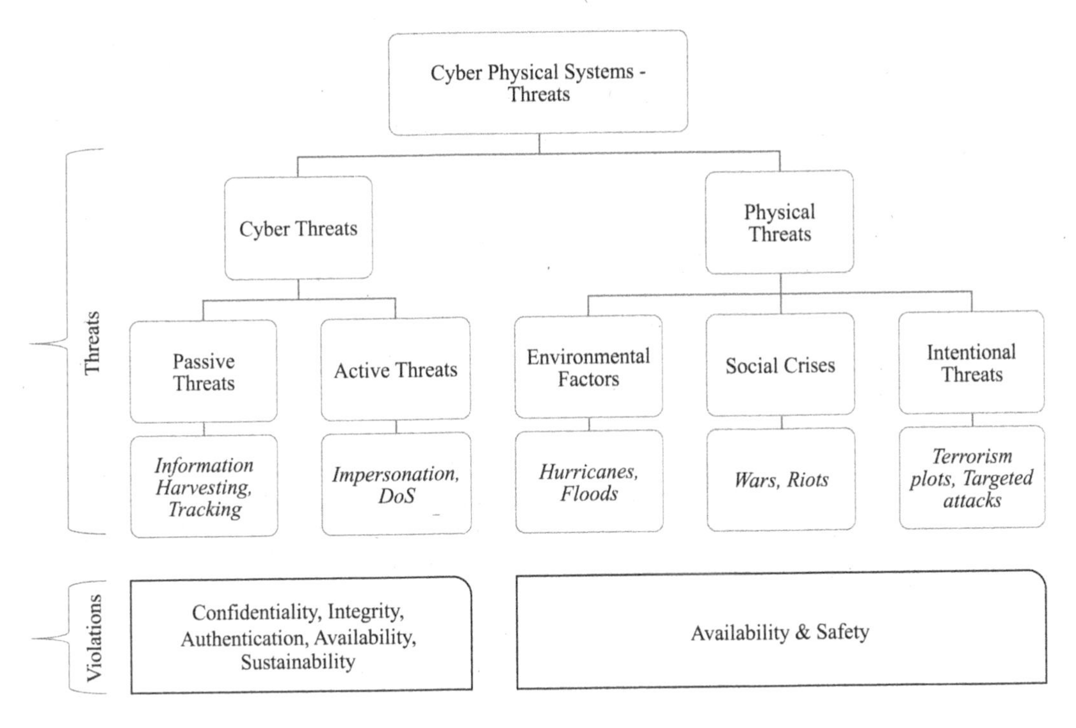

FIGURE 2.5 Threat sources in CPSs and the properties they violate.

Source: AlTawy, Riham, and Amr M. Youssef. "Security tradeoffs in cyber physical systems: A case study survey on implantable medical devices"

TABLE 2.4 Security Threats Against CPS at Each Layer

Layers	Threats	Details
Physical Layer	Direct interventions and damages	
Sensor/Actuator Layer	Node capture	An instrument for mounting counterattacks
	Node destruction	Destruct, extract, or modify node physically
	Power consumption	Quickly drain out limited power of the sensor
	Cryptographic attacks	Crack secret keys with brute force, dictionary, or monitoring
Network Layer	Replay	Forward message to an incorrect destination or with a delay
	DoS	Result in jamming, colluding, and flooding; ill-redirect routing
	Sybil	An adversary illegitimately takes on multiple identities
	Spoofing	Change routing information illegitimately
	Wormhole	Disrupting routing
	Selective forwarding	Disrupting continuity of transmission
Control Layer	Desynchronization	Break timeliness
Information Layer	Privacy	Steal information by eavesdropping and traffic analysis
	Policy	Breach policy by excuse attack and newbie-picking

Source: Han, Song, Miao Xie, Hsiao-Hwa Chen, and Yun Ling. "Intrusion detection in cyber-physical systems: Techniques and challenges"

2.6 HISTORY OF CYBER SECURITY ISSUES IN CPS MEDICAL DEVICES

As mentioned earlier, most of the protocols used in Industrial Control Systems predated the Internet, and they have been adapted to operate over standard Ethernet link layer with UPD or TCP as transports. The networking has resulted in exposing the physical systems that were air-gapped in the past, to cybersecurity threats. The primary focus area for many of the ICS or Medical devices was reliability and not cybersecurity. The industry has been very vigilant about cybersecurity in the last decade, which has led to the discovery of many fundamental vulnerabilities. One of the challenges is the delaying of patching, even after finding the vulnerabilities and fixes. The systems, especially the IMDs (Implantable Medical Devices), are difficult to patch.

Cyber threats have been evolving for decades, and many have hit the headlines because of their impact. The Morris Worm in 1998 was the first worm to hit the Internet and cause massive disruption. The Stuxnet worm in 2010 was the first attack on an Industrial Control System across international borders. The 'Repository for Industrial Security Incidents' (RISI) has a database of incidents on Industrial systems from 1982 onward. The website lists over 180 incidents till 2014, starting with the 'Siberian Gas Pipeline Explosion caused by CIA Trojan virus' in 1982 and 'Union Carbide Chemical Leak in West Virginia' in 1985. Most of the attacks initially were opportunistic and tended to use simpler exploits. Still, with time, the attacks have become more sophisticated with the rise of hacktivism, cybercrime, cyberterrorism, as well as cyberwar. Though most of the attacks originate externally, there are many internal, unintentional incidents also. The unintentional incidents are mostly due to software bugs.

In 2016, Burns et al. published an article titled *A Brief Chronology of Medical Device Security*"[19] that laid out key milestones in the security of medical devices in the past few decades as CPS systems grew by leaps and bounds. The article divides events in the time between 1980 to the present day into four periods based on some inflection points. The first period from the 1980s to the present is the period when complex systems to automate things became more generally available and resulted in many accidental failures. The second period from the 2000s to the present is the period of Implantable Medical devices. The third period from 2006 is the period that introduced the threat of unauthorized access to devices that could cause harm. The fourth and the final period from 2012 to present is the era of the cyber threat to medical device security. Noted below are the highlights from these four periods taken from the article.

2.6.1 Period 1: 1980s–Present – Availability of Complex Systems and Accidental Failures

- **1985 June to 1987 January**: Software glitches in Therac-25 accelerators caused therapists to administer extra radiation, which resulted in six patients receiving harmful levels of radiation.
- **2002 November 13**: A researcher accidentally flooded BIDMC (Beth Israel Deaconess Medical Center) network with data. The critical systems became nonavailable, and the hospital was forced to pull the systems offline and go to a paper-based system.

2.6.2 Period 2: Implantable Medical Devices (2000–Present)

- **1990–2000**: FDA (Food and Drug Administration) recalled over 100,000 ICDs (Implantable Cardiac Defibrillators).
- **2001**: The number of IMDs exceeded 25 million units in the United States.
- **2005**: Death of a cardiac patient became headline news when an ICD short-circuited instead of giving an electric shock.

2.6.3 Period 3: Unauthorized Parties and Medical Devices (2006–Present)

- **2006**: Researches highlighted the reasons for the inability to patch security fixes on embedded devices. The key reasons were intermittent network connectivity and the devices being spread out.
- **2008**: Researchers were able to eavesdrop and even control the FDA approved ICDs.
- **2008**: In the "Reigel vs. Medtronic" case, the US Supreme Court ruled in favor of the medical device manufacturers. The manufacturers were not liable for harm caused if a device was FDA approved.
- **2011**: Researchers discovered vulnerabilities like eavesdropping and impersonation in implantable insulin pumps. Several high-profile incidents of active and passive attacks made it known to the public.
- **2011**: Researchers proposed the "Shield concept" where an RF (Radio Frequency) shield would act as a proxy for communications with IMDs.
- **August 4, 2011**: Jerome Radcliffe, a diabetic patient, exposed a vulnerability in his insulin pump by reverse engineering the communication. He took control of his insulin pump by using the ID of the device by connecting to it wirelessly from a distance of 100 feet wirelessly[10]. He presented his findings at the Black Hat conference in Las Vegas.
- **2011**: The Workshop on Health Security and Privacy at San Francisco where a successful update of custom firmware on CAED (Cardiac Automated External Defibrillator) was done[10]. This was possible because CAED did not verify the authenticity of the update.

2.6.4 Period 4: Cybersecurity of Medical Devices (2012–Present)

- **2012 February**: The main topic of discussion in ISPAB's (Information Security and Privacy Advisory Board) annual board meeting was about the cybersecurity of medical devices.
- **2012 October**: Barnaby Jack made a video presentation at Ruxcon Breakpoint Security Conference in Melbourne, Australia, where he showed the ability to deliver a deadly 830-volt shock to a pacemaker using a laptop 50 feet away[10].

- **2013 June**: The FDA released draft guidance to manage security in medical devices. They gave the final version in 2014.
- **2014**: A survey of all the past incidents revealed that the telemetry interface, especially the radio-based communication channels, has been the center of the threats.

2.7 SECURING MEDICAL MONITORING AND CONTROL SYSTEMS

Sections 4.6 and 4.7 clearly articulate the need for securing CPS. As pointed out by AlTawy and Youssef[10] and Cardenas et al.[20], patching and frequent updates are not well suited for CPS. Cardenas et al.[20] mention the updates in the context of comparing IT security vs. CPS security, while AlTawy and Youssef[10] talk about it in the context of IMDs. They go on to say that there is no systematic way of predicting the effect of a patch on the system or the individual, and many times the patient is better off without the patch since the risk of life is higher than the risk of being unsecure. This means that security must be a focal point as early as requirements and architecture in the product development life cycle. A system that is architected with security in place would be more robust than one without. Seifert and Reza[6] proposed a couple of methods by which existing architectures could be modified to be more secure. Of the different architectures reviewed in Section 4.3, the publish-n-subscribe and blackboard architectures were the most suited for CPS health-care. Seifert and Reza[6] proposed the EventGuard pattern, which was a modification of the publish-n-subscribe model and Secure Blackboard pattern, which was a modification of the Blackboard architecture.

2.7.1 Secure Blackboard Architecture

The Blackboard Architecture consists of a set of knowledge sources that write to a centralized shared device called the blackboard. A controller monitors the blackboard for updates and makes decisions on the actions that need to be taken based on a set of rules. As discussed earlier, this architecture scores well on all the quality metrics, including uncertainty, robustness, scalability, and so forth, since it is built to handle a variety of different redundant sensors. But one drawback in terms of security is that there is no authentication that determines whether a knowledge source can contribute information to the blackboard or not. Going one level beyond the basic authentication, there also needs to be a notion of what type of knowledge is permissible by a category of knowledge source. The Secure Blackboard architecture, as shown in Figure 2.6[6], which rectifies the issues by introducing three new components—an authenticator that validates the legitimacy of the knowledge source, a reference monitor to verify operation type using role-based access control mechanism, and a secure logger to log events.

Seifert and Reza[6] did an end-to-end analysis of the security issues in each of the architectures surveyed, using the STRIDE threat analysis process and DREAD threat ranking system. At each critical entry point in the system, the attack is placed on one of the six STRIDE categories, which stands for Spoofing, Tampering, Repudiation, Information Disclosure, Denial of Service, and Elevation of Privilege. The DREAD model gives a score, generally one to ten,

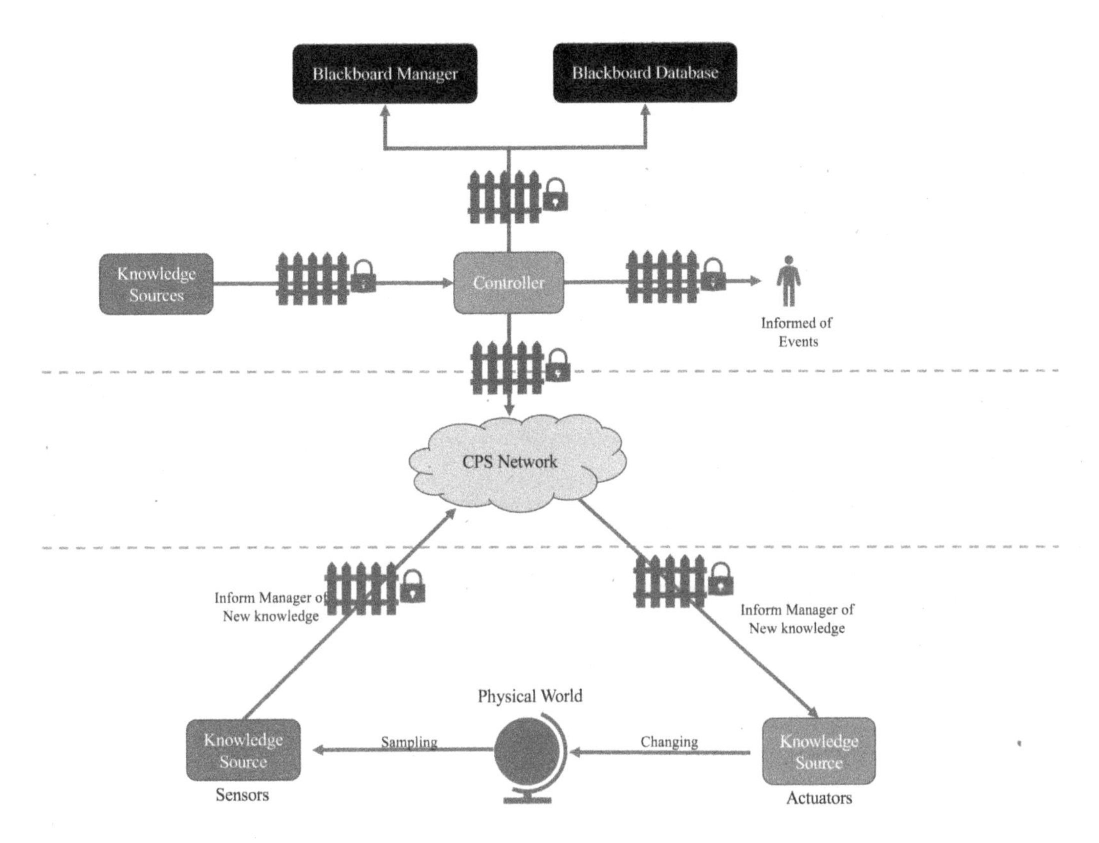

FIGURE 2.6 Secure blackboard architecture.

Source: Seifert, Darren, and Hassan Reza. "A security analysis of cyber-physical systems architecture for healthcare"

based on the following five factors—Damage, Reproducibility, Exploitability, Affected Users, and Discoverability. The conclusion was that the Blackboard architecture had low DREAD scores on all the categories of STRIDE in comparison to other architectures.

2.7.2 Existing Security Solutions in CPS Medical Devices

Čaušević et al.[7] provide a detailed description of security and privacy solution related to CPS healthcare focusing on WBANs (Wireless Body Area Networks). They divide the solutions into academic and commercial solutions. The academic solutions are further categorized as biometric, cryptographic, and IMD (Implantable Medical Devices) solutions.

2.7.2.1 Biometric Solutions

Biometric solutions take advantage of the fact that many functions of the body are like fingerprints, unique to that person. So, parameters like the timing of the heartbeat, gait or walking pattern, voice pattern, heat signature, and so forth or generating a hash using a combination of factors could be used to identify a person uniquely and to check if the readings from a specific device match the expected data.

Some of the examples given by Čaušević et al.[7] are BARI+, a biometric-based distributed key management approach, Wang et al.[42] proposed a biometric solution using ECG to secure communication and HMM (Hidden Markov Model) to recognize feature for authentication.

2.7.2.2 Cryptographic Solutions

Cryptographic solutions are based on using efficient cryptographic techniques to keep data encrypted while at rest and during transit and also using efficient and secure key sharing methodologies. Some of the examples given by Čaušević et al.[7] are—SecMed (Securing User Access to Medical Sensing Information) CRYPE, an attribute-based encryption system that achieves patient-related data access control BANA (body area network authentication scheme by Shi et al.; Adapt-lite, an approach that enables adaptive security mechanisms; MAACE (Mutual Authentication and Access Control scheme based on Elliptic Curve Cryptography), an energy-efficient security scheme based on ECC.

2.7.2.3 Security Solutions for Implantable Medical Devices

IMDs reside inside the body and provide critical functionality like heartbeat and insulin delivery, to name a couple. There are many research publications that have surveyed IMD security. AlTawy and Youssef's[10] research on the security tradeoffs in CPSs with a focus on IMDs provides a review of the security challenges and solutions. They cite the following as threats that IMDS are vulnerable to—Information harvesting, Tracking the patient, Impersonation, Relaying attacks, and Denial-of-service attacks[10]. They cite five challenges in security solutions adopted by IMDs—Critical physical environment, Constrained resources, Legacy compatibility, Bureaucracy, and Emergency authentication[10]. They surveyed the authentication protocols for IMDs and categorized them into four groups—Proximity-based, biometric-based, proxy-based, and hybrid approaches. Figure 2.7 shows these groups and the security solutions within each of them. The chapter discusses the strengths and weaknesses of each of the solutions.

2.7.2.4 Commercial Solutions

There are many existing commercial solutions for health monitoring, and there also has been an increase in startups focused on health monitoring. The success of devices like Fitbit, Apple Watch, Peloton, and so forth points to the direction of health monitoring and analytics being a lucrative area of investment. Čaušević et al.[7] list the following commercial solutions—Google Health by Google, HealthVault by Microsoft, Dossia by a group of companies including AT&T, Intel, and Wal-Mart, eCare Companion by Philips, and Health fitness application by Apple.

2.8 INTRUSION DETECTION IN CPS

2.8.1 Intrusion Detection System in the Context of CPS Security

Control Engineering conducted a survey in 2015[24] and published a cybersecurity report that asked respondents to list the top three resources used to monitor control systems.

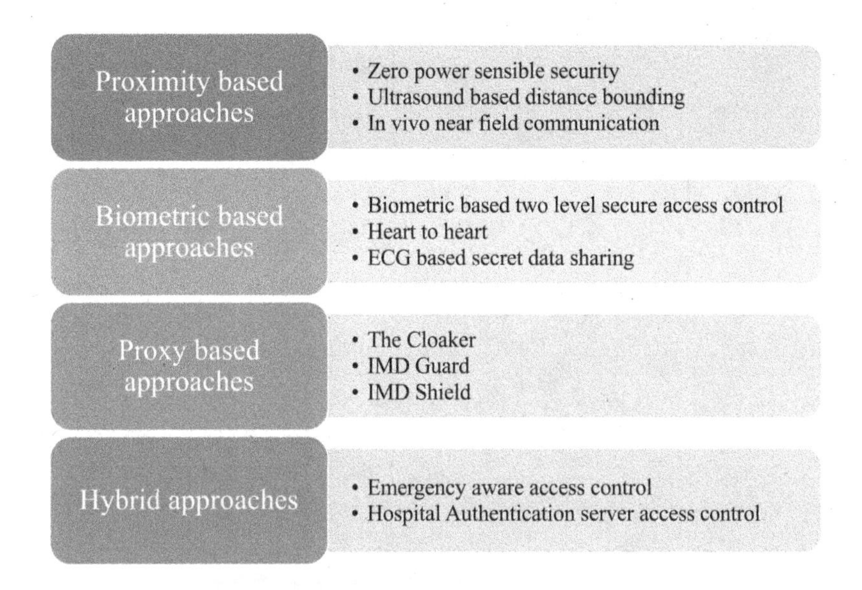

FIGURE 2.7 IMD authentication protocols.

About 99% said that they used Antivirus software, and 84% of the respondents said they used Intrusion detection and prevention systems. Figure 2.8 shows the top five systems used for securing Control Systems.

In simple terms, an IDS is a collection of hardware and software resources that can detect, analyze, and report indications of intrusions in computer systems and networks. According to the respondents, the top three advantages of using an IDS were detection,

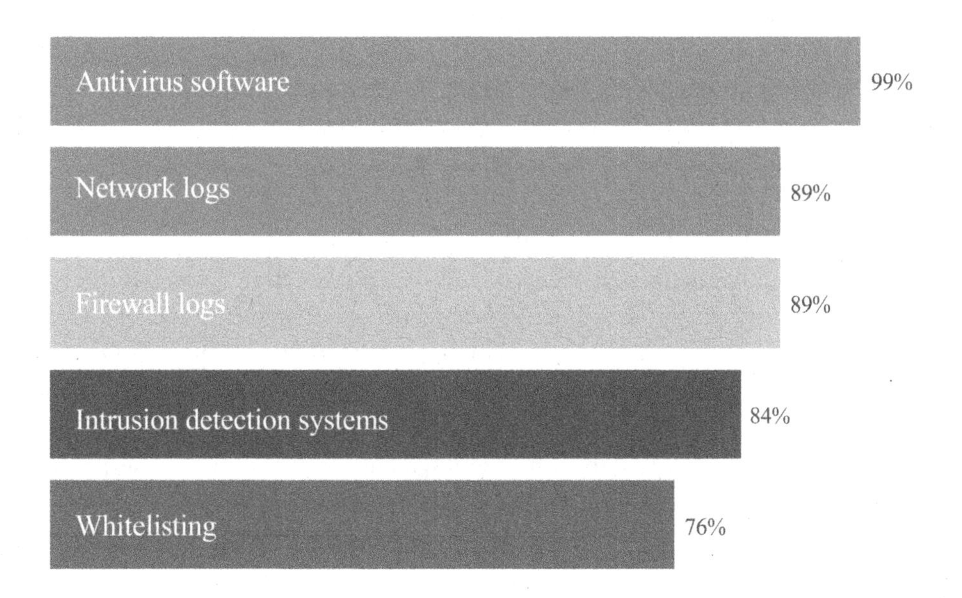

FIGURE 2.8 Resources used to monitor control systems.

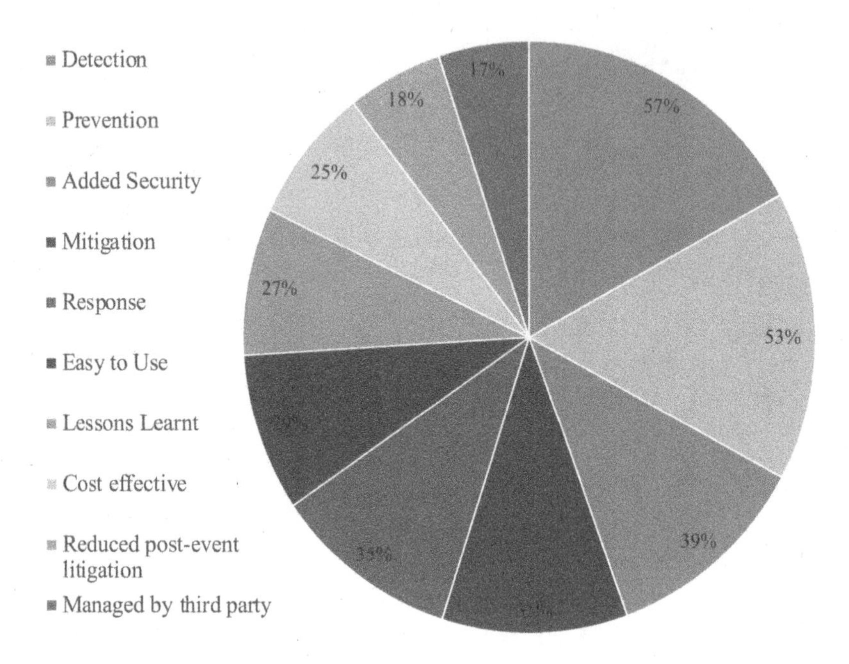

FIGURE 2.9 Advantages of using an intrusion detection system for CPS.

prevention, and added security. Figure 2.9 shows other advantages suggested by the survey.

As discussed in the "Security Threats" section, the effects of a security breach on CPS are catastrophic to life and infrastructure and hence, it is best to secure the systems from intrusions both known and new. Also, the need for quicker responses to signs of potential attacks is critical. The response could be active or passive. In active response scenarios, the system could block traffic in near real time. But the downside of this is that in case of false positives, genuine traffic could get blocked. In passive response scenarios, the system requires someone to investigate the alert and give an approval to go ahead. But this delays the action and if there is a security breach, it could cripple the system quickly.

2.8.2 Challenges of Implementing Intrusion Detection in CPS

In their paper about challenges in securing CPS, Cardenas et al.[20] compare IT security with CPS security, to highlight the difficulties of securing CPS systems, especially patching after a vulnerability is discovered[20]. They go on to mention that when it comes to Intrusion detection, CPS systems may be simpler than enterprise systems, as they exhibit comparatively more straightforward network dynamics—servers changes are rare, the topology is fixed, the user population is stable after deployment, the communication patterns are regular, and the protocols are limited. They also mention that when it comes to intrusion detection, control systems can provide a paradigm shift[20] since it might be possible to detect attacks that are undetectable in regular network traffic scanning, by monitoring the physical system. But the key thing to note is that traditional intrusion detection mechanisms could not be used in CPS without any changes mainly because of the physical aspect of the CPS. The extremely large-scale nature and heterogeneity of CPS require the intrusion detection mechanisms to be hybrid, instead of only one specific technique. There is also the issue of uncertainty since CPS could be deployed in unreliable environments and could get conflicting data from a collection of sensors. Though fast and straightforward detection techniques could be used in resource-constrained systems, in CPSs, the detection needs to be performed in a distributed and online manner simultaneously. If the system is part of multiple organizations, say many hospitals have connected devices that transmit data through the same network, then privacy protection becomes an issue. The data is not only distributed but also hierarchical and multitenant. Modern distributed IDSs lack a scalable privacy protection mechanism[13] and hence there is a need for hybrid mechanisms that combine several best practices. Cardenas et al.[20] proposed a hybrid mechanism, summarizing the countermeasures toward prevention, detection and recovery, resilience, and deterrence[20]. To summarize, the following could be listed as the challenges of implementing Intrusion Detection in CPS.

1. Physical aspects of the system
2. Large scale—CPS could be deployed across thousands of nodes over a large area
3. Heterogeneity—The same system could be deployed in multiple environments
4. Uncertainty
5. Preservation of privacy

2.8.3 Categories of Intrusion Detection Techniques

Han et al.[13] define three broad categories of IDSs[13]:

1. Misuse/Signature-based Intrusion Detection—In this type of IDS, a known threat has a signature, and detection is done if the signature of an event matches with the threat signature.
2. Anomaly-based Intrusion Detection—In this type of IDS, the comparison is between the definition of normal observed behavior versus significant deviations.
3. Stateful Protocol Analysis—In this, predetermined genuine activities of each protocol state are compared against the observed events to identify deviations.

Of the three techniques, Han et al.[13] recommend that Anomaly-based intrusion detection should be used as the major detection technique for CPS[13]. The recommendation is based on two reasons—firstly, anomaly-based detection requires a smaller amount of memory than signature-based since it does not need to store all the signatures; secondly, it is independent of any prior knowledge of the anomaly and thus useful to identify unknown attacks.

Mitchell and Chen also categorize CPS IDS techniques along the same lines but they go into a little more depth and create a hierarchical classification tree to organize the

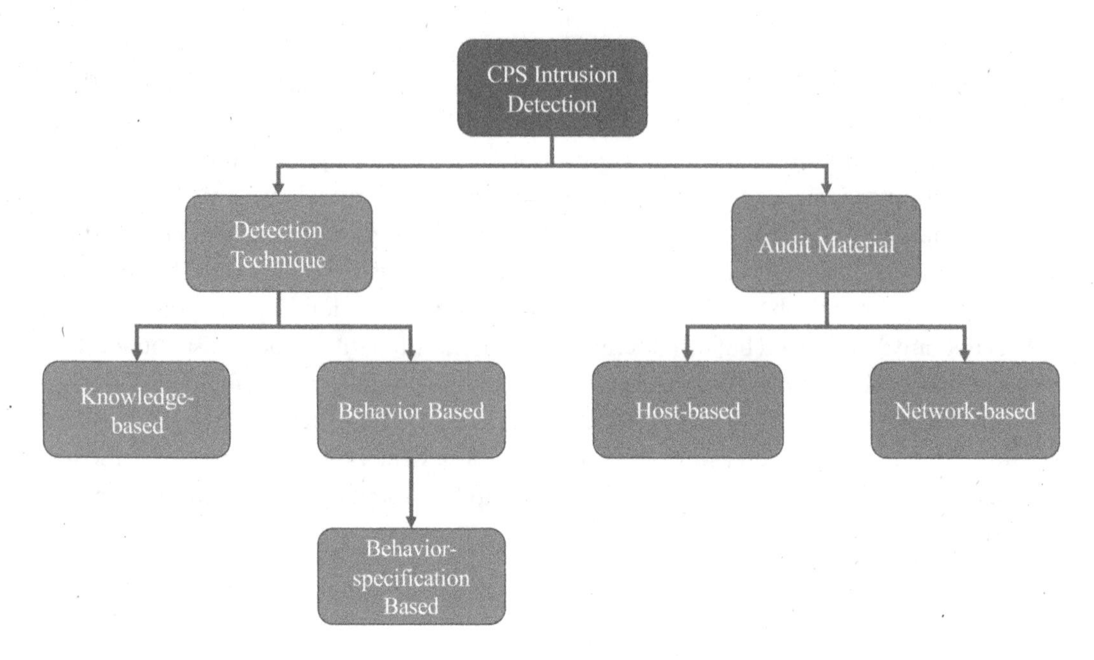

FIGURE 2.10 A classification tree for intrusion detection techniques for CPSs.

techniques[22]. Figure 2.10 shows the classification tree which is based on two dimensions—detection technique and audit material. Knowledge-based detection technique is same as the Signature-based IDS discussed earlier and behavior-based detection technique is same as Anomaly-based IDS discussed above. If data collection is the focus, it could be host-based or network-based. In host-based, logs from nodes are analyzed to see if they are any intrusions. In network-based IDS, the network activity is analyzed to determine if a node is compromised.

2.8.4 Design Outline of an Anomaly-Based Intrusion Detection System

Figure 2.11 shows the outline of an anomaly-based IDS. The training dataset is obtained by gather data from the logs when the system is functioning normally. After feature construction and preprocessing, the system creates a model that represents the average behavior of the system. Then, an outlier is detected if the test instance deviates from the normal profile by a predefined threshold. Based on the response, the system could enter into a fail-stop,

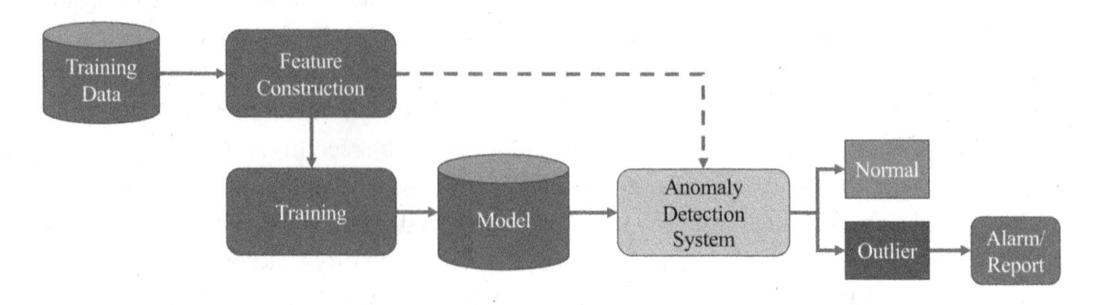

FIGURE 2.11 Design of an anomaly detection system.

fail-safe, fail-loud, or fail-quite modes. There could be a security audit that is done, and the results of that could be fed into the training data.

Han et al.[13] propose a few specific requirements for anomaly-based IDSs in CPS, and it ties in well with the challenges discussed earlier. They are that the system should be able to (a) handle the distributed nature of the deployments and data source, (b) handle the online and real-time nature of the data, (c) handle uncertainty by performing well against known and unknown attacks, (d) should be fault-tolerant, and (e) should not expose privacy[13].

2.9 MACHINE LEARNING TECHNIQUES FOR INTRUSION DETECTION IN CPS

2.9.1 Overview

A key thing to remember is that, it is not enough to just deploy an IDS and sit back, but the system needs to be actively fine-tuned to make sure of its effectiveness. Otherwise, the increase in false-positive percentages would give rise to high operational costs. Since CPS have numerous sensors that are deployed in multiple places, they generate a lot of data in real time. An example of the scale is given in the thesis on "*Reduction of False Positives in Intrusion Detection*" by Donald Burgio. It says that, in 2014, Hewlett Packard's corporate network generated between 100 billion to 1 trillion security events daily[23]. If an IDS deployed to detect intrusion is not effective and generates a lot of false positives, it would be very expensive to handle. In 2017, IBM estimated that yearly cost of handling false positives for organizations is around $1.3 million a year, and it wastes around 21,000 hours of human effort on average[23], and proposed using Artificial Intelligence as a solution. Machine Learning techniques, especially Classification techniques like Decision tree, Rule-based classifier, Bayesian classifier, Artificial Neural Network (ANN), Support Vector Machines (SVM), and so forth, and Clustering techniques like K-means, Hierarchical clustering, DBSCAN, and so forth, are extensively used to analyze CPS data. Also, the distributed nature of the data could mean that specific algorithm to handle specific situations should be used. Han et al.[13] give examples of using kernel density estimator (KDE), support vector machine (SVM), and clustering ellipsoids to handle distributed detection in CPS[13]. Least-squares SVM (LS-SVM) is a technique that could be applied effectively for online data. Feature extraction is another key technique that is important for CPS since the data may contain much noise due to the distributed nature and different types of environments. Burgio lists ML algorithms like ANN, Bayesian statistics, Gaussian Regression, Support Vector Machine (SVM), HMM, Decision Trees, and K-Nearest Neighbor (KNN), specifically for IDS research[23].

2.9.2 Categorization of Machine Learning Algorithms for CPS Security

Critical tasks like "discovering a pattern in existing data", "detecting outliers", "predicting values", and "feature extraction" are the focus areas of CPS security, and various Machine Learning algorithms provide support for those. Han et al.[13] classify the Machine Learning algorithms for IDS into three categories: (a) rule-based; (b) statistical; and (c) data mining/computational intelligence[13]. The rule-based technique, as the name suggests, is based on rules created using prior knowledge. If the incoming data is inconsistent with a rule, then

TABLE 2.5 Classification of Machine Learning Techniques for IDS

Technique Categories	Characteristics	Shortcoming
Rule-based	Fast, simple, online	Low generality, prior-knowledge dependent
Statistical	Average complexity, distributed, average generality	Prior-knowledge dependent or high complexity
DM and CI	Distributed, strong generality, prior-knowledge free	Highest complexity

an alert is created. Rule-based systems are usually fast and are computationally cheaper. But the major downside is the inflexibility around rule changes or lack of handling of incomplete data. Statistical techniques use statistical methodology to predict outliers and have a stronger ability to resist the incomplete and imprecise training data than the rule-based techniques. Lastly, data mining and computational intelligence technique use the methodology of learning the normal profile from multivariate training data with a supervised, semi-supervised, or unsupervised learning procedure. Table 2.5 summarizes the categorization.

Moh et al.[29] categorized machine learning algorithms for IoT based on different use-cases, and they are applicable for CPS security also[5]. Table 2.6 lists machine learning algorithms.

From the point of detecting outliers, the use-cases can be further divided into the following: (a) Malware detection, (b) Intrusion detection, and (c) Data anomaly detection.

Since "anomaly detection" is basically a Classification problem, it follows that the most used machine learning techniques are the ones that are commonly used in Classification, like Decision Tree, Bayesian Networks, Naïve Bayes, Random Forest, and Support Vector Machines (SVM). In many new instances, ANNs have been used. ANNs are generally not used for malware detection since it takes a longer time for training.

TABLE 2.6 Categorization of Machine Learning Solutions for IoT Security[5]

Use Case	Machine Learning Algorithm
Pattern Discovery	• K-means • DBSCAN
Discovery of Unusual Data Points	• Support Vector Machine • Random Forest • PCA • Naïve Bayes • KNN
Prediction of Values and Categories	• Linear Regression • Support Vector Regression • CART • FFNN
Feature Extraction	• PCA • CCA

TABLE 2.7 Categorization of Machine Learning Solutions for Outlier Detection

Use Case	Machine Learning Algorithm
Malware Detection	• SVM • Random Forest
Intrusion Detection	• PCA • Naïve Bayes • KNN
Anomaly Detection	• Naïve Bayes • ANN

The next section reviews examples of machine learning algorithms used for the use-cases in Table 2.7.

2.9.3 Chronology of Research on Machine Learning Techniques for IDS

The following section outlines some of the research as a chronology to show different areas of focus in the years leading up to the present. Bruneau notes that the original idea behind automated Intrusion Detection is credited to James P. Anderson who published a study to detect unauthorized access to computer systems[21]. The first model of a real-time IDS, named Intrusion Detection Expert System (IDES) was developed by Dorothy Denning and Peter Neumann between 1984 and 1986[21]. Burgio made a comprehensive list of research done on Machine Learning techniques used for IDS[23]. Though the current focus is more toward ANN, Deep Learning, and Extreme Learning Machines, over the past couple of decades, almost all the key Machine Learning techniques have been used to do Intrusion Detection at different network layers and systems.

The 1990s saw several Machine Learning techniques being used. Crosbie and Spafford[31] using Genetic Programming, Cannady[32] using ANN, Warrender et al.[33] using HMM are a few of the examples. The period of 2000s also witnessed a lot of research in this area. As more IoT devices came on the market, IDS became a critical component of security. Few examples of that era are—using of Self-Organizing Map by Rhodes et al.[35], Data Mining techniques by Julisch et al.[37], HMM by Ourston et al.[38], and ANN by Wang et al.[42]. 2010s saw a huge uptick in IoT devices and also research in this area. It was also a time big-data analysis and Machine Learning became more accessible due to several tools like Apache Mahout (2009), scikit-learn (2007), Apache Spark (2014), and Google Tensorflow (2015). From Wang et al.[42] who used ANN for intrusion detection, Hutchins et al.[48] who used kill-chain model for analysis of adversary campaigns, Sindhu et al.[49] who used Decision Tree for a light-weight IDS, Alam and Vuong[54] who applied Random Forest algorithm with multiple decision trees to detect malware, Ham et al.[56] who reviewed various approaches for detecting malware in Android devices and showed that Linear SVM demonstrated better performance, Pajouh et al.[61] who proposed a model for intrusion detection based on two-layer dimension reduction using Principal Component Analysis (PCA) and Linear Discriminate Analysis (LDA) and two-tier classification module using Naïve Bayes and K Nearest Neighbors (KNN), Kotenko et al.[58] who combined a Multi-Layered Perceptron Network (MLPN) along with a Probabilistic Neural Network (PNN) to predict the state of an IoT element, Canedo and Skjellum[60] who proposed using Artificial Neural Network (ANN) in gateway and application layers of an IoT network to detect differences between valid and invalid data, to Tama and Rhee[67] who used Boosting for an in-depth experimental study of anomaly detection, there have been significant research done in Intrusion detection in recent time. The following list enumerates most of the research for IDS in the past decades.

1990s

- 1995—Crosbie and Spafford—*Applying genetic programming to intrusion detection.* Used Genetic Algorithm[31].
- 1998—Cannady—*Artificial neural networks for misuse detection.* Used Artificial Neural Network (ANN)[32].
- 1999—Warrender et al.—*Detecting intrusions using system calls.* Used Hidden Markov Model (HMM)[33].
- 1999—Bass—*Intrusion detection systems and multisensor data fusion: creating cyberspace situational awareness.* Used ensemble ML with Situational awareness[34].

2000s

- 2000—Rhodes et al.—*Multiple self-organizing maps for intrusion detection.* Used Self-Organizing Map[35].
- 2001—Jajodia et al.—*Detecting intrusion by data mining.* Used Data Mining techniques[36].
- 2002—Julisch and Dacier—*Mining intrusion detection alarms for actionable knowledge.* Used Data Mining techniques[37].
- 2003—Ourston et al.—*Applications of hidden Markov models to detecting multi-stage network attacks.* Used Hidden Markov Model (HMM)[38].
- 2003—Lin et al.—*Performance analysis of pattern classifier combination by plurality voting.* Used Ensemble Classifiers[39].
- 2004—Wright et al.—*HMM profiles for network traffic classification.* Used Hidden Markov Model (HMM)[40].

- 2005—Depren et al.—*An intelligent intrusion detection system (IDS) for anomaly and misuse detection in computer networks.* Used Self-Organizing Map[41].
- 2007—Wang et al.—*Network security situation awareness based on heterogeneous multi-sensor data fusion and neural network.* Used Artificial Neural Network (ANN)[42].
- 2008—Liang et al.—*A novel stochastic modeling method for network security situational awareness.* Used Hidden Markov Model (HMM)[43].
- 2009 - Tsai et al.—*Intrusion detection by machine learning: a review.* Used Artificial Neural Network (ANN), Genetic Algorithm, Particle Swarm Optimization, Ant Colony Optimization[44].
- 2009—Perdisci et al.—*McPAD: A multiple classifier system for accurate payload-based anomaly detection.* Used Support Vector Machines (SVM)[45].
- 2009—Hu et al.—*A simple and efficient hidden Markov model scheme for host-based anomaly intrusion detection.* Used Hidden Markov Model (HMM)[46].

2010s

- 2010—Wang et al.—*A new approach to intrusion detection using artificial neural networks and fuzzy clustering.* Used Artificial Neural Network (ANN)[47].
- 2011—Hutchins et al.—*Intelligence-driven computer network defense informed by analysis of adversary campaigns and intrusion kill chains.* Used Kill-chain Model[48].
- 2012—Sindhu et al.—*Decision tree-based light weight intrusion detection using a wrapper approach.* Used Decision Tree[49].
- 2012—Koc et al.—*A network intrusion detection system based on a hidden Naïve Bayes multiclass classifier.* Used Naïve Bayes[50].
- 2012—Hoque et al.—*An implementation of intrusion detection system using genetic algorithm.* Used Genetic Algorithm[51]
- 2012—Cheng et al.—*Extreme learning machines for intrusion detection.* Used Extreme Learning Machines (ELM)[52].
- 2013—Zong et al.—*Weighted extreme learning machine for imbalance learning.* Used Extreme Learning Machines (ELM)[53].
- 2013—Alam and Vuong—*Random forest classification for detecting android malware.* Used Random Forest[54].
- 2013—Wong et al.—*Semi-supervised learning BitTorrent traffic detection.* Used K-Means algorithm for classification[30].

- 2014—Creech and Jiankun—*A semantic approach to host-based intrusion detection systems using contiguous and discontiguous system call patterns.* Used Extreme Learning Machines (ELM)[55].
- 2014—Ham et al.—*Linear SVM-based android malware detection for reliable IoT services.* Used Linear SVM[56].
- 2015—Shah et al.—*Analysis of machine learning techniques for intrusion detection system: a review.* Used Multi-Layer Perceptron (MLP)[57].
- 2015—Kotenko et al.—*Neural network approach to forecast the state of the Internet of Things elements.* Used Artificial Neural Network (ANN)[58].
- 2016—Chen et al.—*Anomaly network intrusion detection using hidden Markov model.* Used Hidden Markov Model (HMM)[59].
- 2016—Canedo and Skjellum—*Using machine learning to secure IoT systems.* Used Artificial Neural Network (ANN)[60]
- 2016—Pajouh et al.—*A two-layer dimension reduction and two-tier classification model for anomaly-based intrusion detection in IoT backbone networks.* Used Principal Component Analysis (PCA), Linear Discriminate Analysis (LDA) and a two-tier classification module using Naïve Bayes and KNN[61]
- 2016—Mehetrey et al.—*Collaborative ensemble-learning based intrusion detection systems for clouds.* Used Ensemble Learning, Decision Tree, Collaborative Filtering[28].
- 2016—Moh et al.—*Detecting web attacks using multi-stage log analysis.* Used Bayes Nets (Bayesian Belief Networks) for classification[29].
- 2017—Bamakan et al.—*Ramp loss K-support vector classification-regression; a robust and sparse multiclass approach to the intrusion detection problem.* Used Support Vector Machines (SVM)[62].
- 2018—Blanco et al.—*Tuning CNN input layout for IDS with genetic algorithms.* Used Deep Learning, Convolutional Neural Networks[63].
- 2018—Muna et al.—*Identification of malicious activities in Industrial Internet of Things based on deep learning models.* Used Deep Learning[64].
- 2018—Ross et al.—*Multi-source data analysis and evaluation of machine learning techniques for SQL injection detection.* Used J48 rule-based algorithm, the JRip rule-based algorithm[27].
- 2019—Vinayakumar et al.—*Deep learning approach for intelligent intrusion detection system.* Used Deep Learning[65].
- 2019—Igbe—*Artificial immune system-based approach to cyber attack detection.* Used Negative Selection Algorithm, Dendritic Cell Algorithm[66].

- 2019—Yen and Moh—*Intelligent log analysis using machine and deep learning.* Discusses about Recurrent Neural Networks (RNN), Multilayer Perception (MLP), K-Means algorithms[26].
- 2019—Yen et al.—*Semi-supervised log anomaly detection through sequence modeling.* Proposed CausalConvLSTM for modeling log sequences, used Convolutional Neural Network (CNN) and Long Short-Term Memory (LSTM)[25].
- 2019—Tama and Rhee—*An in-depth experimental study of anomaly detection using gradient boosted machine.* Used Boosting[67].

2.10 CONCLUSION

This chapter presented an overview of CPS, its architectural models, and its security and privacy issues. It reviewed four different architectures and derived the quality requirements to measure the effectiveness of each architecture from the healthcare domain perspective. The chapter also reviewed the communication protocols, and went into the security issues, including the existing CPS health monitoring systems, and threats and challenges for CPS systems. The chapter further described the history of CPS security issues to be aware of the challenges faced in the past and to showcase the evolution. Finally, it dwelled into securing medical monitoring and control systems and described intrusion detection on CPS using various machine learning techniques.

REFERENCES

1. Rajkumar, Ragunathan, Insup Lee, Lui Sha, and John Stankovic. "Cyber-physical systems: the next computing revolution". In *Design Automation Conference*, pp. 731–736. IEEE, 2010.
2. Kurzweil, Ray. *The singularity is near: When humans transcend biology.* Penguin, 2005.
3. "World Internet Users Statistics and 2019 World Population Stats". *Internet World Stats.* Accessed June 30, 2019. https://www.internetworldstats.com/stats.htm.
4. "Facebook Reports Second Quarter 2019 Results". *Facebook.* Accessed July 24, 2019. https://investor.fb.com/investor-news/press-release-details/2019/Facebook-Reports-Second-Quarter-2019-Results/default.aspx.
5. Moh, Melody, and Robinson Raju. "Using machine learning for protecting the security and privacy of Internet of Things (IoT) systems". *Fog and Edge Computing: Principles and Paradigms* (2019): 223–257.
6. Seifert, Darren, and Hassan Reza. "A security analysis of cyber-physical systems architecture for healthcare". *Computers* 5, no. 4 (2016): 27.
7. Čaušević, Aida, Hossein Fotouhi, and Kristina Lundqvist. "Data Security and Privacy in Cyber-Physical Systems for Healthcare". *Security and Privacy in Cyber-Physical Systems: Foundations, Principles and Applications* (2017): 305–326.
8. Eippert, Jill MD. "News: Apple Watch Saves the Day (and a Life)". *Emergency Medicine News:* 41, no. 11 (November 2019): 29. doi: 10.1097/01.EEM.0000604536.28864.70.
9. Harari, Yuval Noah. *Sapiens: A brief history of humankind.* Random House, 2014.
10. AlTawy, Riham, and Amr M. Youssef. "Security tradeoffs in cyber physical systems: A case study survey on implantable medical devices". *IEEE Access* 4 (2016): 959–979.

11. Gunes, Volkan, Steffen Peter, Tony Givargis, and Frank Vahid. "A survey on concepts, applications, and challenges in cyber-physical systems". *KSII Transactions on Internet & Information Systems* 8, no. 12 (2014).
12. Raju, Robinson, Melody Moh, and Teng-Sheng Moh. "Compression of wearable body sensor network data using improved two-threshold-two-divisor data chunking algorithms". In *2018 International Conference on High Performance Computing & Simulation (HPCS)*, pp. 949–956. IEEE, 2018.
13. Han, Song, Miao Xie, Hsiao-Hwa Chen, and Yun Ling. "Intrusion detection in cyber-physical systems: Techniques and challenges". *IEEE systems journal* 8, no. 4 (2014): 1052–1062.
14. Knapp, Eric D., and Joel Thomas Langill. *Industrial Network Security: Securing Critical Infrastructure Networks for Smart Grid, SCADA, and Other Industrial Control Systems.* Syngress, 2014.
15. Henze, Martin, Jens Hiller, René Hummen, Roman Matzutt, Klaus Wehrle, and Jan Henrik Ziegeldorf. "Network Security and Privacy for Cyber-Physical Systems". *Security and Privacy in Cyber-Physical Systems: Foundations, Principles and Applications, H. Song, GA Fink, and S. Jeschke (Eds.)* (2017): 25–48.
16. Bou-Harb, Elias, Nasir Ghani, Abdelkarim Erradi, and Khaled Shaban. "Passive inference of attacks on CPS communication protocols". *Journal of Information Security and Applications* 43 (2018): 110–122.
17. Farahani, Bahar, Farshad Firouzi, Victor Chang, Mustafa Badaroglu, Nicholas Constant, and Kunal Mankodiya. "Towards fog-driven IoT eHealth: Promises and challenges of IoT in medicine and healthcare". *Future Generation Computer Systems* 78 (2018): 659–676.
18. Haque, Shah Ahsanul, Syed Mahfuzul Aziz, and Mustafizur Rahman. "Review of cyber-physical system in healthcare". *International Journal of Distributed Sensor Networks* 10, no. 4 (2014): 217415.
19. Burns, A. J., M. Eric Johnson, and Peter Honeyman. "A brief chronology of medical device security". *Communications of the ACM* 59, no. 10 (2016): 66–72.
20. Cardenas, Alvaro, Saurabh Amin, Bruno Sinopoli, Annarita Giani, Adrian Perrig, and Shankar Sastry. "Challenges for securing cyber physical systems". In *Workshop on Future Directions in Cyber-Physical Systems Security,* 5, no. 1 (2009).
21. Bruneau, Guy. "The history and evolution of intrusion detection". *SANS Institute* 1 (2001).
22. Mitchell, Robert, and Ing-Ray Chen. "A survey of intrusion detection techniques for cyber-physical systems". *ACM Computing Surveys (CSUR)* 46, no. 4 (2014): 1–29.
23. Burgio, Donald A. "Reduction of False Positives in Intrusion Detection Based on Extreme Learning Machine with Situation Awareness". (2019).
24. Pelliccione, Amanda. "2015 Cyber Security Study." *Control Engineering.* (May 2015).
25. Yen, Steven, Melody Moh, and Teng-Sheng Moh. "CausalConvLSTM: Semi-Supervised Log Anomaly Detection Through Sequence Modeling". In *2019 18th IEEE International Conference on Machine Learning and Applications (ICMLA)*, pp. 1334–1341. IEEE, 2019.
26. Yen, Steven, and Melody Moh. "Intelligent log analysis using machine and deep learning". In *Machine Learning and Cognitive Science Applications in Cyber Security*, pp. 154–189. IGI Global, 2019.
27. Ross, Kevin, Melody Moh, Teng-Sheng Moh, and Jason Yao. "Multi-source data analysis and evaluation of machine learning techniques for SQL injection detection". In *Proceedings of the ACMSE 2018 Conference*, pp. 1–8. 2018.
28. Mehetrey, Poonam, Behrooz Shahriari, and Melody Moh. "Collaborative ensemble-learning based intrusion detection systems for clouds". In *2016 International Conference on Collaboration Technologies and Systems (CTS)*, pp. 404–411. IEEE, 2016.
29. Moh, Melody, Santhosh Pininti, Sindhusha Doddapaneni, and Teng-Sheng Moh. "Detecting web attacks using multi-stage log analysis". In *2016 IEEE 6th International Conference on Advanced Computing (IACC)*, pp. 733–738. IEEE, 2016.

30. Wong, R., Teng-Sheng Moh, Melody Moh, and Qurban A. Memon. "Semi-Supervised Learning BitTorrent Traffic Detection". *Distributed Network Intelligence, Security and Applications, Qurban A. Memon (ed.)*, CRC Press-Taylor & Francis Group, USA, 2013.
31. Crosbie, Mark, and Gene Spafford. "Applying genetic programming to intrusion detection". In *Working Notes for the AAAI Symposium on Genetic Programming*, pp. 1–8. Cambridge, MA: MIT Press, 1995.
32. Cannady, James. "Artificial neural networks for misuse detection". In *National Information Systems Security Conference*, 26 (1998): 443–456.
33. Warrender, Christina, Stephanie Forrest, and Barak Pearlmutter. "Detecting intrusions using system calls: Alternative data models". In *Proceedings of the 1999 IEEE Symposium on Security and Privacy (Cat. No. 99CB36344)*, pp. 133–145. IEEE, 1999.
34. Bass, Tim. "Intrusion detection systems and multisensor data fusion". *Communications of the ACM* 43, no. 4 (2000): 99–105.
35. Rhodes, Brandon Craig, James A. Mahaffey, and James D. Cannady. "Multiple self-organizing maps for intrusion detection". In *Proceedings of the 23rd National Information Systems Security Conference*, pp. 16–19. 2000.
36. Jajodia, Sushil, Daniel Barbará, Julia Couto, Ningning Wu, and Leonard Popyack. "ADAM: Detecting intrusions by data mining". In *Workshop on Information Assurance and Security*, 1 (2001): 1100.
37. Julisch, Klaus, and Marc Dacier. "Mining intrusion detection alarms for actionable knowledge". In *Proceedings of the Eighth ACM SIGKDD International Conference on Knowledge Discovery and Data Mining*, pp. 366–375. 2002.
38. Ourston, Dirk, Sara Matzner, William Stump, and Bryan Hopkins. "Applications of hidden Markov models to detecting multi-stage network attacks". In *Proceedings of the 36th Annual Hawaii International Conference on System Sciences*. pp. 10-pp. IEEE, 2003.
39. Lin, Xiaofan, Sherif Yacoub, John Burns, and Steven Simske. "Performance analysis of pattern classifier combination by plurality voting". *Pattern Recognition Letters* 24, no. 12 (2003): 1959–1969.
40. Wright, Charles, Fabian Monrose, and Gerald M. Masson. "HMM profiles for network traffic classification". In *Proceedings of the 2004 ACM Workshop on Visualization and Data Mining for Computer Security*, pp. 9–15. 2004.
41. Depren, Ozgur, Murat Topallar, Emin Anarim, and M. Kemal Ciliz. "An intelligent intrusion detection system (IDS) for anomaly and misuse detection in computer networks". *Expert Systems with Applications* 29, no. 4 (2005): 713–722.
42. Wang, Huiqiang, Xiaowu Liu, Jibao Lai, and Ying Liang. "Network security situation awareness based on heterogeneous multi-sensor data fusion and neural network". In *Second International Multi-Symposiums on Computer and Computational Sciences (IMSCCS 2007)*, pp. 352–359. IEEE, 2007.
43. Liang, Y., H. Q. Wang, H. B. Cai, and Y. J. He. "A novel stochastic modeling method for network security situational awareness". In *2008 3rd IEEE Conference on Industrial Electronics and Applications*, pp. 2422–2426. IEEE, 2008.
44. Tsai, Chih-Fong, Yu-Feng Hsu, Chia-Ying Lin, and Wei-Yang Lin. "Intrusion detection by machine learning: A review". *Expert Systems with Applications* 36, no. 10 (2009): 11994–12000.
45. Perdisci, Roberto, Davide Ariu, Prahlad Fogla, Giorgio Giacinto, and Wenke Lee. "McPAD: A multiple classifier system for accurate payload-based anomaly detection". *Computer Networks* 53, no. 6 (2009): 864–881.
46. Hu, Jiankun, Xinghuo Yu, Dong Qiu, and Hsiao-Hwa Chen. "A simple and efficient hidden Markov model scheme for host-based anomaly intrusion detection". *IEEE Network* 23, no. 1 (2009): 42–47.
47. Wang, Gang, Jinxing Hao, Jian Ma, and Lihua Huang. "A new approach to intrusion detection using Artificial Neural Networks and fuzzy clustering". *Expert Systems with Applications* 37, no. 9 (2010): 6225–6232.

48. Hutchins, Eric M., Michael J. Cloppert, and Rohan M. Amin. "Intelligence-driven computer network defense informed by analysis of adversary campaigns and intrusion kill chains". *Leading Issues in Information Warfare & Security Research* 1, no. 1 (2011): 80.
49. Sindhu, Siva S. Sivatha, Suryakumar Geetha, and Arputharaj Kannan. "Decision tree based light weight intrusion detection using a wrapper approach". *Expert Systems with Applications* 39, no. 1 (2012): 129–141.
50. Koc, Levent, Thomas A. Mazzuchi, and Shahram Sarkani. "A network intrusion detection system based on a Hidden Naïve Bayes multiclass classifier". *Expert Systems with Applications* 39, no. 18 (2012): 13492–13500.
51. Hoque, Mohammad Sazzadul, Md Mukit, Md Bikas, and Abu Naser. "An implementation of intrusion detection system using genetic algorithm". *arXiv preprint arXiv:1204.1336* (2012).
52. Cheng, Chi, Wee Peng Tay, and Guang-Bin Huang. "Extreme learning machines for intrusion detection". In *The 2012 International Joint Conference on Neural Networks (IJCNN)*, pp. 1–8. IEEE, 2012.
53. Zong, Weiwei, Guang-Bin Huang, and Yiqiang Chen. "Weighted extreme learning machine for imbalance learning". *Neurocomputing* 101 (2013): 229–242.
54. Alam, Mohammed S., and Son T. Vuong. "Random forest classification for detecting android malware". In *2013 IEEE International Conference on Green Computing and Communications and IEEE Internet of Things and IEEE Cyber, Physical and Social Computing*, pp. 663–669. IEEE, 2013.
55. Creech, Gideon, and Jiankun Hu. "A semantic approach to host-based intrusion detection systems using contiguousand discontiguous system call patterns". *IEEE Transactions on Computers* 63, no. 4 (2013): 807–819.
56. Ham, Hyo-Sik, Hwan-Hee Kim, Myung-Sup Kim, and Mi-Jung Choi. "Linear SVM-based android malware detection for reliable IoT services". *Journal of Applied Mathematics* 2014 (2014).
57. Shah, Asghar Ali, Malik Sikander Hayat, and Muhammad Daud Awan. *Analysis of Machine Learning Techniques for Intrusion Detection System: A Review*. Infinite Study, 2015.
58. Kotenko, Igor, Igor Saenko, Fadey Skorik, and Sergey Bushuev. "Neural network approach to forecast the state of the Internet of Things elements". In *2015 XVIII International Conference on Soft Computing and Measurements (SCM)*, pp. 133–135. IEEE, 2015.
59. Chen, Chia-Mei, Dah-Jyh Guan, Yu-Zhi Huang, and Ya-Hui Ou. "Anomaly network intrusion detection using hidden Markov model". *International Journal of Innovative Computing, Information and Control* 12 (2016): 569–580.
60. Canedo, Janice, and Anthony Skjellum. "Using machine learning to secure IoT systems". In *2016 14th Annual Conference on Privacy, Security and Trust (PST)*, pp. 219–222. IEEE, 2016.
61. Pajouh, Hamed Haddad, Reza Javidan, Raouf Khayami, Dehghantanha Ali, and Kim-Kwang Raymond Choo. "A two-layer dimension reduction and two-tier classification model for anomaly-based intrusion detection in IoT backbone networks". *IEEE Transactions on Emerging Topics in Computing* 7, no. 2 (2016): 314–323.
62. Bamakan, Seyed Mojtaba Hosseini, Huadong Wang, and Yong Shi. "Ramp loss K-Support Vector Classification-Regression; a robust and sparse multi-class approach to the intrusion detection problem". *Knowledge-Based Systems* 126 (2017): 113–126.
63. Blanco, Roberto, Juan J. Cilla, Pedro Malagón, Ignacio Penas, and José M. Moya. "Tuning CNN input layout for IDS with genetic algorithms". In *International Conference on Hybrid Artificial Intelligence Systems*, pp. 197–209. Springer, Cham, 2018.
64. Muna, AL-Hawawreh, Nour Moustafa, and Elena Sitnikova. "Identification of malicious activities in Industrial Internet of Things based on deep learning models". *Journal of Information Security and Applications* 41 (2018): 1–11.

65. Vinayakumar, R., Mamoun Alazab, K. P. Soman, Prabaharan Poornachandran, Ameer Al-Nemrat, and Sitalakshmi Venkatraman. “Deep learning approach for intelligent intrusion detection system”. *IEEE Access* 7 (2019): 41525–41550.
66. Igbe, Obinna. “Artificial Immune System Based Approach to Cyber Attack Detection”. PhD dissertation, The City College of New York, 2019.
67. Tama, Bayu Adhi, and Kyung-Hyune Rhee. “An in-depth experimental study of anomaly detection using gradient boosted machine”. *Neural Computing and Applications* 31, no. 4 (2019): 955–965.

CHAPTER 3

The Future of Privacy and Trust on the Internet of Things (IoT) for Healthcare

Concepts, Challenges, and Security Threat Mitigations

Anastasios N. Bikos and Nicolas Sklavos

CONTENTS

3.1 INTRODUCTION

The Internet of Things (IoT) is interconnecting more and more devices every day, and we are headed for a cyber-world that will have 24 billion IoT smart gadgets by 2020. As defined by the Casagras European Union Framework[1], the context of the Internet of Things is "a global network infrastructure, linking physical and virtual objects through the exploitation of data capture and communication capabilities. This infrastructure includes existing and involving Internet and network developments. It will offer specific object-identification, sensor, and connection capability as the basis for the development of independent cooperative services and instrumentations. These will be characterized by a high degree of autonomous data capture, event transfer, network connectivity, and interoperability". Specifically, the IoT is a novel technology trend that is vastly gaining significant popularity based on modern wireless telecommunications. The essential nucleus idea of this concept is the pervasive presence around in our daily world of a numerous of things, or objects—such as Radio-Frequency IDentification (RFID) tags, sensors, actuators, mobile phones, etc.—that can interconnect, communicate, and collaborate for standard functionality. Nowadays, IoT links the Internet with sensors and a multitude of devices, mostly using internet protocol (IP)-based connectivity. Unquestionably, the main benefit of the IoT idea is the high impact it will have on several aspects of everyday life and behavior of potential users, from the private and public sector to our individual life domain[17]. One such area of exploitation is Healthcare.

In the healthcare industry, IoT provides options for remote monitoring, early prevention, and medical treatment for disabled patients. Inside the scope of IoT context, people or objects can be equipped with sensors, actuators, RFID tags, etc. Such wearable devices can facilitate the access by patient's caregivers and hospital facilities. For instance, RFID tags of patient's medical devices can be readable, geolocatable, and controllable by IoT software frameworks. IoT truly enables a wide range of smart applications and services to tackle with challenges that individuals or healthcare sector faces. Since there is, undoubtedly, increasing consciousness and reliability of consumers with regards to their health standards, demand for remote patient applications, digital healthcare along with better quality healthcare emerges as more popular today than ever before. Nowhere is this more exact than the use of such devices in the health and fitness industry. Personal fitness bands, Internet-connected insulin pumps, and patient tracking throughout an individual's hospital journey are just a few of the numerous examples of how IoT is currently used in the healthcare sector. One the one hand, however, this is only an early beginning for this industry. Still, on the other, the potential for further innovation and IoT instances in this domain is vast. Some analysts even predict that the sector will be worth some $117 bn (or £93.2 bn) by 2020. To that direction, a more integrated and IoT-enabled eHealth approach proves essential in all these areas. So far, there have been deployed hundreds of applied technologies for a more integrated and mature IoT-enabled eHealth reality. Concurrently, together with the emergence of IoT and Machine-to-Machine Type Communications, we are about to witness very soon, probably by 2020, the official globally available roll-out of 5G, or 5th Generation of mobile telecommunications network from 3GPP (3GPP 2019). 5G, due to its extremely high bandwidth (10 Gbps/sec), ultralow latency (1 msec) and dense connectivity (100 million devices per square km), will provide significant benefits to drive eHealth industry, for instance, remote medical operations and robotic surgery arms[28,29].

Within the overall interconnected healthcare and eHealth perspective, more integrated approaches and benefits are aimed for the so-called Internet of Healthcare Things (IoHT) or the Internet of Medical Things (IoMT). The period from 2017 to 2022 will be crucial in this transition, with significant changes before 2020, for example, 5G emergence. The applicability area of the previously mentioned specialized IoT in the medical industry is endless. The reason: healthcare is such a vast ecosystem, especially if we start considering personal healthcare, the pharmaceutical industry, healthcare insurance, RTHS, healthcare building facilities (hospitals), robotics, intelligent biosensors, smart pills, everything remote and the various healthcare specializations, activities, and even treatments of) diseases, with help of Machine Learning and Artificial Intelligence. Two things are currently for sure: (1) the main IoT use case in healthcare for the time being is (remote) health monitoring, certainly from an IoT financial perspective and (2) the IoT will soon be ubiquitous in healthcare and health-related activities and processes on various relevant levels. Nonetheless, there is significant growth ahead in a more Industrial IoT context, whereby healthcare providers, such as hospitals, IoT, in conjunction with scenarios and technologies in the areas of robotics, artificial intelligence, and Big Data.

Remote health monitoring and telehealth is perhaps the crucial IoT use case scenario on IoHT. In other words: today's primary use case from an IoT spending perspective is outside the setting considerations of a hospital or other healthcare facility. Remote health monitoring, which is very realistic thanks to the Internet of Things, also considerably helps solve the rise of chronic diseases, among others, due to an aging population. More advanced, currently deployable, eHealth scenarios include Big Data mining and processing of huge medical information for cure disease, virus prevention, and genetic research. Remote health monitoring is also ideal when patients or elders live in remote areas. There is a broad range of specialized wearables and biosensors that help to enable the previous benefits and potentials. The connected and "smart" hospital is another use case of IoT. Upcoming trends such as "smart" beds, the aggregation and real-time availability of data from healthcare devices and assets regarding specific patients, and the appearance of robots in a hospital environment for routine tasks, are some typical examples. Finally, there is growing intuition that eventually the Internet of Things will meet robotics in healthcare. In research by IDC, by 2019 there will be a 50% rise in the use of robots to deliver medications, supplies, and food throughout the hospital premises. Such infostructure will demand a fully interconnected and interoperating environment with critical information handling through sensors and intelligent devices, something that could be provided only by the Internet of Things technology.

This information-intensive industry inevitably gives birth to several challenges. According to the previously mentioned 2017 research by Aruba Networks on the area of IoT in healthcare, the leading IoT use case in healthcare organizations is monitoring and maintenance (73% of respondents), led by remote operation and control (50% of respondents). Connecting IoT devices is indeed an important goal to exploit the benefits from IoT, with 67% of survey respondents planning to connect their IoT devices using conventional Wi-Fi protocol. Interconnecting such devices, sharing life-critical or patient sensitive data, or even the handling of such devices that human life depends on their nominal functionality creates the possibility of malefic actions, data exploitation, or cybersecurity concerns. Besides the previously mentioned benefits raised from the proper usage of IoT in medicine, several "security" threats could be defined, as well. According, again, to the same previous research, 89% of IoT device users have suffered an IoT-related security breach, 49% recognize security malware as a significant threat for IoT, whereas 22% are afraid of Denial-Of-Service attacks on such gadgets.

The security-related domain is the one factor addressed from the deployment of IoT in the medical world. Indeed, there is a massive list of security issues associated with all mentioned networked medical devices. Currently, due perhaps to industry immaturity, or need for more profit by the companies, there is a commercial circulation of medical devices with untested, unpatched, or defective software and firmware. These risks can significantly lead to a lack of service and care delivery, sensitive data breach, loss, or destruction of medical data, or even cause effects from virtual harm to physical harm. Besides the theft or damage of expensive networked medical devices, there is currently known to exist a considerable lack of standards and reliable security protocols able to offer layers of durable protection to the IoT in health. As a result, security breaches are often reported, with unauthorized device setting changes, reprogramming, or infection via malware and denial-of-service attacks. As discussed earlier, the usage of Wi-Fi technologies to share, distribute and access patient data, monitoring systems, and implanted medical devices, could impose an obvious target for malefic attackers to compromise mobile health devices. These vulnerabilities, which primarily have not been addressed in healthcare before, in turn, can pose potential harm to patients. It is ubiquitous, like nowadays, that having passwords, encryption schemes, and the latest versions of hardware and software on a typical medical device is something not entirely apparent. Furthermore, securing IoT devices means more than merely defending the actual devices themselves. Due to the interconnected and super scaled nature of the IoT, there is the need to build security into software components and network connections that link to those devices, as well[4–6].

The privacy-related perspective is the second most significant issue for IoT in health. Undoubtedly, the raw amount of medical data that IoT devices can generate in hospitals and patient care centers is staggering. The massive data creation, distribution, and access gradually disregard the outlook of conventional information mining and gives way to the Big Data concept. "Big Data" is related to the aggregate collection of vast and complicated data sets, which overpasses available computational methods or systems. In healthcare, big data can help analyze several such data for prediction and medication purpose toward the well-being of patients. It also helps to be aware of the people's health for disease control and prognostic examination. One the other hand, this massive data availability creates more entry points for hackers and leaves sensitive information vulnerable. In a more recent study, it has been reported that 80% of damages for Cyberattacks involve a breach of privacy-related details, with 21% of subcases to belong to personal health information. Apart from the private and ethical concerns of stealing sensitive medical data from human patients, there is substantial reputational damage to companies and organizations in the field from these privacy breaches. It is evident, therefore, that Trust from consumers will be crucially challenged in the IoT eHealth commercial arena.

As a result, there is a straightforward demand not only for strong IT countermeasures to confront IoT security and privacy challenges, but also to well establish law legislations and data security laws in a national and global perspective. Better security standards, secure authentication or the most recent two-factor authentication system (2FA) should be installed to access patient records when a user is to provide auxiliary information to sign in (e.g., a retinal scan, phone text code, DNA sample, fingerprint, etc.), not just the login and password credentials. If this point is successfully implemented, hackers are granted fewer chances for malefic activities. To further minimize the risks of data breaches, robust encryption schemes should be adopted. Unlike software-based encryption solutions that exploit valuable computational device resources, hardware-level encryption, as a rule, does not have a tremendous impact on performance. Data transmission encryption, however, is no less significant. To minimize the chances of sensitive data being stolen, hospitals should control the boundaries of data access for gadgets by allowing them to use the internal Wi-Fi network (not an external one) and avoiding transmitting data via the cellular network. A final security solution, from a standardization perspective, is to build robust and flexible security frameworks to be adapted for the custom environments needs of the various medical centers. Together with training and cybersecurity awareness among hospital staff and patients, the security and privacy issues from the engagement of IoT in practical Healthcare could be tremendously mitigated.

This manuscript underlines and discusses the state of the art of the Security and Privacy challenges related to the IoT in the healthcare industry. While the list of IoT devices and "smart" software frameworks in medical applications are limitless, the authors contribute by depicting the most recent security vulnerabilities or lack of adequate protection that such industrial, medical products come along with. Security attack models and malefic activities performed by hackers against IoHT data and gadgets are also outlined. Furthermore, a strong emphasis is given for establishing secure protocols and standards for such devices, rather than depend on traditional generic wireless communication networks, together with an illustrative comparison of the most novel security frameworks that IoT in medical care are beginning to encompass. Finally, practical and theoretical security recommendations and policies are being proposed to minimize, as much as possible, the security "gaps" among IoT hardware and software modules in the Healthcare industry. The rest of the paperwork is organized as follows: Section 3.2 presents the most recent and widespread application frameworks (software, hardware, and middleware) that are currently deployed in IoT based healthcare. The authors depict examples of the "smart" IoT systems that hospitals and care centers are utilizing now, together with their benefits for the health system. In the next section, Section 3.3, a full list of significant security flaws and vulnerabilities of such devices and software environments is being projected, together with the most sophisticated attack models that hackers and malefic users aim to deploy. In Section 3.4, the authors perform a full separation of the Security concerns, on the one hand, and the Privacy issues, on the other, that could potentially affect human patients, hospital staff, and generally the health system, in case of malefic activities. Section 3.5 provides all possible security recommendations and countermeasures to minimize the effect of the previous risks, to build a more secure and reliable eHealth. Finally, Section 3.6 concludes this chapter.

The circumstance of smart Healthcare and IoT has met on-going effort from the research literature. Qi et al.[31] have explored various applicability scenarios on the IoT for eHealth, including blood-pressure monitoring, monitoring of oxygen saturation, and heartbeat performance. Finally, the authors have reviewed research work in the infrastructure and networking technologies to drive IoT for eHealth. Baker et al.[32] described a model of future IoT healthcare systems, specific components that could potentially operate into their order as well as short- and long-range communication protocols for smart eHealth. Mahmoud et al.[33] investigated Cloud of Things (CoT) platforms and how to deploy them into eHealthcare in terms of energy efficiency, portability, and performance. Dhanvijay and Patil[34] focused on IoT-based healthcare management systems that can enable data transmission and reception and studied the results of their research in terms of energy, power as well as security and privacy.

In particular, the contribution of this manuscript is to deliver a full context review analysis of the Internet of Everything for Healthcare and smart medical devices in terms of the security classes of *Security, Privacy, and Trust.* Future research directions and horizons for building a more secure IoT for eHealth are being illuminated.

3.2 INSIDE THE HEALTH IoT WORLD: HOW IT IS APPLIED TODAY AND BENEFITS FOR HEALTHCARE

3.2.1 The Vision of IoT

By browsing through the literature, an interested reader might have considerable difficulty to fully understand what IoT means and what its impact will be on our everyday lives in the upcoming decades. The social, economic, and technical implications from the full deployment of IoT, remain unclear. The term, IoT, by itself, is composed of two terms. The first leads toward a network-oriented vision of the IoT, whereas the second move the focus into generic and intelligent objects to be fully integrated into a common framework. In practical terms, "Internet of Things" semantically means "a world-wide network of interconnected objects uniquely addressable, based on standard communication protocols". This implies a considerable number of (heterogeneous) objects involved and meshed in the process[1]. From an architecture point of view, IoT consists of several layers and sublayers starting from the edge technology layer at the bottom to the application layer at the top, as shown in Figure 3.1. The two lower layers contribute to data capturing, while the two higher layers are mainly responsible for data utilization in software[7].

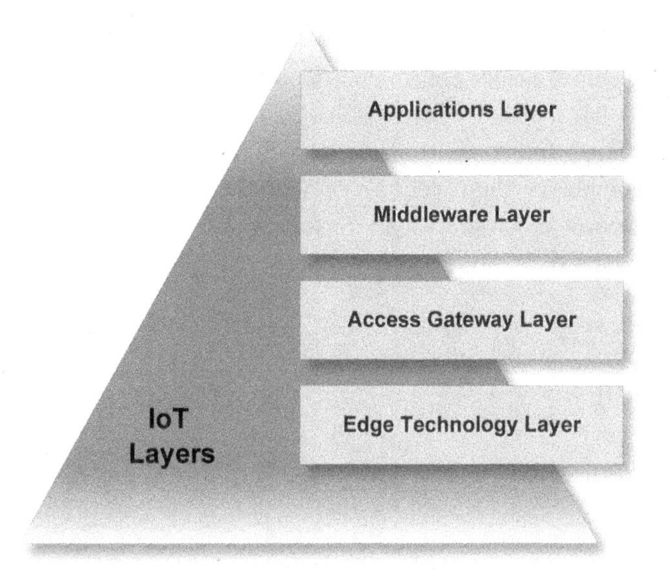

FIGURE 3.1 The IoT architecture layers.

The functions of the layers (from the bottom up) are as follows:

1. *Edge technology layer (or perception layer)*: This is basically the hardware layer, which includes data collection components such as wireless sensor networks (WSNs), RFID systems, cameras, intelligent terminals, electronic data interfaces (EDIs), and global positioning systems (GPS). These hardware components offer identification and information storage (i.e., via RFID tags), information collection (e.g., via smart sensor networks), information processing (i.e., via embedded edge processors), communications, control, and actuation (i.e., via robots).

Most common IoT technologies are:

1. *RFID systems:* They are the most critical components of the IoT system. They enable data transmission by an ultraportable device called RFID tag. The RFID reader obtains the tag and converts it into raw data, able to be manipulated by a specific software component[1].
2. *Wireless sensor networks (WSNs):* A WSN may consist of many sensing nodes, which report the sensing data results to special nodes (sinks).

   Today, most of the commercial wireless sensor network solutions are based on the IEEE 802.15.4 standard, which designs the physical and MAC layers for low-power, low bit-rate communications in wireless personal area networks (WPAN)[1].
3. *Access gateway layer (network layer or transport layer):* This layer is responsible for data handling, including data transmission, and message routing. It sends to the middleware layer all information received from the edge layer, using conventional communications technologies such as Wi-Fi, Li-Fi, Ethernet, GSM, WSN, and WiMax.

4. *Middleware layer:* The middleware is a software layer, or a set of sublayers interposed between the technological and the application levels. Its primary purpose is to perform information abstraction and to hide. The IoT technology often follows the Service Oriented Architecture (SOA) approach, inside its operations. The adoption of the SOA principles allows for decomposing complex and monolithic systems into software modules consisting of an environment of more straightforward and well-defined components[1].

5. *Applications layer:* This is the top layer. It is responsible for the delivery of various applications to different IoT users, with concurrent Quality of Service (QoS), cloud-computing technologies, data processing, machine-to-machine (M2M) services, along with any necessary program user interface.

Figure 3.2 illustrates the exact association and deployability of the previous IoT architecture, inside the scope of eHealthcare.

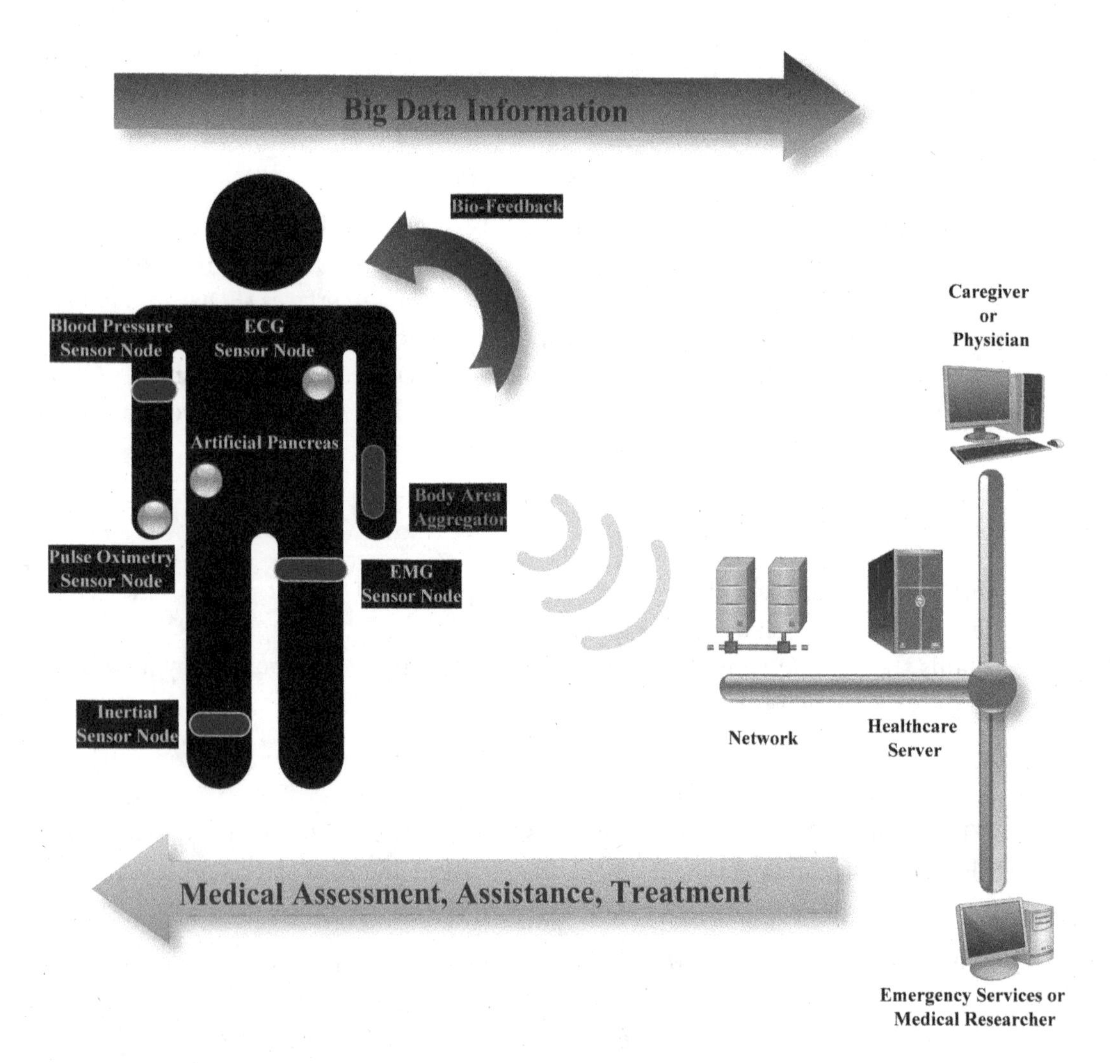

FIGURE 3.2 A typical IoHT monitoring system.

3.2.2 Enabling Technologies for IoHT

The IoT adoption in Healthcare, nowadays, could not be more facilitated than ever before; the reason is that IoT gadgets are embedded computing devices that exhibit many following qualities, like unique identity, ability to wirelessly communicate, ability to sense, ability to be controlled remotely, and possibility offering for massive data analysis and collection. Also, what drives the even further growth of the IoT world is the tremendous budget cut of the sensor's product cost, the emergence of social media and cloud computing, as well as several other IT innovations. Thus, the applicability ratio of IoT in Health is vast: typical examples of connected health and medical devices range from pacemakers, insulin pumps, continuous glucose monitors, hearing aids, heart rate patches and wireless scales for monitoring congestive heart failure, baby monitors for temperature and heart rate control, as well as patient identification and tracking in hospitals. The introduction of IoT in the Health industry also aids sensitive social groups. IoT plays a significant role in a wide range of IoT deployable scenarios for disabled users, as well, from preventing diseases and disabilities at one end of the spectrum to managing chronic conditions and limitations at the other.

A significantly relevant direction is cloud-assisted eHealth services. As an emerging patient-centric model, cloud-based personal health record (PHR) systems allow end-users to securely store their medical data collected from smart medical devices and platforms on the semitrusted cloud service providers and offer selective access to third parties, that is, pharmacists, or medical insurance companies. The manner this is being performed is through *Private Information Retrieval (PIR)* cryptographic techniques, to obtain access to encrypted private data only, while outsourcing the least information possible to the cloud server. For instance, if the outsourced data in a (cloud) database query contains encrypted patient records, the semitrusted cloud provider might learn the patient-client has been diagnosed with a disease when it sees a medical specialist has retrieved the patient records. Thus, the secure management of sensitive online patient's data is a trivial task.

Finally, Figure 3.2 illustrates the typical architecture of an IoT system in healthcare[8]. Notably, the specific figure depicts the previously mentioned applicability scenario of IoHT, where medical big-data gathered from "smart" wearable devices on a patient's body can be transmitted to the cloud, processed, analyzed to later on assess, assist, and treat medical conditions.

3.3 SECURITY VULNERABILITIES IN IoT MEDICAL DEVICES AND SECURITY THREATS

3.3.1 The Risks of Connected Healthcare Devices

When it comes to the healthcare industry, hospitals, care providers, and other facilities have a "patients" first mentality. Offering patients the best possible quality of medical services is the number one priority. Unfortunately, this does not always account for the online version of such therapeutic frameworks that are beginning to become embraced with the Internet of Things. According to security experts, the healthcare field is 200% more likely to suffer from malefic data manipulation and theft and undergo 340% more security breaches than any other industry. The underlying reason for this security outburst is, most certainly, the network connectivity of such devices and software applications, together with their fundamental dependability on the Internet and vast providers' networks[2,3].

The security vulnerabilities of IoT medical equipment affects many different types of in-hospital infostructure including diagnostic devices [i.e., MRI (Magnetic Resonance Imaging) machines and CT (Computerized Axial Tomography) scanners], therapeutic equipment (e.g., infusion pumps and medical lasers), life support equipment (e.g., heart support machines), and Internet-connected devices for monitoring patients' vital signs (e.g., thermometers, glucometers, blood pressure cuffs, and wearables.). These sophisticated medical devices are utilized in conjunction with mobile terminals (i.e., tablet computers and smartphones), which enable health professionals to configure them, observe them, and visualize their data. Furthermore, several IoT applications integrate RFID tags as a method to uniquely identify and associate with each other devices, doctors, nurses, and medicine all in a fully meshed network. Security risks in IoT medical gadgets constitute a significant concern, as it does not only mean compromising patient's data confidentiality; security vulnerabilities can have life-threatening implications. Since the configuration and control commands are transmitted to the medical IoT devices wirelessly, hackers can invade the wireless network to gain total control of such appliances and send unauthorized commands with fatal results. Typical examples, for instance, would be a malicious attack against an insulin pump that could lead to a wrong dose to a diabetes patient. As another example, the hacking of an electrical cardioversion device could trigger an unnecessary shock to a patient. There is a massive list of "built-in" security vulnerabilities and flaws inside IoT medical devices that could potentially create a series of malefic activities against them.

1. *Password hacking:* It is widespread for medical equipment to be protected by a weak user or login authentication passwords that equivalently could be hacked quite easily. This is the case when the built-in passwords provided by the device industry vendors are maintained by default. Hackers can quickly discover such passwords to penetrate to the device configuration information. Thus, a series of more sophisticated attacks could then take place.

2. *Poor security patching:* Some medical devices come out of the "box" poorly or inadequately patched with updated firmware against security flaws and malware. Typical reasons are because the patch has not been applied from the manufacturer before, or the device itself runs an older version of the operating system. Poorly patched tools are more vulnerable to malware and other attacks, which makes them an easy target for hackers.

3. *Denial of service (DoS) attacks:* Medical devices are usually lightweight, and resource-constrained, which makes them prone to DoS attacks. The transmission of multiple simultaneous requests to the device can cause it to stop, disconnect from the network, or even become malfunctioned, causing life risk to the patient.

4. *Unencrypted data transmission:* It is quite usual for the malefic attackers to keep monitoring the network to eavesdrop and steal passwords. The online transfer of unencrypted data can, therefore, facilitate their efforts to gain access to the device to extract information or to exploit the device by sending malefic commands.

One exterior, yet critical, vulnerability factor for IoHT medical equipment is their reliance upon conventional wireless protocols, which many of them provide weak and unprotected communication links. Most of the widespread medical devices are Wi-Fi-enabled, which renders Wi-Fi the technology that carries most of the traffic that is transmitted between medical devices. However, some Wi-Fi enabled networks can further increase the security vulnerabilities of IoT devices due to the presence of weak or deprecated security standards. For example, the WEP (Wireless Encryption Password) mechanisms that are encompassed inside Wi-Fi security are vulnerable, as WEP passwords can be easily stolen. This can adjacently lead hackers to launch security attacks based on the sniffing of unprotected traffic. To confront WEP problems, IEEE and the Wi-Fi community have specified and prototyped more modern Wi-Fi standards and protocols [i.e., WPA2, WPA2-PSK (TKIP/AES)], with much more robust encryption capabilities. However, not all medical gadgets vendors provide firmware support for these updated standards, putting the security operation and interoperability with other devices at risk[7].

3.3.2 Security Attacks in Healthcare System

As mentioned in the previous subsection, both IoT health devices and networks are vulnerable to security attacks due to the increased attack surface. The IoT paradigm keeps evolving, and several new IoT health devices and services are expected. Therefore, an attacker may invent different types of security threats to degrade the security levels of both existing and future IoT medical devices and networks. Next, the authors aim to build a complete threat model that includes all known sophisticated adversary attacks against eHealthcare. Accurately, a *two-dimensional* attack taxonomy is depicted to classify existing and potential new threat types best. The first-dimension parameter of this threat allocation would be the service-oriented types of IoHT compromise, namely, information-, host-, and network-specific compromise. The second dimension is the attack types based on each specific IoT architecture layer.

1st DIMENSION OF ATTACK TYPE(S): SERVICE-ORIENTED ATTACKS

1. *Attacks Based on Information Disruption*

   Ready to be processed and stored health data can be obtained, manipulated, or altered by a malefic attacker to provide wrong information and remove information integrity. Such attacks include the following[10]:

   Interruption: An adversary initiates a series of denial-of-service (DoS) attacks to cause communication links to become lost or unresponsive. This attack-type explicitly threatens healthcare availability, standard network functionality, and device responsibility.

   Interception: An adversary eavesdrops on medical data included in transmitted messages to threaten data privacy and confidentiality.

   Modification: An adversary gains unauthorized access to health data and performs tampering attacks to create confusion and false identities in the IoT health network.

Fabrication: An adversary forges messages by injecting false information to threaten message integrity and authenticity.

Replay: An adversary performs replay attacks to threaten message freshness, and cause network confusion.

2. *Attacks Based on Host Properties*
   Three possible types of attacks can be launched based on host properties[10]:

   User Compromise: An adversary compromises the user's health devices and networks by cheating or stealing cryptographic primitives. This type of attack reveals sensitive information, such as passwords, cryptographic keys, and user data.

   Hardware Compromise: An adversary tampers with physical devices and components and may extract device program-codes, keys, and data. Furthermore, an attacker may reprogram compromised devices with malicious codes.

   Software Compromise: An attacker exploits several software (i.e., operating systems, system software, and applications) vulnerabilities and forces IoT health devices to malfunction states (i.e., buffer overflow and resource exhaustion).

3. *Attacks Based on Network Properties*
   This type of attack comes in two necessary forms: protocol- and layer-specific security compromise.

   Standard Protocol Compromise: An attacker utilizes standard protocols (software and networking protocols) and acts maliciously to threaten service availability, message privacy, integrity, and authenticity.

   Network Protocol Stack Attack: As shown in Figure 3.3[10], this attack category involves a series of threats that aim to exploit security flaws that belong to different layers of the IoT protocol stack, as defined by the IETF working group. More analytically, these attack types are explained inside the scope of the second attack taxonomy dimension, next.

2nd DIMENSION OF ATTACK TYPE(S): IoT PROTOCOL LAYER-ORIENTED ATTACKS

The second dimension of attack types classifies security threats against the eHealthcare system, each belonging to a different layer of the IoT architecture plane, according to Figures 3.1 and 3.4. They are accurately categorized as follows[9]:

1. *Attacks at Data Collection Level*
   These malefic security incidents can cause several threats to the IoHT data collection layer, such as modifying information, excluding some critical data, or resending data packets. These attack vectors apply for the 1st layer (perception layer) and 2nd layer (network or transport layer) of the IoT (see Figure 3.1). Equivalently, for the IETF

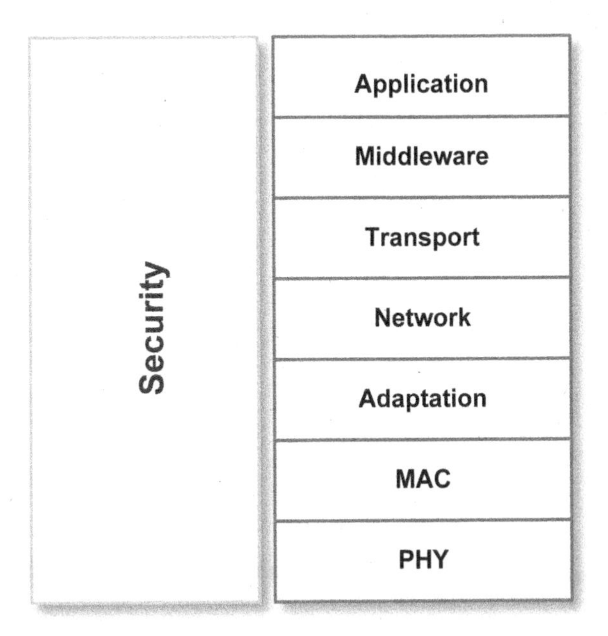

FIGURE 3.3 How security issues affect IoT in each architecture layer.

protocol stack, these attack types are associated with the lower four layers (Physical, MAC, Adaptation, and Network) (see Figure 3.4).

Jamming Attack: The attacker performs radio interference with the signal frequencies of the BAN (Body Area Networks). This results in isolating and preventing the various sensor nodes within the range of the attacker signals to give or receive any messages among the affected nodes if the jamming continues.

Data Collision Attack: It takes place when two or more nodes attempt to transmit simultaneously. Furthermore, it refers to jamming attacks when a malefic user tries to deliberately generate extra collisions by sending repeated messages on the channel. When the frame header is altered due to a crash, the error checking mechanism at the receiving end detects that as the presence of an error and rejects received data. Thus, a modification in the data frame header is a threat to data availability in the BAN.

Data Flooding Attack: The attacker can repeatedly broadcast many requests to the victim node for connection, using all its power resources until they reach a maximum limit, causing a flooding attack.

Desynchronization Attack: In this attack, the attacker tampers messages between sensor nodes by duplicating it many times, using fake sequence numbers, something that leads to the Wireless BAN to an endless cycle state. The latter causes sensor nodes to transmit messages again and again and waste their energy.

Spoofing Attack: The malefic user targets the routing information to perform several disruptions such as spoofing, altering, or replaying the routing information, leading to the creation of routing loops to the network.

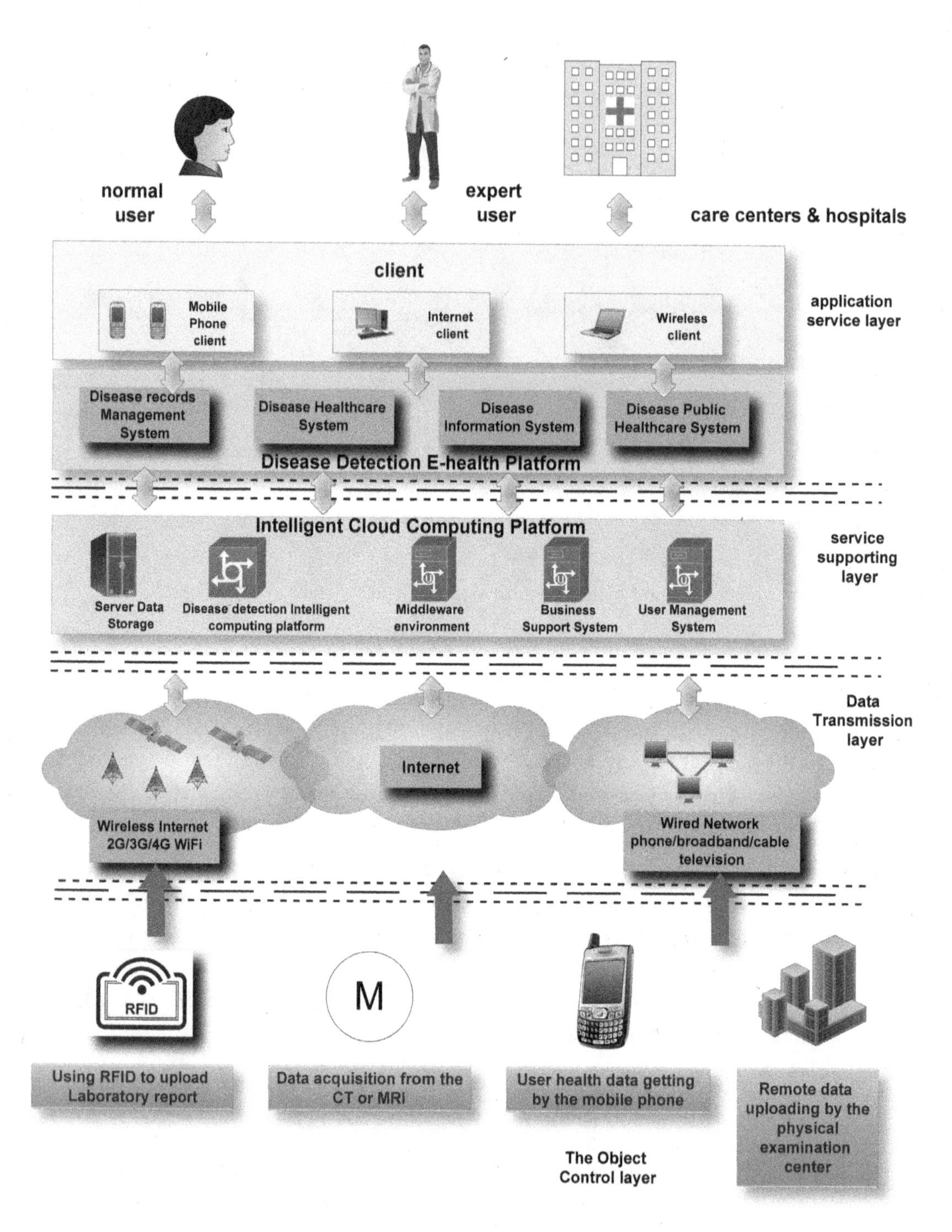

FIGURE 3.4 The "Internet of Healthcare Things" typical architecture plane.

Source: S. M. R. Islam, D. Kwak, M. H. Kabir, M. Hossain, and K. S. Kwak, "The Internet of Things for Health Care: A Comprehensive Survey"

Selective Forwarding Attack: It takes place when the attacker's malicious node in a data flow path forward selected messages and excludes others. The damage is reported to become worse, mainly when the malicious nodes are located close to the base station.

Sybil Attacks: In the Sybil concept, the malicious attacker node represents more than one genuine identity in the network. It is a particularly present type of attack in IoT networks and health applications where the location information is required to be exchanged between the nodes and their neighbors to route the geographically addressed packets efficiently. Such type of attack is quite challenging to trace due to the high mobility of the working nodes.

2. *Attacks at Transmission Level*

These attacks can cause severe threats to transmission level like spying, altering information, interrupting communication, sending more signals to jam the base station, as well as networking traffic. They are mainly associated with the MAC, transport, and network layers of each protocol stack (see Figures 3.1 and 3.4)[9].

Eavesdropping of Patient's Medical Information: Health monitoring systems will record the patient's health data from BANs and sensors to be transmitted to the healthcare providers. Unauthorized users can easily exploit the vulnerabilities of wireless protocols to spy on this data. Thus, the network protocols and IoHT applications must be designed with security precautions to protect against breaches of patient's sensitive data.

Man in the Middle Attacks: The attacker intercepts a communication between two parties and exchanges messages between them. The communication link is entirely under the control of the attacker, making him able to read, insert, and alter the intercepted messages.

Data Tampering Attack: A tampering attacker can damage and replace encrypted data by authorized network nodes.

Scrambling Attacks: It is a type of jamming attack on radio frequencies for short time intervals against the control and management information of the WiMAX protocol frames to affect the standard functionality of the network. The results are communication interruption and message availability issues.

Signaling Attacks: Before a patient's smartphone starts transmitting data, there is some initial signaling operation that needs to take place with the serving base station. Such signaling services contain authentication, key management, registration, and IP-based connection establishment. The attacker can launch a signaling attack on the serving base station by padding other state signals. Thus, the excessive load on the base station results in DoS attacks, causing availability issues.

Unfairness in Allocation: It can degrade the network performance by interrupting the Medium Access Control (MAC) priority schemes.

Message Modification Attack: Here, the attacker can capture the patient wireless communication channels and extract the patient medical data to be tampered later, which can mislead the involved parties (doctor, nurse, or family).

Hello Flood Attack: This type of security attack makes the malefic user able to fool the network nodes to select the attacker for routing their messages. The manner this is being performed is by sending a hello message with high-powered radio transmission to the network.

Data Interception Attack: Here, an interception can take the place of the patient's information by the attacker during their interaction between computer terminals of healthcare systems, through hospital LANs.

Wormhole Attack: This specific type of attack is silent, yet severe because it copies the packet at one location and replays it at another site in the network without any changes in the content. The results can damage the network flow and topology.

3. *Attacks at Storage Level*
   These attack categories may cause several threats to storage level, such as modifying patient medical information or altering the configuration of system monitoring servers. They are explicitly applied to the application layer of the IoT stack[9].

Inference Patient's Information: Malefic users aim to combine authorized information with other available data, which leads them to identify sensitive patient data, such as diseases and human medical profiles.

Unauthorized Access of Patient Medical Information: This type of security compromise can take place by unauthorized individuals without valid authentication; thus, as a result, patient's data can be accessed, modified, and damaged.

Malware Attack: It is a malicious software program able to perform harmful actions. This type of attack can infect and propagate to the whole hospital server, something that can cause unavailability and disruption.

Social Engineering Attacks: In this type of attack, a third-party attacker can obtain access to the system by impersonating either the patient or authorized user(s) to access secret information. Here, authorized users can also outsource patient's data to concerned parties such as Health Insurance Company for unethical personal gains.

Removable Distribution Media Attack: This type of security attack involves stealing or losing a computer or storage medium, something that can be used to take information or spread malware software to the hospital's healthcare framework.

3.4 SECURITY AND PRIVACY CHALLENGES IN IoT HEALTHCARE

It is generally recognized that privacy and data protection and information security are complementary requirements for IoT services, especially when they applied and deployed inside the Health industry. Information security is regarded as retaining the confidentiality, integrity, and availability (CIA) of information. It is also considered that information security is perceived as a fundamental requirement in the provision of IoT services for the Healthcare industry, both to ensure information protection for organizations, patients, and care centers, but also the health benefits of the citizens as well. It could be noticed that regarding information security, the general information security demands should apply for IoT; however, since IoT (and IoHT) is a particular case and more of a vision rather than a concrete technology (see Section 3.2.1), it is well understood that it is quite complicated to define all the requirements yet accurately. Still, from various already performed studies for identifying risks in such highly interconnected environments, we can spot the following *Security and Privacy* issues[7,10].

3.4.1 Security Challenges in eHealth

eHealth faces many security challenges; it has been reported that during the year 2016, significant cyberattacks against hospitals grew 63%. According to an Intel security survey[11], at least 19 care centers in the United States were infected with ransomware in Q1 and Q2 of 2016. In contrast, a related group of Q1 attacks on hospitals generated $100,000 in ransom payments. Further results from these types of malware attacks have been shutting down of operations at health centers. We could explicitly identify and classify the following most critical issues as related to the *Security factor* of IoT in Healthcare[7]:

1. *Integrating RFID into IoT:* RFID technology has been considerably utilized in IoT applications for disabled users, and patients, to track, identify, and associate with medical data. However, RFID systems are more vulnerable to security attacks due to their resource constraints. There are, indeed, several security attacks that correspond to those components, such as eavesdropping, the man in the middle attack, service degradation, spoofing, cloning, and abuse of tag. Thus, securing, and restricting access to data from RFID tags is vital.
2. *Integrating WSNs into IoT:* Wireless Sensor Network technologies possess limited computational and energy resources. They are mostly battery-powered and connected through lossy links. Thus, they are more prone to be attacked by an adversary. For example, the physical attack of the node can exploit critical information, such as security protocols, source code, and other cryptographic keys. Another critical security issue is how to secure the communication channel between a sensor node and the Internet. WSNs have been utilized in many sensitive instances in the IoHT, thus if the security of a WSN is compromised, that attack might result in human life consequences and loss of resources.
3. *Unauthorized Data Access/Access Control:* Different client users are assigned for different IoHT applications, and each software application will likely have a significant number of user subscribers. Therefore, effective authentication technology should be encompassed to prevent unauthorized user involvement. Moreover, program-specific access control is essential to avoid illegal entities from accessing to system's resources (i.e., data, services, and hardware). As a result, access control mechanisms play a vital

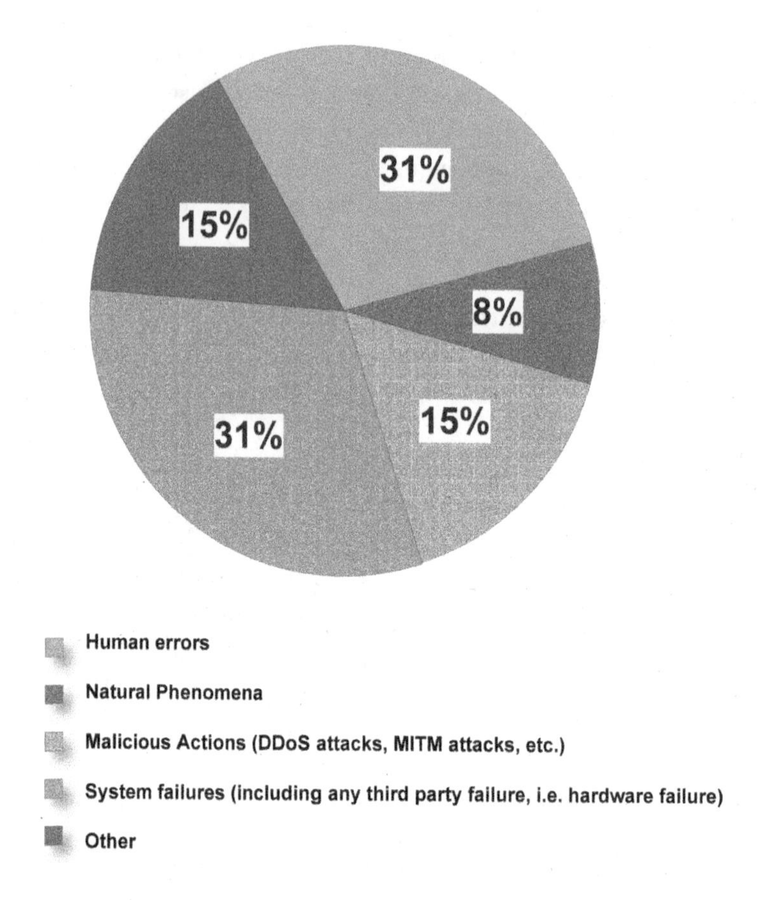

FIGURE 3.5 Common root causes of eHealth security incidents.

Source: D. Liveri, A. Sarri, and C. Skouloudi, "Security and Resilience in eHealth Security Challenges and Risks"

role in ensuring that breaches of security do not take place in IoT, and life-threating conditions will not occur.

Incidents Management: Incident's administration is a significant security challenge in eHealth security. Although most security experts implement security policies in their eHealth systems and infrastructures, there are incidents than can neither be neglected nor be avoided. Security incidents root causes, as depicted in Figure 3.5[12], include human errors, natural phenomena, malicious cyber-actions, and system failures. System failures and human errors account for most of the reported incidents. Natural causes play the least effects, whereas social influences could be deliberate (malicious activities and attacks by adversaries), or nondeliberate (incorrect security practices by hospital personnel that may result in security incidents)[12].

4. *Security Issues for Communication Technologies:* In Table 3.1[7], the security issues for the leading Wireless Protocol communication technologies used in IoHT are listed and described. All these protocols, despite having provided authentication and encryption schemes, possess substantial security pitfalls, that is, most are prone to Denial-of-Service attacks, fundamental exploitation techniques, and eavesdropping.

TABLE 3.1 Security Features of IoHT Wireless PAN/LAN/WAN Communications Technologies

Features	Zigbee	Bluetooth	Li-Fi	Wi-Fi	LTE/LTE-A
Standard	IEEE 802.15.4	IEEE 802.15.1	IEEE 802.15.7	IEEE 802.11	LTE: 3GPP Rel.9 LTE-A: 3GPP Rel.10 5G-NR (New Radio): 3GPP Rel.13[27]
Healthcare Applications	Remote sensing and control: disease monitoring, personal wellness monitoring	Machine to machine (M2M): provide a wireless connection between devices	Control medical equipment, automate procedures, and robotic surgeries	Networking and Internet access	M2M services; monitoring, and tracking patients and devices
Provided Security	128 AES encryption	64/128 AES encryption	1s and 0s encryption	WEP, WPA, WPA2 encryption, access control, authentication, confidentiality	Authentication, 128 AES encryption
Security Issues	Exploiting the critical exchange process, and battery lifetime	Denial of service, scalability and secure pairing	Reliability and network coverage; data in light "stream".	Eavesdropping and malicious attackers	A man-in-the-middle attack, user identity threats, and sequence number synchronization

Source: Wassnaa AL-mawee. "Privacy and Security Issues in IoT Healthcare Applications for the Disabled Users a Survey"

5. *Security issues for Health Cloud Computing:* Several cloud computing concerns are related to security in the healthcare field[7].

a) *Secure Data Access:* Adversaries can attack PHRs in the cloud, and that can lead to loss of sensitive personal data, failure of medical care, and theft of PHRs. Therefore, it is crucial to managing secure data access in the storage cloud. Another issue would be for a malicious insider attack aiming to modify the patient's data.

1. *Data Protection:* Data stored in the cloud resides in a shared environment. The patient medical records and various sensitive information from groups of elder or disabled persons are collected and stored in the cloud, together with other application data. The big question is how to classify the best cloud data security between sensitive and nonsensitive (public) data. Finally, backup plans of encrypted cloud data are a necessity to avoid PHRs loss.

2. *Big Data:* Inevitably, the term "Big Data" well applies inside the context of IoHT. In healthcare, Electrocardiogram (ECG) data are collected, processed, and stored as big data.

For example, when such a system collects ECG signals from one person at 240 Hz (240 data elements per second), it collects 864,000 data per hour. Therefore, data will be considered huge and, thus, called big data if even collected from 10,000 persons. The massive scale of Big Data, together with its database manipulation inside the cloud computing, undoubtedly, raises many security issues; encryption and decryption should scale and be fast enough for this data amount, whereas functional integrity and authentication methods should be applied.

3.4.2 Privacy Issues in IoHT Applications

In IoT applications for hospital or house-kept patients, massive amounts of health data need to be gathered and analyzed. Reasonable questions arise as to who owns the medical data for a disabled, elder, or regular patient, who has the right to access it, and where the data is stored. In this section, the most common issues, and implications in IoHT systems, from a *Privacy* perspective, are being discussed.

3.4.2.1 Risks of Patients' Privacy Exposure

As mentioned earlier, a PHR is *"an individual electronic record of health-related information that conforms to the nationally recognized interoperability standards. PHR can be drawn from multiple sources while being managed, shared, and controlled by the individual"*. PHRs are directly reported to the relevant eHealth center, whereas the primary security and privacy issues are to keep the patient's PHR strictly confidential. It is well understood that the risk of privacy exposure of the PHRs is accountable; the distinction between sensitive (private) and nonsensitive (public) patient data is challenging to accomplish, whereas there are many legislation issues based on access rights assigned by the data owner.

3.4.2.2 Threats of Cyber-Attacks on Privacy

Cyber-attacks can potentially inject false data into a healthcare system, causing critical damage in IoT components. It is, therefore, fundamental to offer an adequate level of protection against cyber-attacks in "smart" home IoHT applications, or "smart" hospitals. However, the resource-constrained nature of many IoT devices present in an eHealthcare environment does not allow to implement the standard security solutions.

3.4.2.3 Data Eavesdropping and Data Confidentiality

Generally, the patient's health data, including the disabled and elders, are retained under the legal obligations of confidentiality and made available only to the authorized health caregivers. It is crucial to prevent stealing data from storage environments or to eavesdrop on them while they circulate over the wireless communication links. For example, a widespread IoHT-based disabled glucose monitoring and insulin delivery system utilizes wireless communication links, which are often used to launch privacy attacks.

3.4.2.4 Identity Threats and Privacy of Stored Data

Loss of a patient's privacy, especially her identity data, may lead to significant physical, financial, and emotional consequences to the patient. Besides identity theft, the disclosure of stored PHRs can cause further losses of patient's privacy.

3.4.2.5 Location Privacy

Most IoHT systems and devices contain geo-location information of their patient users for purposes such as tracking, identification, and remote monitoring on them. Location privacy is concerned with location privacy threats and eavesdropping on a user's current location. An adversary might launch attacks to precisely identify the current patient's location and derive privacy-related issues from him.

3.5 TOWARD TRUSTED eHEALTH SERVICES

3.5.1 Security Policies and Law Enforcement for IoHT

The IoT holds big promises for healthcare, especially in proactive eHealth and personal remote monitoring. To fulfill these promises, significant challenges must be overcome, particularly regarding privacy and security. We could generally identify and classify the following problems which need to be tackled in the world of pervasive devices[14]:

- Management, scalability, and heterogeneity of devices
- Networked knowledge and context
- Privacy, security, and trust will have to be adapted to both accessories and data

This will involve the development of highly efficient cryptographic algorithms and more reliable protocols that provide basic security properties such as confidentiality, integrity, and authenticity, as well as secure implementations for the various kinds of computational resource-constrained devices. Besides, vulnerability management and threat identification, it is crucial for health industries, organizations, and care centers to fully encompass security legislation policies and risk assessment strategies in a national and multinational domain. Individually, the following security steps should be taken into consideration, for companies and health-care experts:

1. *Culture assessment:* Evaluate the organization's security behavior and develop the structure of High Resilient Security Programs.
2. *Policy compliance:* Audit adherence with regulatory obligations and assess compliance with a unified framework and industry standards for healthcare.
3. *Security metrics:* Involves collecting evidence-based operational security metrics to enable quality decision making and to track adequate progress.
4. *Cyber-Security Awareness campaigns:* Structure, launch, and manage effective awareness programs across the enterprise, for example, Social Engineering/Phishing.

The upper technology solutions can be proven as robust and reliable steps toward the evaluation of security solutions, recommendations, and implementation of chosen techniques to detect potential cyber-attacks and enforce the security policy. Law enforcement is another necessary target domain. Currently, all health insurers are excluding electronic data loss and privacy breaches from the liability policy; thus, a separate system for data privacy breach of insured patients is required. As previously discussed, PHRs are meant to be strictly confidential for each patient and individual involved in the healthcare context. Unfortunately, data breach and privacy exploitation also take place; thus, the legislation of what should be defined as sensitive private data, how it should be accessed and manipulated by health insurance companies is a controversial legal issue. It is worth depicting at this point that IoT and IoHT are still in a premature context; it remains more like a vision rather than a concrete technology. That applies to its legal dimension and law enforcement.

3.5.2 Security Defense Countermeasures for Healthcare IoT

Next, we consider a real-life eHealth use-case scenario to practically examine the best possible security countermeasures for IoHT and recommended mitigations that need to be applied from a three-fold IoT approach: (1) *Security in the IoHT Landscape*, (2) *Endpoint Devices Security*, and (3) *Network Security*. The scenario is as follows:

Consider a patient with a blood pressure monitoring device, which takes a blood pressure reading every 15 minutes and the device itself or another local device, which has a collector function, stores the readings. Once a day, the device or collector and the medical facility's application server communicate with each other to transfer the reading to the healthcare server. The collector function could be in the network gateway. The readings may be summarized, or some other data processing technique performed, and then the medical staff reviews the results[13].

There are several things to be considered in this example. From the (1) perspective, the working environment in which the monitoring device operates is highly essential. If the method is only used in a home or a medical facility, then it may be assumed that it is a trusted environment; thus, encryption of transmitted data is not required. However, in any situation that is not considered trusted, then the communication needs to be encrypted for data confidentiality. Currently, a generic insulin pump vendor provides wireless connectivity between their insulin pump and a USB device plugged into the patient's PC. One might consider this a safe and secure execution environment. However, the communication protocol is generic and not secured. Recently it had been shown that this insulin pump could be hacked into and controlled if one is in the proximity. The malefic user could then program the device with custom commands, something that would have fatal consequences. The second biggest issue from the device side is that IoT/IoHT devices do not have enough computational resource capabilities to perform encryption operations inside them[16]. Most current devices have limited memory and processing power to handle standard security protocols (such as AES 128 and others). Also, energy resources and battery life are other critical factors. Furthermore, the wireless network offers some degree of security, although encryption over conventional GSM/GPRS networks is not safe. A strong security countermeasure, for added security, is to use an encryption mechanism that is layered—128-bit AES and then deploy the A-5 encryption on the GPRS channel. The adversary would not be able to break the AES encryption.

Lightweight cryptography (LWC) is defined as crypto algorithms suitable for a limited resource-constrained environment, like medical sensors, RFID tags, and portable healthcare monitoring devices. Many LWC crypto primitives are developed for data security purposes. However, not all of them are practically trusted for medical applications. We could refer to several (AES-based) tested algorithms for IoHT platforms, such as *International Data Encryption Algorithm (IDEA), Camellia, Leak Extraction (LEX), Light Encryption Device (LED), Salsa20,* and *SIMON* ciphers[24]. Authors in Ref. 24 evaluate ten candidate crypto algorithms for medical applications in terms of speed and memory comparison results. All ten algorithms are considered LWC, yet not all of them are suitable for IoHT, as derived from the hardware evaluations. Based on the benchmarking done by the authors, two cipher algorithms, the *SPECK,* and the *SIMON,* are both anticipated to become primary in many IoHT practical scenarios. Their main advantage is their interoperability, fast performance in software, and secureness of operations.

Inside the (2) perspective, it is quite clear that management of long-term diseases and chronic conditions will, therefore, require several wearables and embedded devices, constrained by their specific purposes in their communication and information exchange. The vision of IoT will be that a typical device will act as the collector and processor node of this information operating at the heart of an autonomous yet connected system. A strong security recommendation here should be a distributive security architecture approach; rather than imposing a single architecture and security scheme on every device, it makes more sense to choose security attributes that are appropriate to each device, unifying these to a central policy from the collector as part of a free computing domain. Further security enhancements should be the deployment of parallel virtual machines whenever several identical collection agents might be required—for example, two or more family members with a different medical condition, or a single person with multiple information sources where a hard separation between them is required. This parallel machine usage has the benefit of moving the intensive, power-draining computing functions away from the sensor. Device-specific security algorithms and protocols need to be employed on the sensor, though. Finally, the use of prevention and detection software in the endpoint devices is also critical. This software shall include data encryption capabilities, firewall, antivirus, intrusion detection, antispam, remote diagnostic, and auditing software.

Finally, in the (3) point of view, the eHealth network security architecture should involve multiple layers of prevention, detection, and response controls as the network span through different types of networks. These include wireless, wired, enterprise, private, and public networks. Vital security recommendations for this IoT infrastructure domain should be robust communication security: the proposed security protocols should ensure the smoothness transitions connections among different edge networks, strong authentication (RSA with SHA-256 for digital signature), and confidentiality services, together with authorized user access. Access control prevents identity theft by denying unauthorized parties in the network. Access control matrix or access list is used to provide various accesses to various levels of users. Hashing can be used for preventing data modification by third party. Finally, the encryption of medical data includes confidentiality. Figure 3.6[13] summarizes the whole eHealth building blocks security stack.

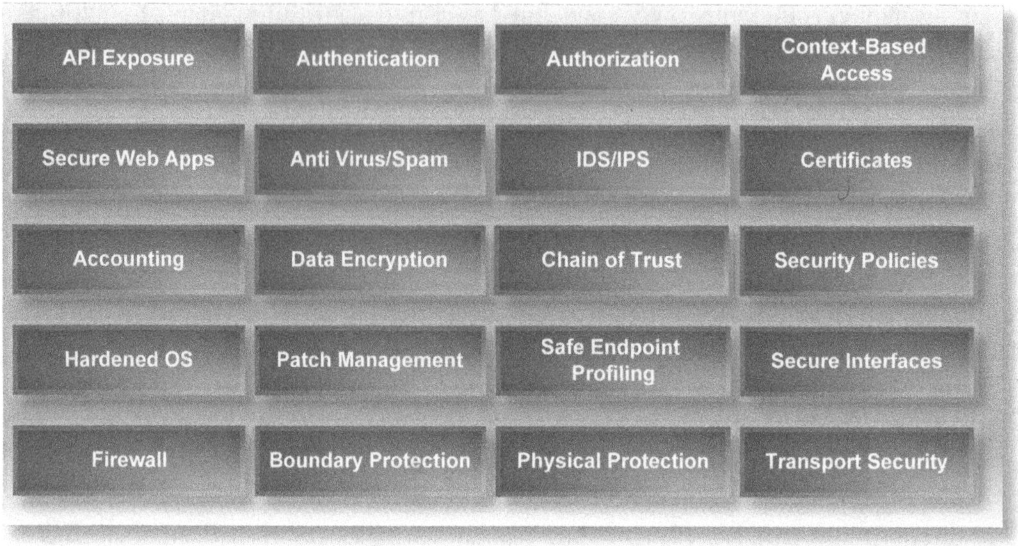

API Exposure	Authentication	Authorization	Context-Based Access
Secure Web Apps	Anti Virus/Spam	IDS/IPS	Certificates
Accounting	Data Encryption	Chain of Trust	Security Policies
Hardened OS	Patch Management	Safe Endpoint Profiling	Secure Interfaces
Firewall	Boundary Protection	Physical Protection	Transport Security

FIGURE 3.6 The “Internet of Healthcare Things” security stack.

Source: D. Lake et al. “Internet of Things: Architectural Framework for eHealth Security”

To mitigate the security and privacy risks, secure network security infrastructures for short or long-range communication are needed. Thus, further discussion and comparison of related work should be focused, concerning the efficient authentication and authorization security architectures for IoT-based healthcare systems. The several research proposal solutions, however, although they contain delegation-based structures, they tend to suffer from lack of scalability and reliability. Specifically, most of their architecture cannot be extended to be employed for multidomain infrastructures, like in-home/hospital domains. Also, their proposed architecture suffers from a considerable network transmission overhead that could result in long transmission latencies, in case of DoS attacks. Thus, there have been many efforts in designing secure gateways for one or several specific applications and architectural layers of IoHT, as an alternative research direction. Finally, in an efficient and reliable authentication and authorization architecture for IoHT, protection against Distributed Denial of Service (DDoS) attacks must be ensured. Table 3.2 compares the various already studied security architecture frameworks, inside the scope of eHealth and IoT, for merely user authentication purposes.

Integrated and layered perspective on Security, Trust, and Privacy can potentially deliver a security framework-based input to address protection issues in the IoHT. As suggested in Ref. 14, the authors have selected a cube structure mechanism to model security, trust, and privacy in the IoHT. The cube has three intersected dimensions, which is an ideal modeling structure for depicting the convergence of safety, trust, and privacy among all involved entities and infrastructure of IoT in health.

From Figure 3.7, the type and structure of information required to grant/reject any user access request for IoHT services are complicated. They should address the following IoT notations: security (authorization), trust (reputation), and privacy (respondent).

TABLE 3.2 Comparative Study of IoHT Security Architectures

Security Frameworks	Features	Benefits	Limitations	Potential Improvements
CodeBlue[18]	A wireless communications infrastructure for critical care environments	• Provides routing, naming, discovery, and security for wireless medical sensors • Designed to scale across a wide range of network densities • Operates on a variety of wireless devices • Designed by Harvard	• Security aspects remain unaddressed • Prone to a DoS attack, snooping attack, gray-hole attack, Sybil attack, and masquerading attacks[21]	• Elliptic Curve Cryptography (ECC)[19] • TinySec[20] • Key generation and symmetric encryption solutions
Datagram Transport Layer Security (DTLS)[25]	A delegation-based security architecture	• Relies on an off-path centralized delegation server • Establishes interoperable network security between end-peers from independent network domains	• Architecture lacks scalability and reliability • Architecture cannot be extended to be employed for multidomain infrastructures • Considerable network transmission overhead	• Should adapt multidomain networks functionality • Resistance against DoS attack
SwissGate[22] *6LoWPAN (IPv6 over Low power Wireless Personal Area Networks)*[26] *UT-GATE*[23] *Secure and Efficient Authentication and Authorization (SEA) Architecture*[15]	Smart eHealth Gateway(s)	• Autonomously perform local data storage and processing, to learn, and to make a decision at the edge of the network • Provide preliminary results and reduce the redundant remote communication to cloud servers • Bring intelligence into IoT-based ubiquitous healthcare systems	• DoS attacks or compromises of the delegation server	• Should perform authentication and authorization more securely and efficiently using a distributed approach • Should act on behalf of medical constrained devices • An exclusive smart eHealth gateway can handle authentication and authorization of remote end-points

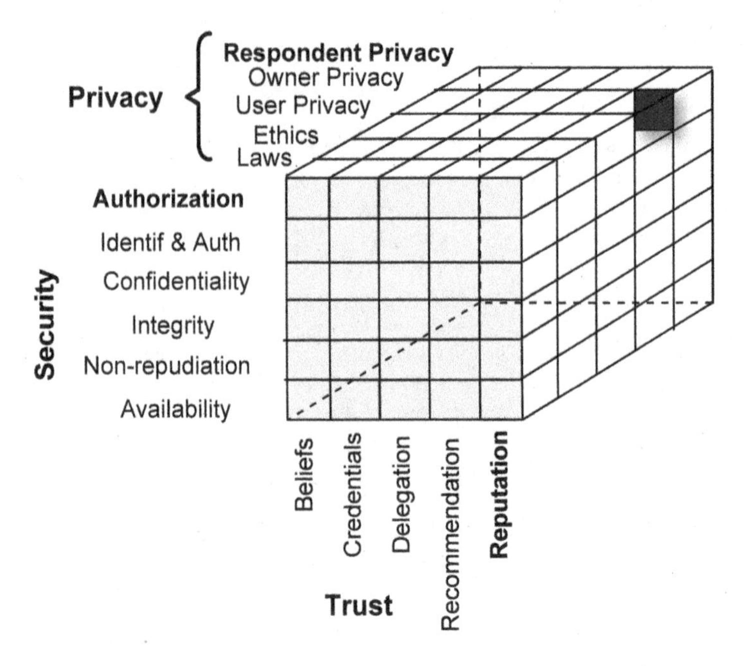

FIGURE 3.7 A security framework model for IoHT.

3.6 CONCLUSIONS AND OUTLOOK

The recent widespread development and deployment in the paradigm of the "IoT" offered an excellent promise for providing state-of-the-art technological solutions for disabled people, elders, and human patients. Current IoT-driven healthcare services and systems not only manage to increase life expectancy and quality but also offer a huge marketing possibility for the industry. However, there are many privacy and security challenges and concerns derived from IoT healthcare applications. Security network attacks against such "smart" wearable equipment and hospitals have significantly risen lately with life-threatening implications and loss of medical service availability. Securing and protecting sensitive medical records and human patient personal traits is another critical concern. This chapter surveys the newest security attacks against IoHT while providing a complete two-dimensional attack taxonomy based on the architectural approach of IoT. Furthermore, it aims to separate and emphasize the Security issues on the hand that emerges from IoHT applicability, and Privacy concerns for personal data, on the other. For more in-depth insights into industry trends and upcoming technologies, this chapter offers a broad view on how recent and on-going advances in sensors, devices, Internet applications, and various other techniques have pushed the IoT-based health services to their full potential, coming together, however, with several security consequences. To better understand IoT Healthcare protection laws and legislations, this work discusses IoT policies and law enforcement scenarios for the benefit of various stakeholders interested in assessing IoT-based eHealth technologies. Finally, security countermeasures, possible communication protocol enhancements, and frameworks are depicted to mitigate security vulnerabilities on the Internet of Things Healthcare.

GLOSSARY

3GPP: Third-Generation Partnership Project
CoT: Cloud of Things
IoHT: Internet of Things (IoT) for Healthcare
IoMT: Internet of Medical Things
IoT: Internet of Things
PHR: Personal Health Record
RFID: Radio-Frequency Identification
RTHS: Real-Time Health Systems
TKIP/AES: Temporal Key Integrity Protocol/Advanced Encryption Standard

REFERENCES

1. L. Atzori, A. Iera, and G. Morabito. "The Internet of Things: A Survey". Computer Networks 54, no. 15 (2010): 2787–2805. https://doi.org/10.1016/j.comnet.2010.05.010
2. S. Zeadally, A. K. Das, and N. Sklavos, "Cryptographic Technologies and Protocol Standards for Internet of Things", Internet of Things (2019): 100075. https://doi.org/10.1016/j.iot.2019.100075
3. N. S. Bechtsoudis. "Aiming at Higher Network Security Through Extensive Penetration Trees", IEEE Latin America Transactions 10, no. 3 (2012).
4. M. B. Tamboli and D. Dambawade, "Secure and Efficient CoAP Based Authentication and Access Control for Internet of Things (IoT)", In IEEE International Conference on Recent Trends in Electronics Information & Communication Technology (RTEICT), Banglore, India, pp. 1245–1250, 2016.
5. E. Al Akeem, C. Y. Yeun, and M. J. Zemerly, "Security and Privacy Framework for Ubiquitous Healthcare IoT Devices", In 2015 10th International Conference for Internet Technology and Secured Transactions (ICITST), London, pp. 70–75, 2015.
6. L. M. R. Tarouco et al. "Internet of Things in Healthcare: Interoperability and Security Issues", In 2012 IEEE International Conference on Communications (ICC), Ottawa, ON, pp. 6121–6125, 2012.
7. Wassnaa AL-mawee. "Privacy and Security Issues in IoT Healthcare Applications for the Disabled Users a Survey", Western Michigan University, Master's Theses, 12-2012.
8. J. Lach, J. Aylor, H. Powell, and A. Barth, "Body Area Sensor Networks", INERTIA-Team Research Faculty, University of Virginia. https://sites.google.com/a/virginia.edu/inertia-team/engineering-research/body-area-sensor-networks
9. A. M. Altamimi. "Security and Privacy Issues In eHealthcare Systems: Towards Trusted Services". (IJACSA) International Journal of Advanced Computer Science and Applications 7, no. 9 (2016).
10. S. M. R. Islam, D. Kwak, M. H. Kabir, M. Hossain, and K. S. Kwak, "The Internet of Things for Health Care: A Comprehensive Survey", IEEE Access 3 (2015): 678–708.
11. A. Drougkas | Officer in NIS, The NIS Directive and Cybersecurity in eHealth, Belgian Hospitals Meeting on Security| Brussels | 13th October, European Union Agency for Network and Information Security 2017.
12. D. Liveri, A. Sarri, and C. Skouloudi, "Security and Resilience in eHealth Security Challenges and Risks", ENISA (European Union Agency for Network and Information Security), 2015.
13. D. Lake et al. "Internet of Things: Architectural Framework for eHealth Security". Journal of ICT 3–4 (2014): 301–328.
14. S. Babar, P. Mahalle, A. Stango, N. Prasad, R. Prasad. "Proposed Security Model and Threat Taxonomy for the Internet of Things (IoT)". Communications in Computer and Information Science 89 (2010): 420–429. 10.1007/978-3-642-14478-3_42.

15. S. R. Moosavi et al. "SEA: A Secure and Efficient Authentication and Authorization Architecture for IoT-Based Healthcare Using Smart Gateways", Procedia Computer Science 52 (2015): 452–459.
16. N. Sklavos, "Cryptographic Hardware & Embedded Systems for Communications", In Proceedings of the 1st IEEE-AESS Conference on Space and Satellite Telecommunications, Rome, Italy, October, pp. 2–5, 2012.
17. St. Limnaios, N. Sklavos, and O. Koufopavlou, "Lightweight Efficient Simeck32/64 Crypto-Core Designs and Implementations, for IoT Security", In 2019 27th IFIP/IEEE International Conference on Very Large Scale Integration (VLSI-SoC), Cusco, 2019.
18. D. Malan et al. "CodeBlue: An Ad hoc Sensor Network Infrastructure for Emergency Medical Care". In WIBSN'04, London, 2004.
19. N. Koblitz. "Elliptic Curve Cryptosystems". JAMS 48: (1987): 203–209.
20. C. Karlof et al. "TinySec: A Link-Layer Security Architecture for Wireless Sensor Networks". In ICENSSS, Baltimore, MD, pp. 162–175, 2004.
21. G. Kambourakis et al. "Securing Medical Sensor Environments: The CodeBlue Framework Case". In ARES'07, Vienna, Austria, pp. 637–643, 2007.
22. R. Mueller et al. "Demo: A Generic Platform for Sensor Network Applications". In MASS'07, Pisa, Italy, pp. 1–3, 2007.
23. Rahmani et al. "Smart e-Health Gateway: Bringing Intelligence to IoT-Based Ubiquitous Healthcare Systems". In CCNC'15, Las Vegas, NV, 2015.
24. N. Alassaf, B. Alkazemi, and A. Gutub. Applicable Light-Weight Cryptography to Secure Medical Data in IoT Systems. Journal of Research in Engineering and Applied Sciences (JREAS) 2 (2017): 50–58.
25. E. Rescorla and N. Modadugu. "Datagram Transport Layer Security (DTLS) Version 1.2". In RFC 5238, 2012.
26. W. Shen et al. "Smart Border Routers for Healthcare Wireless Sensor Networks". In WiCOM'11, Wuhan, China, pp. 1–4, 2011.
27. M. T. Ahad and K. A. Yau, "5G-Based Smart Healthcare Network: Architecture, Taxonomy, Challenges and Future Research Directions", IEEE Access 7 (2019): 100747–100762.
28. M. Pike, N. M. Mustafa, D. Towey, and V. Brusic, "Sensor Networks and Data Management in Healthcare: Emerging Technologies and New Challenges", 2019 IEEE 43rd Annual Computer Software and Applications Conference (COMPSAC), Milwaukee, WI, pp. 834–839, 2019.
29. D. Zhang, B. Xiao, B. Huang, L. Zhang, J. Liu, and G. Yang, "A Self-Adaptive Motion Scaling Framework for Surgical Robot Remote Control", IEEE Robotics and Automation Letters 4, no. 2, (2019): 359–366. DOI: 10.1109/LRA.2018.2890200
30. 3rd Generation Partnership Project (3GPP), 3rd Generation Partnership Project; Technical Specification Group Services and System Aspects; Release 16 Description; Summary of Rel-16 Work Items (Release 16), Updated October 2, 2019.
31. J. Qi, P. Yang, G. Min, O. Amft, F. Dong, and L. Xu, "Advanced Internet of Things for Personalized Healthcare Systems: A Survey", Pervasive Mobile Computing 41 (2017): 132–149.
32. S. B. Baker, W. Xiang, and I. Atkinson, "Internet of Things for Smart Healthcare: Technologies Challenges and Opportunities", IEEE Access 5 (2017): 26521–26544.
33. M. M. E. Mahmoud et al. "Enabling Technologies on Cloud of Things for Smart Healthcare", IEEE Access 6 (2018): 31950–31967.
34. M. M. Dhanvijay and S. C. Patil, "Internet of Things: A Survey of Enabling Technologies in Healthcare and Its Applications", Computer Networks 153 (2019): 113–131.

CHAPTER 4

Toward the Detection and Mitigation of Ransomware Attacks in Medical Cyber-Physical Systems (MCPSs)

Alberto Huertas Celdrán, Félix J. García Clemente and Gregorio Martinez Perez

CONTENTS

4.1 INTRODUCTION

Nowadays, hospitals are adapting their devices, procedures, and mechanisms to the new information and communication technologies (ICTs). The revolution we are experiencing is provoking an improvement of clinical environments, allowing them to be more powerful, accessible, effective, and precise. The main aspects considered by experts for the coming years can be summarized in the following list [1]:

- Digitalization and learning from patient experience.
- Personalization of medicine through noninvasive technologies.
- Improvements in surgical operations.
- Improvement of human quality of life within the medical environment.
- Support diagnosis of diseases through the use of artificial intelligence.

As can be seen, some of the previous aspects are closely related to new technologies. Among all of them, we highlight the development of advanced monitoring systems in mobile units and the use of digitalization and artificial intelligence to improve diagnostics. In this context, the combination of some of the previous aspects is enabling the improvement of current hospitals and their evolution to those of the future.

The next generation of hospitals will have Medical Cyber-Physical Systems (MCPSs) [2], which are composed of medical devices and algorithms that achieve continuous and high-quality care for patients. An MCPS combines the physical world, in which interoperable medical devices monitor patients' data, with the cyber world, where algorithms process and analyze the previous data. The processing of such data is aimed at improving the care offered to patients. Nowadays, MCPS deployed in testbed clinical environment are supervised by medical professionals for proper functioning. At this point it is important to emphasize that the elements that make up an MCPS should be interoperable to communicate perfectly, they should focus on patient data, and they should consider the cybersecurity of the entire system and data. To make possible the MCPS vision, the ASTM F2761 standard [3] proposes a patient-centric architecture for Integrated Clinical Environments (ICE) that enables the open coordination and interoperability of heterogeneous medical devices and applications. The ICE standard specifies the needed characteristics to integrate medical devices and other equipment, through a network interface, in a single high precision medical system for patient care. This standard proposes a medical system that tends to be more resistant to errors and more interoperable.

Despite the benefits of MCPS and ICE, they open the door to new cyberattacks in clinical environments. In this context, the Ponemon Institute surveyed in 2016 stating that 64% of organizations reported a successful attack targeting medical records (9% more than the previous year) and almost 90% of attacks on health care organizations caused data breaches [4]. Besides, a 2018 Verizon data breach report [5] stated that ransomware attacks [6] are the 85% of all malware in medical care, and more than 70% of attacks confirmed disclosure data. Finally, more than 20% of medical devices in this environment have suffered some kind of ransomware attack. As demonstrated, ransomware is one of the most harmful and recent cyber threats. Especially, ransomware is a kind of malware that encrypts data and asks for a rescue or ransom to recover them. Usually, ransomware works in two phases, one of spreading and another of encryption. During the spreading phase, the ransomware looks for vulnerabilities of other computers located in the same network to exploit and infect them. The encryption phase cyphers files and data requesting a ransom, usually monetary. If the ransom is paid, the decryption key is provided (or not). In such a context, the medical devices used in MCPS, being interoperable and using network interfaces to enable their communication, present important vulnerabilities. Two of the most famous ransomware in recent years are WannaCry [7] and Petya [8]. In this context, in January 2018, the Hancock Health Hospital (US) paid attackers $55,000 to unlock systems following a ransomware infection [9]. Previous outbreaks, such as the infamous NotPetya and WannaCry cases in 2017, also affected hospitals worldwide and allegedly forced some of them (16 in the case of the British NHS) to shut down services, send patients to other hospitals and even postpone scheduled surgeries [10].

Once the impact of ransomware on medical scenarios has been highlighted, it is important to note that the ICE standard has not been designed to protect the elements that form an MCPS from a cybersecurity point of view. In this context, existing protection mechanisms like antiviruses, intrusion detection systems (IDSs), or firewalls are not suitable to detect cyberattacks performed by novel or unseen ransomware. It is because their functioning relies on having metadata like, for example, signatures of known cyberattacks. In addition, not all the traffic patterns generated by ransomware and malware are distinguishable from the normal traffic patterns generated by medical devices and systems with networking capabilities. In this sense, both a malware encrypting a shared folder and an application compressing the same files have a similar traffic pattern. Similarly, normal changes in the clinical environment can be misinterpreted as attacks if the detection mechanisms do not adapt properly. Finally, achieving an acceptable balance between detection and false alarm rates is a difficult task. A high false alarm rate can be rather frustrating for the administrator and a low detection rate can make the system ineffective.

To improve the previous challenges, we propose a system capable of detecting and mitigating automatically and in real time the spreading phase of ransomware. The proposed solution is generic enough to detect different families of ransomware. In this context, as proof of concept to validate our solution we have focused on two of the most important ransomware, WannaCry and Petya. Regarding the detection capabilities, the proposed solution monitors continuously the communications of the devices belonging to the MCPS environment. During the monitoring, real-time processing to create network flows is carried out, after that we group the flows in time windows, and they are automatically analyzed based on rules predefined by the network administrator. If any flow belonging to ransomware is detected, it is identified and classified according to the rules. Once the type of ransomware has been identified, its mitigation process begins. The mitigation depends on the type of ransomware detected, and this can range from a simple rule in a firewall to cut the ransomware traffic, to the replacement of the infected software of a medical device, through virtualization techniques. Finally, once the mitigation is carried out, it is verified that the network flows are normal again and that there is no ransomware in the MCPS. To demonstrate the system usefulness during the detection and mitigation processes, we have implemented a proof of concept in which we perform all these steps and obtaining satisfactory results in terms of precision and time.

This chapter is organized according to the following structure. In Section 4.2 we provide an overview of the Integrated Clinical Environment (ICE). Section 4.3 presents the needed background to understand our contribution and the state of the art regarding similar works oriented to detect and mitigate ransomware attacks. Section 4.4 presents the design details of our solution, which can detect and mitigate ransomware attacks in an automatic and real-time fashion. Section 4.5 defines the proposed clinical scenario, the implementation of the proposed solution and the experiments carried out to demonstrate its usefulness. Finally, in Section 4.6 we show the conclusions and the work that we plan for the future.

4.2 INTEROPERABLE CLINICAL ENVIRONMENT

With the aim of facilitating the interoperability of the elements that make up the hospital rooms of the future and making the MCPS vision possible, the ASTM F2761 standard [3] proposes a patient-centered ICE architecture that allows the open coordination of devices and heterogeneous medical applications. Figure 4.1 shows the most relevant framework components:

- ICE equipment interfaces: They are connected to medical devices to enable their network properties.
- ICE supervisor: It is focused on hosting medical applications that receive and control vital signs of patients.
- ICE network controller is responsible for enabling communications between the supervisor and the interfaces of ICE equipment, as well as managing and maintaining the discovery of medical devices and their information.
- Data Logger is focused on problem-solving and forensic analysis.
- External interface allows communication with external resources of the hospital, such as electronic health records.

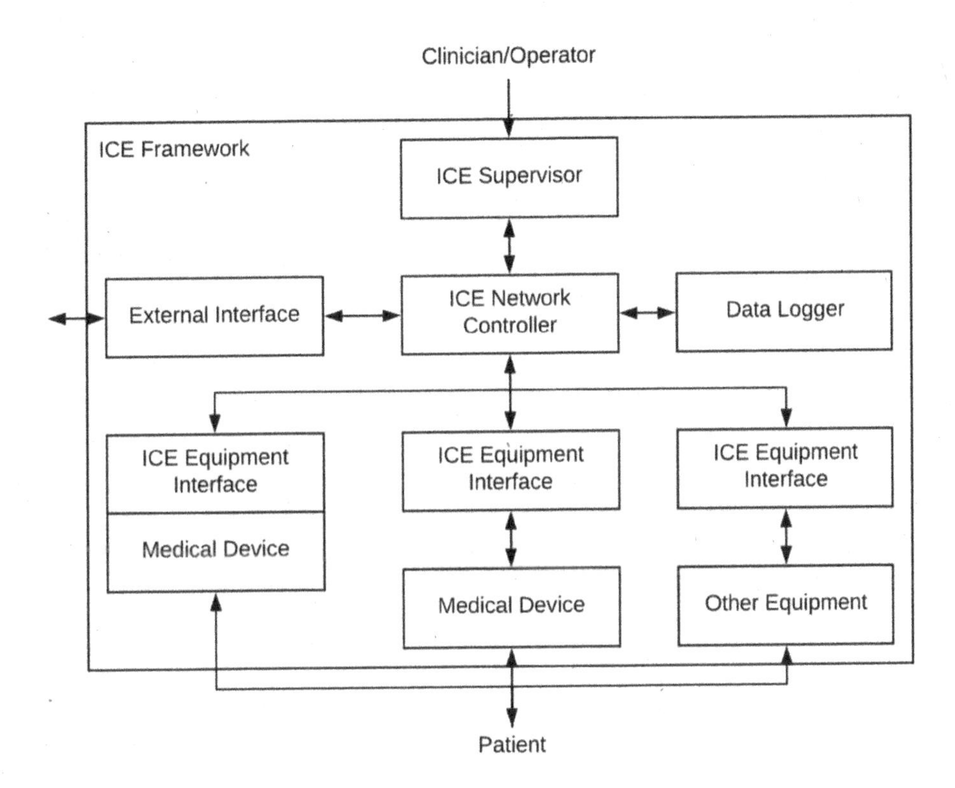

FIGURE 4.1 ICE framework.

4.3 MALWARE AND RANSOMWARE

Classic forms of attacks such as spam, phishing, denial of service (DoS), botnets, trojans, or worms depend on the code used to infect computers and exploit their vulnerabilities. This code is known as malware. Malware is defined as software that meets the malicious objectives of an attacker [11]. These objectives are usually intended to obtain access to systems or resources without the consent of the owners. Different types of malware can work together; they are complementary, so the damage is greater and more difficult to detect. As stated in the introduction, the elements defined by the ICE standard are vulnerable to different types of attacks, among which we highlight malware. Also, nowadays, any user who connects to the Internet faces different security problems. Now the attacks can be very sophisticated using complex techniques or attacks that do not require much knowledge; instead, they use techniques such as social engineering. These attacks also usually depend on the environment in which the target software is located. It is estimated that every day there are more than 1 million new threats [12].

One of the most famous types of malware is known as ransomware. Ransomware is a malware that encrypts the files of infected devices, affecting the availability of data and/or devices, and requests a ransom to decrypt the files and prevent loss of information. Recent ransomware attempts to exploit system vulnerabilities to spread rapidly across networks and affect as many devices or machines as possible. This propagation process is much more harmful when the number of devices is significantly higher. Today, there are several ransomware families with important differences in terms of encryption and dissemination behaviors.

In recent years, ransomware has gained relevance as a devastating method of cybercrime in medical settings. Locky [13], SamSam [14], and more recently, WannaCry and Petya are known families of ransomware that have affected hospitals around the world and forced. Some of them close the services, send patients to other hospitals, and even postpone scheduled surgeries. In this work, we focus on WannaCry and Petya, two of the most important and dangerous ransomware.

WannaCry is a ransomware that automatically spreads across the network exploiting a vulnerability in the Server Message Block (SMBv1) of MS Windows. When a machine has been infected with WannaCry, it tries to connect to a predefined domain. If the site was registered, the ransomware does not run; acting as a kill switch. In contrast, if the connection fails, the ransomware begins with the following two phases:

- Spreading phase: WannaCry tries to spread through the network to infect vulnerable medical devices and computers. To that end, WannaCry uses EternalBlue, an exploit developed by the NSA that attacks a vulnerability in the MS Windows SMBv1 protocol. After successful exploitation, the DoublePulsar payload is sent to execute remote code and infect other devices and computers connected to the network.
- Encryption phase: It can be done before, in parallel, or after the spreading phase (depending on the ransomware version and family). During this phase, the data files are encrypted. Once the process is complete, WannaCry requests a ransom to decipher the files. The ransomware that belongs to the WannaCry family generates and saves a new RSA key pair, which is used to encrypt the medical database and destination data files. Once the previous files are encrypted, WannaCry removes the originals and communicates with an Onion server using a Tor server to transfer the encryption keys. When the ransom is paid, WannaCry obtains the RSA private key decrypted from the Onion server and decrypts the files.

4.3.1 Malware Analysis

When analyzing malware, two types of analysis can be performed: static and dynamic. The static analysis uses reverse engineering techniques and focuses on the analysis of the program instructions. This analysis can be used to subsequently perform a heuristic detection, explained in the following section. Instead, dynamic analysis is used for behavior-based detection. It runs the malware to see malware behavior and carry out its objectives. Normally, static analysis is performed before dynamic analysis, since the latter demands many more resources. In addition, static analysis can serve as input for the dynamic, complementing each other.

Static Analysis of a malicious file can be done in two ways. The first does not demand too many resources and consists of carrying out an analysis of the static characteristics of the analyzed file such as metadata, certificates, or character strings, among others. Instead, the second requires more knowledge and experience regarding the type of file. For example, when analyzing a portable executable [15], it is important to know the executable structure because the sample assembly code provides the intentions and purpose of the file. Likewise, specific static tools are used to perform reverse engineering processes, which consists of passing the assembly language to a higher-level language and identifying the purpose of the file functions. One of the problems that this type of analysis faces is the obfuscation of the code, a method used by the creators of malware to increase the complexity of the previous process.

Dynamic Analysis executes the malware in a controlled environment, to observe its behavior. During the analysis, the interaction that malware has with the system is monitored. The actions intended to be identified in the Windows operating systems [16]:

- Creation, deletion, reading and writing of system files and keys in the Windows registry.
- Network connections detailing the protocol, origin, destination, port, among others.
- Creation and completion of processes with their parameters and breakdown of the functions used and calls to the Windows operating system.

It is essential to consider what type of analysis should be performed in a controlled environment to prevent any malicious activity from spreading. Usually, dynamic analysis performs the following six processes:

1. First, a snapshot of the system must be created before the analysis, the image must then be used to reverse the state of the machine to study more samples or to compare the differences between the initial state and the state after the analysis.
2. The next step is to run the monitoring tool to detect the changes made by the malware in the system.
3. The third step consists of executing the malware.
4. Once all the processes generated by the have finished, the monitoring process is stopped.
5. Then all the relative information is collected, and a report is generated.
6. Finally, the system is reverted to its initial state.

4.3.2 Malware Detection

The detection of different types of malware has always been a very important aspect of data integrity in any environment. In this section, we will see the current state of these detection systems. Detection systems have been improving their techniques to deal with the new types of malware. Detection techniques are categorized, depending on the approach used [17]. The main categories are as follows.

4.3.2.1 Signature-Based Approach

This detection technique is used mainly by antivirus, because it allows to quickly detect malware. Traditional methods of detection involve identifying malware by comparing code in a program with the code of known virus types that have already been encountered, analyzed, and recorded in a database. The most common signatures for this system are the hashes of the files. Solutions based on signatures compare the hash of the file being analyzed with the hashes of known malware, this technique is fast and effective as long as the malware is in the database and the analyzed file has not undergone modifications. The difficulty arises because a constant update is needed to obtain the generated signatures; otherwise, it becomes a vulnerable system. In addition, new malware and ransomware cannot be detected using this technique.

4.3.2.2 Heuristic-Based Approach

The operation of this detection is based on the heuristic analysis of the file through rules and/or algorithms that look for some type of pattern indicating that the file is potentially malware. In this context, the heuristic model was specifically designed to spot suspicious characteristics that can be found in unknown, new viruses and modified versions of existing threats as well as known malware samples. It is possible that if the heuristics are good, signature checks are not necessary, although as a general rule both detection methods are used [18]. Remarkably, these techniques usually use Machine Learning techniques [19] to learn how the behavior of the malware evolves and thus have more power to detect them [20]. Heuristic analysis is ideal for identifying new threats, but to be effective heuristics must be carefully tuned to provide the best possible detection of new threats but without generating false positives on perfectly innocent code. For this reason, heuristic tools are typically deployed along with other methods of detection, such as signature analysis and other proactive technologies.

4.3.2.3 Behavior-Based Approach

Behavior-based techniques execute the malware in a controlled environment to know the actions performed during the infection process. Among other advantages, in some cases, behavior-based detection solutions become the only means to detect and protect from a threat such as fileless malware. Behavioral solutions benefit from Machine Learning (ML) based models on the endpoint to detect previously unknown malicious patterns in addition to behavior heuristic records. Collected from different sources, system events are delivered to the ML model. After processing, ML model produces a verdict if the analyzed pattern is malicious.

4.3.3 Ransomware Detection and Mitigation in Interoperable Medical Settings

This section gives an overview of the cybersecurity concerns found in the current state of ICE and MCPS solutions. Historically, medical devices have been developed as standalone systems without communication capabilities. However, the MCPS vision is emerging to provide interoperability, safety, and security to clinical environments. In this context, key characteristics in the search for appropriate technology, regulations, and ecosystems to ICE are highlighted in [21]. Nowadays, OpenICE [22] is a commonly adopted implementation of the ICE framework. OpenICE is a distributed patient-centric architecture that implements the components defined by the ICE framework. On one hand, the equipment interfaces can run on computers with limited resources (e.g., BeagleBone Black, Raspberry Pi, etc.), which are physically attached to medical devices to provide network capabilities [23]. On the other hand, the communications between the interfaces and the supervisor are managed by the external Data Distribution Service (DDS) middleware [24], which covers partially the cybersecurity of OpenICE.

In the literature, there are different solutions oriented to analyze and mitigate the security and privacy issues of healthcare and MCPS [25, 26]. In this context, one of the most recent works is presented in [27], where the authors review the major security techniques available in the state of the art and study their applicability and utility for the design of MCPS. After that, the authors define an abstract architecture for MCPS to demonstrate various threats. In [28], it is developed a prototype to protect the communications of solutions based on the ICE framework through the security mechanisms provided by the OMG Data Distribution Service (DDS) standard [24]. Specifically, the DDS middleware adds support for authentication, authorization, access control, confidentiality, integrity, and nonrepudiation of the exchanged information. After several experiments, the authors state that transport-level security (TLS) solutions may not provide sufficient resilience against insider attacks, while DDS potentially addresses or mitigates disturb, eavesdrop, and DoS attacks. Despite the important outputs of this proposal, it is not clear how DDS is able to mitigate DoS attacks. Another work, oriented to protect the security and privacy of solutions based on ICE, is proposed in [29]. The authors of this work propose a cloud-based secure logger that receives the information sensed by ICE interfaces attached to medical devices. The proposed logger relies on standard encryption mechanisms to maintain a secure communication channel even on an untrusted network and operating system. This solution is effective against replay, injection, and eavesdropping attacks. However, any behavior that does not lead to message alteration is not detected. In [30], it is designed and implemented an authentication framework for ICE-compliant interoperable medical systems. The proposed framework is composed of three layers, allowing it to fit in the variety of authentication requirements coming from different ICE entities and networking middleware services. The performed experiments demonstrate that the proposed authentication framework protects to device replacement and impersonation attacks to the OpenICE framework.

In addition to the previous security mechanisms proposed in ICE, in the literature we can find solutions in heterogeneous scenarios that leverage Software Defined Networking (SDN) to detect/mitigate attacks [31, 32]; some of them using different ML techniques for detection purposes. However, to the best of our knowledge, none of them has integrated the combination SDN and Network Function Virtualization (NFV) with the ICE standard to be able to replace infected elements such as ICE Equipment Interfaces or ICE Supervisor in a few seconds. Moreover, the detection of the ransomware spreading in ICE with encrypted traffic has not been addressed yet.

Among the ransomware mitigation solutions based on SDN [33], use SDN redirection capabilities together with a blacklist of proxy servers to check if the infected device is trying to connect to one of them to obtain the public encryption key. The mitigation consists in establishing a flow filter to impede this communication and, thus, the encryption of the files. The main drawback of this proposal is that it needs to keep updated a blacklist of proxy servers. These servers must be identified by means of behavioral analysis of known malware, thus making impossible to detect new campaigns. When compared with our solution, our goal is not to prevent the encryption of the files. We attempt to detect the ransomware spreading by using the characteristic traffic patterns generated during that stage. Additionally, our mitigation procedure restores the ICE system to a clean state.

Another relevant and recent paper related to SDN and ransomware is [34]. In this work, the authors use deep packet inspection to identify HTTP POST messages, defining as feature vector the lengths of each three consecutive HTTP POST messages. Then, a classifier is trained by computing both the centroid of the feature vectors belonging to each ransomware class, and a maximal distance to be considered from that class. They obtain a false positive rate (FPR) of about 4% using this method. The main difference with our proposal is that the latter one does not inspect the payload, thus, being suitable even with encrypted traffic. Additionally, our combination of anomaly detector and classifier reaches an FPR of less than 1%. Unfortunately, these results are not comparable due to the context of each solution.

In [35], the authors recognize that malware developers are increasingly using encrypted traffic to avoid payload inspection. Therefore, they propose using SDN to obtain flows and compute a feature vector made up of a combination of interarrival times, packet ratios, and burst lengths. They train a random forest to detect the traffic exchanged between the infected device and the C&C server, obtaining around 10% of FPR. Our solution aggregates the flows belonging to a given time window; thus, obtaining more expressive features. Although our FPR is virtually zero, the results are not comparable due to the different contexts.

Regarding other types of malware, the authors of [36] proposed a system able to detect it using ML classifiers. They focused on analyzing the network traffic and selected four categories of features: basic information, content based, time based, and connection based. To perform the system evaluation the authors used two different datasets, concluding that Bayes network and random forest classifiers produced more accurate outputs than other ML solutions like multilayer perceptron. Similarly, in [37], deep learning techniques are proposed to detect botnet attacks by analyzing network flow patterns from real botnet traffic captures. This is carried out in the context of a MEC oriented system based on NFV/SDN that provides a dynamic management of the resources involved in the detection process. Another solution was presented in [38], where the authors performed a ML analysis of ransomware affecting Windows operating system. They considered the network traffic to achieve an acceptable detection rate. A dataset created from conversation-based network traffic features was used to achieve a true positive detection rate of 97% using the Decision Tree (J48) classifier. The proposal presented in [39] also was focused on ransomware detection. Unlike the previous solutions, this approach extracted relevant features from the API call history produced by ransomware attacks. These features were used along with a Support Vector Machine (SVM) to detect unknown ransomware. The authors demonstrated a correct detection ratio by considering 276 ransomware samples in a Sandbox. The authors of [40] developed a network intrusion detection model based on big data and netflows. In addition, six feature selection algorithms were combined to achieve a better accuracy in terms of classification. On the other hand, in ref. [41] authors presented a methodology for trusted detection of ransomware in a private cloud. The authors considered volatile memory dumps of virtual machines to create features. The results showed that the proposed system was able to detect anomalous states of a virtual machine, as well as the presence of both known and unknown ransomware.

The authors of [42, 43] proposed an automatic, intelligent and real-time system to detect, classify, and mitigate ransomware in ICE. The proposed solution is fully integrated with the ICE++ architecture and makes use of ML techniques to detect and classify the spreading phase of ransomware attacks affecting ICE. Additionally, NFV and SDN paradigms are considered to mitigate the ransomware spreading by isolating and replacing infected devices.

From the industry point of view, Cisco has developed a project called Umbrella [44] that provides a cloud security platform to protect their systems against cyberthreats. Umbrella has been tested at the University of Kansas hospital, one of the most important hospitals in the United States, which has the priority of protecting sensitive data of patients. This solution exposes the importance of this type of software obtaining promising results in terms of detection of malware and specifically of the family ransomware. Of 100,000 attacks received daily, the 30% is ransomware and once the system is installed, almost 100% of them are detected. In addition, this software is capable of tracking attacks in order to correlate users with the events generated by them. This software is efficient and quite complete since it is also capable of mitigating ransomware family attacks, a 75% reduction in response to this case. However, one of the disadvantages Umbrella is that it is an online system, since it uses the cloud. This implies that connectivity issues may cause a significant risk to the clinical environment, in this document we propose a similar solution but considering as a fundamental point its persistence.

Juniper Networks is another company that has taken part in this race of detection and mitigation of malware in interoperable medical environments [45]. The solution proposed by Juniper is a platform that they call software-defined secure networks, SDSN. The platform has the following characteristics: advanced detection technologies for the ransomware family; adaptive policies that communicate beyond firewalls to switches and routers; and ecosystem beyond firewalls, covering switches and routers. In addition, in 2016 Juniper Networks presented its cloud based Skyit Threat Prevention, which combines cutting-edge technologies to detect and deter potential cyber threats. A cloud-based system implies that a service outage can be a disaster for the system, as in other cloud-based systems that we have seen in different documents, we see that the service is always working. Therefore, our solution proposes a generic detection and mitigation system with total availability.

The previous solutions improve in a proactive and reactive way the cybersecurity challenges of clinical and heterogeneous environments. However, they do not consider the real-time and autonomic detection and mitigation of ransomware during its spreading phase. Additionally, these solutions are not integrated in an adaptive and flexible management architecture, which is able to manage the security mechanisms according to the contextual information (e.g., when the ICE system is under a given attack).

4.4 PROPOSAL FOR RANSOMWARE DETECTION AND MITIGATION IN MEDICAL CYBER-PHYSICAL SYSTEMS

This section explains the details of the proposed solution responsible for automatically detecting and mitigating the diffusion phase of ransomware attacks in real time. As proof of concept, we focus the explanation on the behavior of the WannaCry and Petya ransomware. However, it is important to know that the proposed solution is generic enough to detect and mitigate other ransomware families.

Figure 4.2 shows the components of the three main modules making up the proposed solution, which are briefly introduced as follows:

- **Monitor:** This module is in charge of monitoring in real time the network traffic of medical devices and generating network flows according to the network packets.
- **Analyzer:** The component of this module analyses the network flows to find behavior patterns of clean and infected medical devices belonging to the clinical environment.
- **Decision-maker:** The components of this module allow the network administrator to define network management rules able to detect and mitigate ransomware attacks in real time.

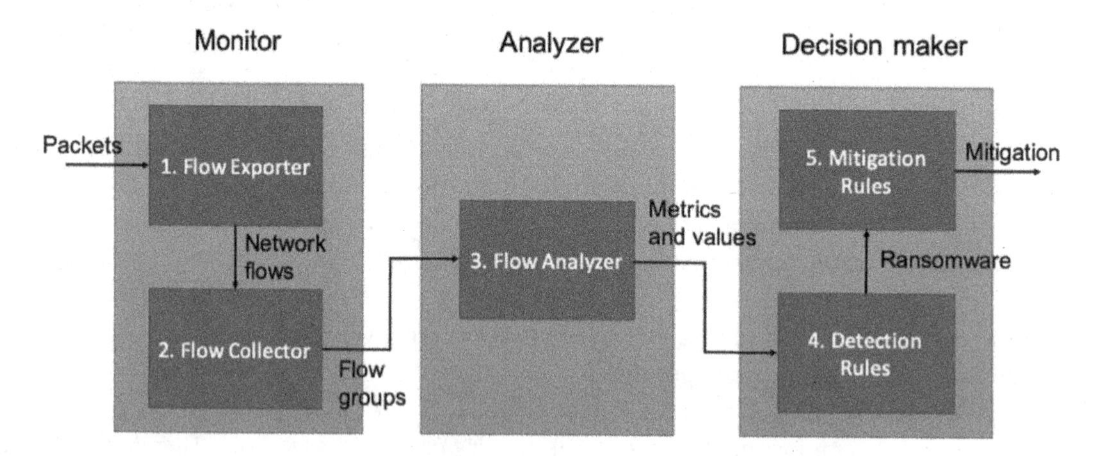

FIGURE 4.2 Design of the proposed solution.

4.4.1 Monitoring, Generation, and Processing of Network Flows

The first step to secure interoperable medical environments is to monitor the traffic produced by medical devices belonging to their networks. The main function of this first module is to collect in real time the network packets generated by the medical devices of the clinical scenario and generate the network flows as the composition of the monitored packets. Once the flows have been generated, we group them into configurable time windows.

The *Flow Exporter* component aims to collect all packets that go from any device belonging to the clinical network to any IP destination. This is possible thanks to the constant monitoring of the network interface that all ICE devices in the medical environment have and in which they are interconnected with each other in the same subnet.

With the goal of generating network flows, we consider the number of packets exchanged between source/destination IPs, the source/destination ports, and the protocols. Network flows are generated by grouping the packets of the level 3 transport layer of the OSI model (see Figure 4.3). Each network flow contains the number of level 3 packets have been sent exchanged between two IPs and ports using a given protocol. This work is made by the *Flow Collector* component, which gathers the information processed by the Flow Exporter and generates network flows based on the previous three aspects (source/destination IP, destination port, and protocol). After that, this component groups the flows in configurable time windows. This procedure facilitates monitoring with different time intervals. The grouping of the flows will be passed to the next module, in charge of analyzing them to find behavior patterns that will be used to detect ransomware of the same family. This decision is directly related to the analysis process, which is explained in Section 4.4.

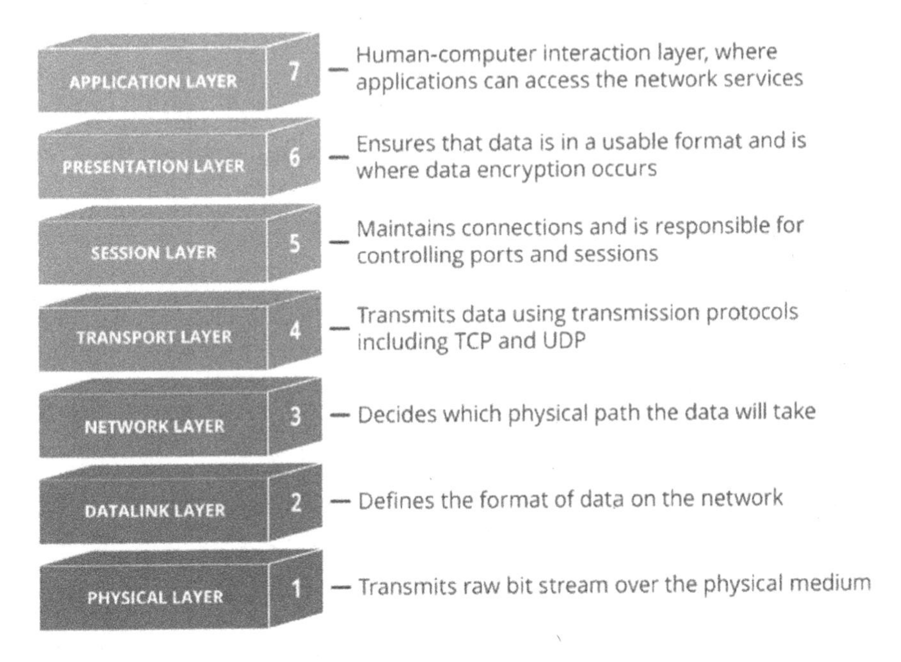

FIGURE 4.3 Layers making up the OSI model.

4.4.2 Network Flow Analysis

The *Flow Analyzer* component of the proposed solution allows the network administrators to analyze the clinical network status through the use of metrics oriented to the network flows. In particular, the metrics taken into account by this module are:

- Number of packets per flow that use a specific protocol and two Internet protocols (IPs)/ports.
- Number and type of flows exchanged in each time window.
- Frequency of exchanged Network flows and packages.
- Addresses, ports, and protocols used by medical devices.

Based on the previous metrics and the study of the network flows acquired during the monitoring phase in different situations (clean and under attack), the network administrator is able to establish particular thresholds able to detect each kind of ransomware.

To explain how the analysis phase is done, we will focus on the spreading phase of WannaCry and Petya, two of the most recent and dangerous ransomware affecting clinical environments. At this point, it is important to note both ransomware exploit the vulnerability EternalBlue. As a starting point, WannaCry and Petya use ARP requests to discover the active devices of the clinical network. After that, the devices responding to the previous ARP requests are those that attempt to attack by exploiting the vulnerability EternalBlue, explained in Section 4.2. The network packets in charge of exploiting EternalBlue can be identified because they use the Transmission Control Protocol (TCP) and the destination port 445. In this context, by considering the previous metrics the proposed solution allows the network administrator to detect an anomalous amount of ARP and TCP packets. Based on the metrics and network flows the proposed solution can dictate the result of the analysis. After performing the analysis, which is detailed in Section 4.4, the main difference between Petya and WannaCry is based on the number of exchanged TCP packets with destination port 445. Once the analysis has been performed, the administrator is in charge of defining network managing rules, which are explained in the following section.

4.4.3 Decision-Making for Attack Detection and Mitigation

This module is responsible for making decisions to detect and mitigate the diffusion phase of ransomware. These decisions are made according to predefined network management rules that contain metrics and thresholds established by the network administrator after the network flow analysis process, explained in Section 4.3.2.

The rules are composed of an antecedent and a consequent. The consequence of the rule is enforced when all conditions of the antecedent are met, which may be one or more. The antecedent conditions are established according to metrics, which will be focused on protocols, ports, number of network packets, and flows exchanged between the devices (as explained in Section 4.4.2). In addition, each metric has a threshold defined by the network administrator according to the study performed during the analysis phase. As indicated, once the antecedent of a rule has been fulfilled, the consequent of the same is executed, which implies the enforcement of one or more specific actions. This consequence will also be defined by the network administrator and will focus on ransomware detection and mitigation.

4.4.3.1 Detection

Since the analysis of WannaCry and Petya has shown that ARP and TCP messages are key, the Detection rules component of the proposed solution relies on them when defining the rules to detect WannaCry and Petya diffusion patterns. To do this, the network administrator establishes for each metric the threshold fitting into each attack. When the thresholds defined for ARP and TCP metrics are accomplished, it means that there is a ransomware attack affecting the clinical environment.

As an example, the next rule establishes that when the number of ARP packets of a device, which IP address is A, contained in one or more flows of a given time windows is higher than a preestablished threshold called *ARP_threshold* and the number of SMBv1 packets of a device, which IP address is A, contained in one or more flows of a given time windows is higher than a preestablished threshold called *SMBv1_threshold* then the machine with IP_A is infected by a ransomware. This composes the background of the ransomware detection rule, as we see below:

If (number of ARP packets per flow from IP_A > ARP_threshold) & (number of SMBv1 packets per flow from IP_A > Threshold) → Ransomware has infected IP_A

As an example, it is important to highlight that WannaCry has a much larger number of TCP message exchange on port 445 than Petya. These values are used by the proposed solution to distinguish different ransomware and to detect different ransomware belonging to the same family. The values of these metrics are detailed in Section 4.5.

4.4.3.2 Mitigation

Once the type of ransomware that attacks the environment has been detected, the Mitigation rule component provides the network administrator with management rules to mitigate the attack. Mitigation rules have a similar structure to those focused on the detection. The main difference is the consequence. Instead of detecting one particular ransomware, they indicate the actions to enforce its mitigation. Rules can perform different actions, depending on the threat detected and what process it is in. In our case, since it is a reactive detection (once the attack has occurred), the proposed system will replace the infected software by using virtualization techniques. However, the proposed rules allow the administrator to define other types of mitigations, such as firewall rules to block the communications between infected devices.

In our proof of concept, we have a configured, clean and ready copy of the machine allocating the medical device, so when we have to replace one in a mitigation process, we proceed as follows: (1) turn off the infected machine, (2) destroy machine to release resources, (3) deploy the clean machine, and (4) start the medical device controller.

4.5 IMPLEMENTATION AND EXPERIMENTAL RESULTS

This section presents the implementation details of the proposed solution and the proof of concept developed to demonstrate the usefulness of the proposed solution.

4.5.1 Scenario Description

As proof of concept, we have designed and deployed a hospital room of the future that uses MCPS. The scenario is composed of a medical supervisor, two different medical devices, and a distributed database. On the one hand, we have a supervisor, in charge of collecting information on medical devices that are connected to the patient. On the other hand, we have medical devices, such as an electrocardiogram or a blood oxygen meter, responsible for monitoring the patient's vital signs. In addition to the above, the MCPS has a medical database distributed in three machines that are part of the scenario. This database is responsible for carrying out readings and writings in a pseudo-random way to simulate the behavior of a real database where medical professionals make queries to obtain patient information.

Figure 4.4 shows how the aforementioned elements are deployed in three virtual machines. These machines are hosted on a physical host, which acts as a staging router

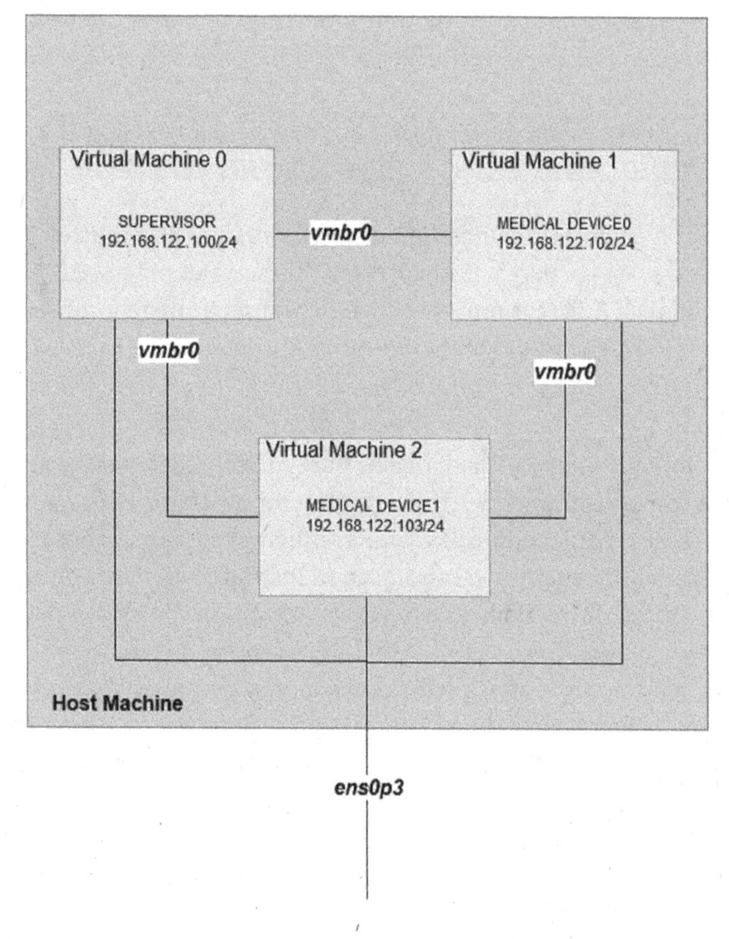

FIGURE 4.4 Integrated clinical environment.

through the network interface ens0p2. This output is controlled through a definition of iptables to control access to the environment. Virtual machines have Windows 7 as an operating system, OpenICE as an implementation of the ICE standard, and SMBv1 as a data exchange protocol for the medical database. We use OpenICE to emulate interoperable medical devices on the machines themselves. The two medical devices are connected to the supervisor thanks to the interoperability offered by the OpenICE protocols. These devices are emulated with OpenICE, which creates one device per virtual machine. The network used for in our proof of concept has been the 192.168.122.0/24 and each MV has a fixed IP address, being as follows:

- Supervisor: 192.168.122.100
- Medical device 0: 192.168.122.102
- Medical device 1: 192.168.122.103

Comparing the proposed scenario with Figure 4.1, it is important to indicate that both the *data logger* and the *ICE network controller* are located on the machine that houses the medical supervisor. The other devices are connected to each other through the *vmbr0* interface.

Based on this clinical scenario and network topology, we have performed two different infections. First, manually we have infected the medical device 1 with WannaCry and it has spread across the network to infect the medical device 0. After that, we have destroyed the environment and created a new one, clean. In this second scenario, we have reproduced the same behavior with Petya. Below we show the monitoring, analysis, and decision-making processes performed by our solution according to the clean and infected scenarios.

4.5.2 Monitoring

To deploy the *Monitor* module and its two components (described in Section 4.5.1), our solution uses the Argus tool [46]. Argus captures network traffic in layer 2 and generates network flows from existing network packets. It allows the monitor module to obtain the number of packets that have been exchanged between the medical devices belonging to the clinical scenario previously defined. In addition, it also generates the active network flows belonging to different connections between the devices of the clinical environment.

To perform the monitoring process, Argus acquires all traffic that passes through the *vmbr0* interface, which is where all medical devices are connected to the supervisor. After that, the traffic is aggregated in network flows according to a windows time and other parameters explained in the following text. Finally, the network flows are saved in an Argus file according to predefined time window used by our solution to detect ransomware attacks. The file generated by Argus with the network flows is read by a function defined as Read Argus (RA). The following command is used by our solution to read the Argus files: *ra -nr file.argus -t 2019/06/11.20 -s stime proto saddr sport dir daddr dport pkts spkts dpkts – "arp".*

TABLE 4.1 Example of Network Flows Generated with RA

Time	Prototype	SrcAddr	SrcPort	Dir	DstAddr	DstPort	TopPkts	SrcPkts	DstPkts
19:15:35.905872	tcp	192.168.122.102	49161	<?>	192.168.122.100	139	3	2	1
19:15:51.135001	tcp	192.168.122.102	49161	<?>	192.168.122.100	139	3	2	1
19:15:53.960477	tcp	192.168.122.103	49158	<?>	192.168.122.102	445	3	2	1
19:15:54.649711	tcp	19:15:54.649711	49165	<?>	192.168.122.103	445	1	1	0
19:15:55.657230	tcp	192.168.122.100	49166	<?>	192.168.122.103	139	61	31	30
19:16:06.146977	tcp	192.168.122.102	49161	<?>	192.168.122.100	139	5	3	2

The options used in the previous command are the following ones and Table 4.1 shows, the RA output when reading from a file that is constantly monitored.

- *-n*, translates the name of the service to the port of that same service.
- *-t*, indicates the time window we want to read from that file in the example, we show all the network flow corresponding to the Argus file on June 11, 2019 at 8pm. If we wanted to change the window and pass it from 1 h to 1 min, we added the minutes to the filter: 2019/06/11.20: 30. The format of this filter is as follows: [[[yyyy/] mm/] dd.] HH [: MM [: SS]]. There is also the possibility of establishing all types of ranges over time, as well as using exclusion, inclusion, and denial among others.
- -s, indicates what will be shown of the Argus trace when it is shown on the screen, we decided that the most important parameters to show in order of appearance in the order are: flow start time, protocol used, source IP, source port, flow direction, destination ip, destination port, number of packets exchanged in the flow, packets sent by the source, and packets sent by the destination.
- Finally, the parameters that follow the two scripts are the filters managed by RA. For this case we put "ARP" to show us only the ARP exchange flows.

The network flows are generated according to the desired time windows. This particular part is done when reading from the Argus file, classifying those flows according to the time windows set to the routine call: ra -nr file.argus -t cx-y -s. In the -t option, the first c indicates that we are going to group all the flows that are within the xy range, these being the seconds elapsed from the window, that is, we convert the date on which we are working in seconds (x) and add the desired time window (x + window = y). In this way, we have a time window in which the flows are grouped.

4.5.3 Network Flow Analysis

After monitoring the network flows of the clinical environment in three different situations (clean, infected by WannaCry, and infected by Petya), this section explains how the selected metrics are useful to show how the network packages and flows are affected by the different attacks.

Before starting with the analysis of the network flows, we have made an experiment to choose the most suitable time windows to detect the ransomware attacks. In this context,

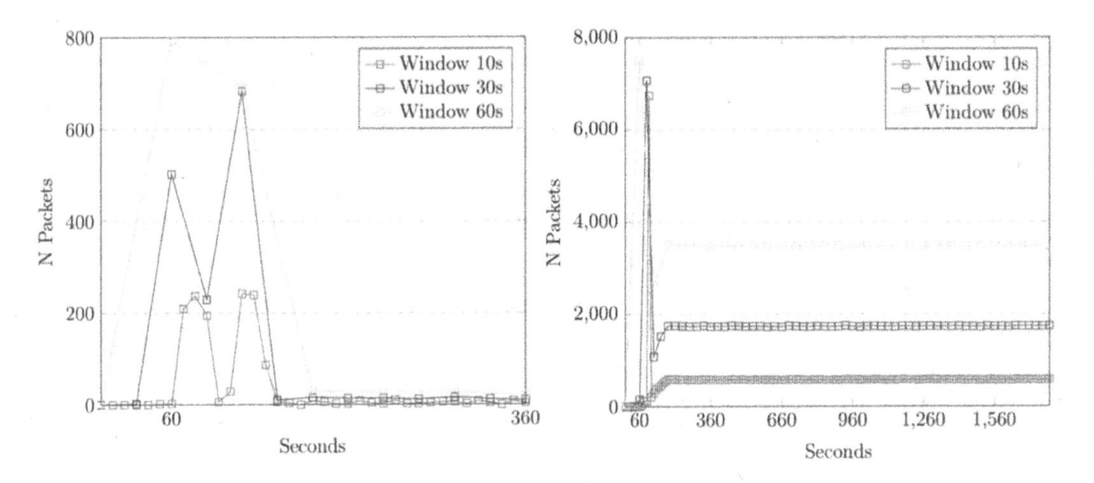

FIGURE 4.5 WannaCry affecting the number of ARP (left side) and TCP (Port 445) packets (Right Side) according to different time windows used to generate the network flows.

we have considered three different time windows of 10, 30, and 60 sec to aggregate the ARP and TCP (port 445) network flows.

In this context, the left side of Figure 4.5 shows the behavior of ARP packets when a medical device has been infected with WannaCry. In addition, the right side shows the same but in terms of TCP packets in port 445. On the other hand, the left side of Figure 4.6 shows the behavior ARP packets along the time when a medical device has been infected with Petya. Finally, the right side of Figure 4.6 shows the behavior for TCP packets on port 445. As can be seen, the behavior of both ransomware attacks is different. In WannaCry the ARP and TCP flows are generated just at the beginning of the attack. In contrast, Petya tries to discover and infect the medical devices of the clinical network continuously. In terms of time windows, we can see that despite the windows of 10 sec and 30 sec are fine to

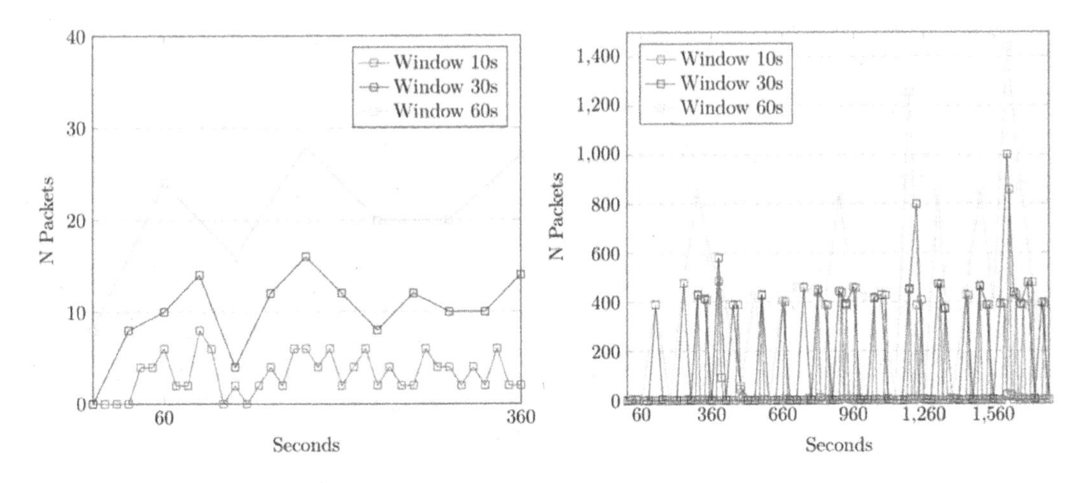

FIGURE 4.6. Petya affecting the number of ARP (left side) and TCP (Port 445) packets (right side) according to different time windows used to generate the network flows.

detect WannaCry, for Petya, 60 sec is the best approach to reduce false positives in the TCP traffic. According to these results, we have chosen a time windows of 60 sec to perform the following experiments of this section.

Once the time window has been selected, the next step is to define the thresholds associated with the ARP and TCP metrics defined in Section 4.4 for both ransomware attacks, WannaCry and Petya. With that goal Figure 4.7 shows the impact of both attacks in terms of ARP and TCP flows generated in the clinical environment. In each plot, we show the number of ARP packets generated by an infected device, the behavior of a clean device, and the behavior of two medical devices of our environment before and after their infections with WannaCry and Petya.

Figure 4.7 shows that the behavior of WannaCry and Petya is quite different. WannaCry discovers new devices to spread itself just at the beginning of the attack. It is achieved by a significant increment of the number of ARP messages (as can be seen in the top-left graph). Furthermore, in the bottom-left graph, we see that WannaCry

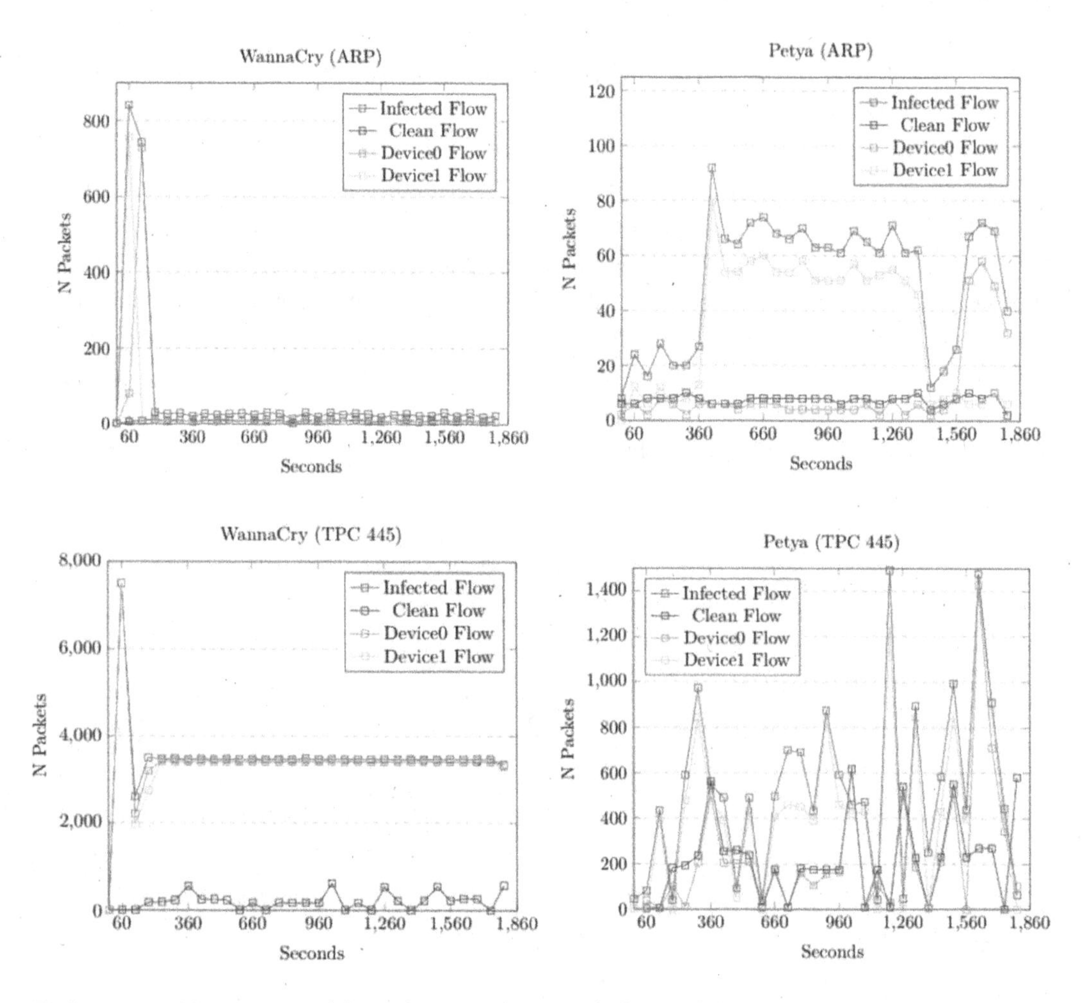

FIGURE 4.7. How ARP and TCP (Port 445) network flows of different devices are affected by WannaCry and Petya.

infects the device 0 and devices 1 during the first 100 sec. I contrast, the spreading phase of Petya is not so fast, but it is more constant in terms of ARP packets sent along the time. In this case, Petya infects the device 1 around the 360 sec and device 0 in the second 1.560. It can be inferred because when both devices are infected, the number of ARP packets sent from these devices increases in a significant way. This is because they are trying to discover more devices to spread themselves. The right side of Figure 4.7 clearly shows the changes in devices 0 and 1 when they are infected. Finally, after analyzing Figure 4.7, the network administrator is able to establish the thresholds used by the detection rules to detect WannaCry and Petya. This last step is explained in the next section.

4.5.4 Network Flow Detection

Once analyzed the behavior of WannaCry and Petya in terms of ARP and TCP communications during their spreading phase, this section is focused on implementing the management rules in charge of detecting and differentiating both ransomware in an automatic fashion. Below we show how the decision-maker module (presented in Figure 4.2) implements the rules by using a script that we have developed as a proof of concept in our medical environment, explained in Section 4.5.1. This script is executed on the host machine that is responsible for virtualizing the medical devices and services composing our integrated clinical environment. First of all, it is important to comment that the script able to detect and distinguish the ransomware attacks receives the following parameters:

- *ip_file*, with the IP addresses that we want to monitor, in our case it will be the IP of our clinical network;
- *argus file* is the file with the network flows, this file is continuously updated with new flows;
- the date of the sample, in the format *yyyy/mm/dd*; and
- the minutes that the monitoring will take.

When that time is up, *cron.d* automatically puts it back into operation. This ensures continuous monitoring, analysis, and detection. After that, we control the number of parameters received. Then we initialize the array with the IP address of the clinical network to analyze and select the start time of the trace, both the one that we are going to analyze and the one of the clean trace. Next, we work on the time window. In this case, the time is 60 sec, and we take out the number of packets exchanged in that flow and store them in different variables for later comparison. After that, we get the network flows per minute and how many packets they have. The goal is to count the current number of packets per minute and see if they exceed twice (the established threshold) the number of packets per flow of a clean trace. If this last condition is met, we consider that the trace is infected, and we will proceed to replace the software of the medical device.

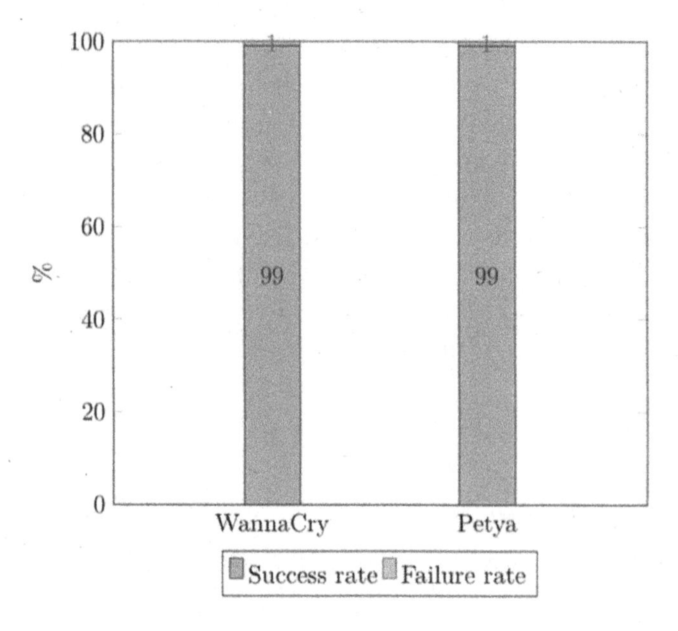

FIGURE 4.8 Detection success rate of the proposed solution.

We have taken several measures to check the final performance of this test with the previous rules obtaining satisfactory results. Figure 4.8 shows 99% of success when detecting WannaCry and Petya ransomware. We consider the failures when a WannaCry flow detects that it is Petya since the threshold of WannaCry is higher than the Petya one.

4.5.5 Mitigation

Once we have detected an infected network flow through the previous rules, the last step is to replace the software of the medical device by using virtualization. The mitigation rules are defined by the administrator and they can be flexible or rigorous depending on the scenario conditions. Our solution has a clone virtual machine of each one of the devices making up the scenario, so when we detect an abnormal network flow through the previous policies, we turn off the medical device and replace its software controller with a clean copy.

As we observed in the previous code, when the two defined thresholds are exceeded and an infected network flow has been detected, we proceed to stop the machine. This machine has the format id ip.clone so we can identify it with its IP. Once the virtual machine has stopped, the next step is to remove the infected virtual deploying the new one. Once copied with the configuration of that device, we start it up and after a few seconds, the VM is reconnected, having the device running again and having eliminated the previous infection. This generic and efficient solution allows us to replace an interoperable medical device with a completely new threat-free one.

Finally, we show an average of the copy and deployment times of the virtual machines once the mitigation has been carried out. Figure 4.9 shows this information divided into two parts. The first one is the time it takes to copy the state of a virtual machine, it should

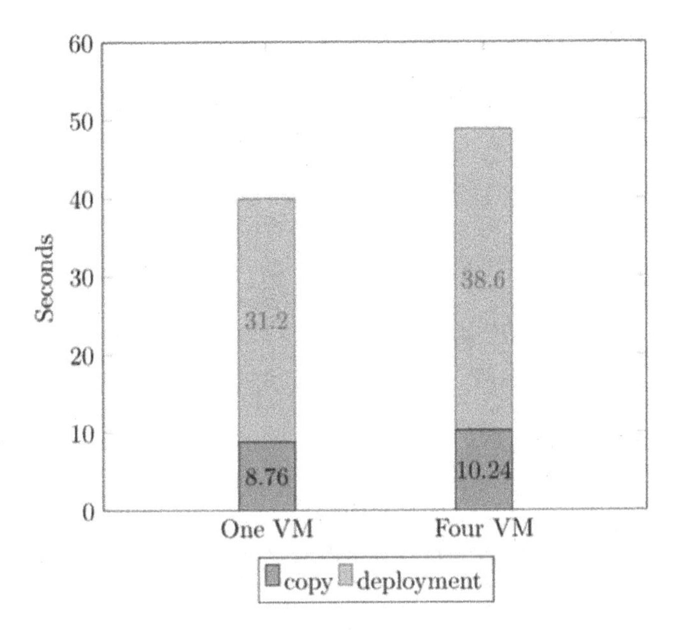

FIGURE 4.9 Time used by our solution to mitigate a ransomware attack in the proposed scenario.

be noted that in this proof of concept they were moving from an SSD disk to another SSD with a transfer rate of 3.61 GB/sec. The second one is the time required by our solution to deploy a single machine. We have performed the measurement of both times when there are one and four machines running in the clinical scenario. As can be seen in Figure 4.9 the increment of time when the number of existing virtual machines is higher increases but in an acceptable way.

4.6 CONCLUSIONS AND FUTURE WORK

Hospitals are evolving to interoperable clinical environments by implementing Medical Cyber-Physical Systems. Despite the benefits of these systems, they open the door to new cybersecurity challenges. Nowadays, ransomware attacks are one of the most dangerous things that affect hospitals due to their significant problem due to the criticality of the actions they take and the data they manage. After a study of the ransomware that affects these environments, it has been detected that they have specific behaviors that can be detectable by real-time systems. In this sense, this work presents an automatic solution able to monitor, detect, classify, and mitigate ransomware attack. As proof of concept and to demonstrate the usefulness of the proposal, we have virtualized an integrated clinical environment and applied two of the most dangerous and novel ransomware attacks, WannaCry and Petya.

The proposed solution monitors the network flows generated by the medical devices belonging to the clinical environment. Those flows are aggregated according to configurable time windows, after that our solution allows the network administrator to analyze the flows by using predefined metrics. Finally, the administrators are in charge of define network management rules to detect and mitigate the ransomware attacks in real time. In our proof of concept, we have shown how to detect WannaCry and Petya and how to mitigate them by using virtualization techniques. The experiments performed in terms of success rate and mitigation times demonstrate the validity of the proposed solution.

As future work, we plan to deploy the proposed solution in other scenarios, not only the clinical one, managing sensitive data that needs security and privacy mechanisms. With more study and tools, the system can be improved to detect other types of cyberthreats, being a good starting point to secure the data and protect them properly. To conclude, we also plan to expand the proposed solution to detect and mitigate other ransowmare attacks such as BadRabit, Locky, GranCrab, or Satana, and similar families. Finally, we consider the use of automatic communication management techniques such as SDN to mitigate the attacks.

ACKNOWLEDGMENTS

This work has been supported by the Irish Research Council under the government of Ireland post-doc fellowship (grant code GOIPD/2018/466), and Spanish Ministry of Science, Innovation and Universities, through the SAFEMAN project (grant code RTI2018-095855-B-I00). Special thanks to Jesús Salinas Pérez-Hita for his work on the design and implementation phases.

REFERENCES

1. Marcos Pablos S, García Holgado A, and García Peñalvo FJ, "Trends in European research projects focused on technological ecosystems in the health sector", In Proceedings of the Sixth International Conference on Technological Ecosystems for Enhancing Multiculturality, Salamanca, Spain, pp. 495–503, 2018.
2. Stankovic JA, "Research Directions for Cyber Physical Systems in Wireless and Mobile Healthcare", ACM Transactions on Cyber-Physical Systems, vol. 1, no. 1, p. 12, 2016.
3. Medical Devices and Medical Systems – Essential Safety Requirements for Equipment Comprising the Patient-Centric Integrated Clinical Environment (ICE) Part 1: General Requirements and Conceptual Model, ASTM International: West Conshohocken, 2013.
4. Sixth Annual Benchmark Study on Privacy & Security of Healthcare Data, Technical Report, Ponemon Institute, USA, 2016.
5. Data Breach Investigations Report, Technical Report, Verizon, USA, 2018.
6. Brewer R, "Ransomware Attacks: Detection, Prevention and Cure", Network Security, no. 9, pp. 5–9, 2016.
7. Chen Q, Bridges RA, "Automated Behavioral Analysis of Malware: A Case Study of WannaCry Ransomware", In Proceedings of the 2017 16th IEEE International Conference on Machine Learning and Applications (ICMLA), Cancun, Mexico, pp. 454–460, 2017.
8. Aidan JS, Verma H K, Awasthi LK, "Comprehensive Survey on Petya Ransomware Attack", In Proceedings of the 2017 International Conference on Next Generation Computing and Information Systems, Jammu, India, pp. 122–125, 2017.
9. Osborne C, "US Hospital Pays $55,000 to Hackers After Ransomware Attack", 2018. Available online: https://www.zdnet.com/article/us-hospital-pays-55000-to-ransomware-operators/ (accessed on 29 September 2019).
10. Mohney G, "Hospitals Remain Key Targets as Ransomware Attacks Expected to Increase", 2018. Available online: https://abcnews.go.com/Health/hospitals-remain-key-targets-ransomware-attacks-expected-increase/story?id=47416989 (accessed on 29 September 2019).

11. Willems C, Holz T, Freiling F, "Toward Automated Dynamic Malware Analysis Using CWSandbox," IEEE Security & Privacy, vol. 5, no. 2, pp. 32–39, 2007.
12. Trend Micro, "Malware: 1 Million New Threats Emerging Daily", 2015. Available online: https://blog.trendmicro.com/malware-1-million-new-threats-emerging-daily/ (accessed on 29 September 2019).
13. Cabaj K, Gregorczyk M, Mazurczy W, "Software-Defined Networking-Based Crypto Ransomware Detection Using HTTP Traffic Characteristics", Computers & Electrical Engineering, vol. 66, pp. 353–368, 2018.
14. Mukesh SD, "An Analysis Technique to Detect Ransomware Threat", In Proceedings of the International Conference on Computer Communication and Informatics, Coimbatore, India, pp. 1–5, 2018.
15. Ero Carrera. PEFILE, a Python module to read and work with PE files, 2018. Available online: https://github.com/erocarrera/pefile (accessed on 29 September 2019).
16. Bazrafshan Z, Hashemi H, Fard SMH, Hamzeh A, "A Survey on Heuristic Malware Detection Techniques", In The 5th Conference on Information and Knowledge Technology, Shiraz, Iran, pp: 113–120, 2013.
17. Idika N, Mathur AP, "A Survey of Malware Detection Techniques", Purdue University, vol. 48, pp. 2–48, 2017.
18. Department of Computer Science Virtual University of Pakistan Rabia Tahir. "A Study on Malware and Malware Detection Techniques, 2018". Available online: http://www.mecs-press.org/ijeme/ijeme-v8-n2/IJEME-V8-N2-3.pdf (accessed on 29 September 2019).
19. McGregor A, Hall M, Lorier P, Brunskill J, "Flow Clustering Using Machine Learning Techniques", Lecture Notes in Computer Science, vol. 3015, 2004.
20. Ye Y, Li T, Adjeroh D, Iyengar SS, "A Survey on Malware Detection Using Data Mining Techniques", ACM Computing Survery, vol. 50, no. 3, pp. 1–41, 2017.
21. Hatcliff J, King A, Lee I, Macdonald A, Fernando A, Robkin M, Vasserman E, Weininger S, Goldman J M, "Rationale and Architecture Principles for Medical Application Platforms", In Proceedings of the 2012 IEEE/ACM Third International Conference on Cyber-Physical Systems, Beijing, China, pp. 3–12, 2012.
22. Arney D, Plourde J, Goldman J M, "OpenICE Medical Device Interoperability Platform Overview and Requirement Analysis", Biomedical Engineering/Biomedizinische Technik, vol. 63, pp. 39–47, 2017.
23. Rajkumar S, Srikanth M, Ramasubramanian N, "Health Monitoring System Using Raspberry PI", In Proceedings of the International Conference on Big Data, IoT and Data Science (BID), Pune, India, pp. 116–119, 2017.
24. Köksal Ö, Tekinerdogan B, "Obstacles in Data Distribution Service Middleware: A Systematic Review", Future Generation Computer Systems, vol. 68, pp. 191–210, 2017.
25. Huertas Celdrán A, Gil Pérez M, García Clemente FJ, Martínez Pérez G, "Preserving Patients' Privacy in Health Scenarios through a Multicontext-aware System", Annals of Telecommunications, vol. 72, pp. 577, 2017.
26. Coventry L, Branley D, "Cybersecurity in Healthcare: A Narrative Review of Trends, Threats and Ways Forward", Maturitas, vol. 113, pp. 48–52, 2018.
27. Almohri H, Cheng L, Yao D, Alemzadeh H, "On Threat Modeling and Mitigation of Medical Cyber-Physical Systems", In Proceedings of the IEEE/ACM International Conference on Connected Health: Applications, Systems and Engineering Technologies, Philadelphia, PA, pp. 114–119, 2017.
28. Soroush H, Arney D, Goldman J, "Toward a Safe and Secure Medical Internet of Things", IIC Journal of Innovation, vol. 2, pp. 4–18, 2016.
29. Nguyen H, Acharya B, Ivanov R, Haeberlen A, Phan L T X, Sokolsky O, Walker J, Weimer J, Hanson W, Lee I, "Cloud-Based Secure Logger for Medical Devices", In Proceedings of the IEEE First International Conference on Connected Health: Applications, Systems and Engineering Technologies, Washington, DC, pp. 89–94, 2016.

30. Cheng L, Li Z, Zhang Y, Zhang Y, Lee I, "Protecting Interoperable Clinical Environment with Authentication", SIGBED Review, vol. 14, pp. 34–43, 2017.
31. Huertas Celdrán A, Gil Pérez M, García Clemente F. J, Martínez Pérez G, "Enabling Highly Dynamic Mobile Scenarios with Software Defined Networking", IEEE Communications Magazine, vol. 55, no. 4, pp. 108–113, 2017.
32. Huertas Celdrán A, Gil Pérez M, García Clemente F. J, Martínez Pérez G, "Sustainable Securing of Medical Cyber-Physical Systems for the Healthcare of the Future, Sustainable Computing: Informatics and Systems, vol. 19, pp. 138–146, 2018.
33. Cabaj K, Mazurczyk W, "Using Software-Defined Networking for Ransomware Mitigation: The Case of CryptoWall", IEEE Network, vol. 30, pp. 14–20, 2016.
34. Cabaj K, Gregorczyk M, Mazurczy W, "Software-Defined Networking-Based Crypto Ransomware Detection Using HTTP Traffic Characteristics", Computers & Electrical Engineering, vol. 66, pp. 353–368, 2018.
35. Cusack G, Michel O, Keller E, "Machine Learning-Based Detection of Ransomware Using SDN", In Proceedings of the 2018 ACM International Workshop on Security in Software Defined Networks & Network Function Virtualization, Tempe, AZ, USA, pp. 1–6, 2018.
36. Narudin FA, Feizollah A, Anuar NB, Gani A, "Evaluation of Machine Learning Classifiers for Mobile Malware Detection", Soft Computing, vol. 20, pp. 343–357, 2016.
37. Fernández Maimó L, Huertas Celdrán A, GilPérez M, García Clemente F J, Martínez Pérez G, "Dynamic Management of a Deep Learning-Based Anomaly Detection System for 5G Networks", Journal of Ambient Intelligence and Humanized Computing, vol. 10, no. 8, pp. 1–15, 2018.
38. Alhawi OMK, Baldwin J, Dehghantanha A, "Leveraging Machine Learning Techniques for Windows Ransomware Network Traffic Detection", Cyber Threat Intelligence, vol. 70, pp. 93–106, 2018.
39. Takeuchi Y, Sakai K, Fukumoto S, "Detecting Ransomware Using Support Vector Machines", In Proceedings of the 47th International Conference on Parallel Processing Companion, Boston, MA, USA, no. 1, pp. 1–6.
40. Wan Y, Chang J, Chen R, Wang S, "Feature-Selection-Based Ransomware Detection with Machine Learning of Data Analysis", In Proceedings of the 3rd International Conference on Computer and Communication Systems, Nagoya, Japan, pp. 85–88, 2018.
41. Cohen A, Nissim N, "Trusted Detection of Ransomware in a Private Cloud Using Machine Learning Methods Leveraging Meta-features from Volatile Memory", Expert Systems with Applications, vol. 102, pp. 158–178, 2018.
42. Huertas Celdrán A, García Clemente F J, Weimer J, Lee I, "ICE++: Improving Security, QoS, and High Availability of Medical Cyber-Physical Systems Through Mobile Edge Computing", In 2018 IEEE 20th International Conference on e-Health Networking, Applications and Services (Healthcom), Ostrava, Czech Republic, pp. 1–8, 2018.
43. Fernández Maimó L, Huertas Celdrán A, Perales Gómez Á L, García Clemente F J, Weimer J, Lee I, "Intelligent and Dynamic Ransomware Spread Detection and Mitigation in Integrated Clinical Environments", Sensors, vol. 19, no. 5, pp. 1114, 2018.
44. Cisco Systems CA USA, "Ransomware: What Every Healthcare Organization Needs to Know". Available online: https://www.cisco.com/c/dam/en_us/solutions/industries/docs/ransomware-what-every-healthcare-organization-needs-to-know.pdf (accessed on 29 September 2019).
45. Himss Media USA, "Cyberthreats, Ransomware, Malware: Is Your Healthcare Organization Prepared?", 2016.
46. Holman RA, Stanley J, "The History and Technical Capabilities of Argus", Coastal Engineering, vol. 54, no. 6–7, pp. 477–491, 2007.

CHAPTER 5

Security Challenges and Requirements for Industrial IoT Systems

Valentin Vallois, Ahmed Mehaoua
and Fouad Amine Guenane

CONTENTS

5.1 INTRODUCTION

In recent years, the rise of the Internet of Things (IoT) has brought many smart devices to the Internet [1]. As for the industry sector, the integration of IoT devices initiated the fourth industrial revolution [2]. The concept of IoT isn't new for industries, they already used sensors linked to Supervisory Control And Data Acquisition (SCADA) for monitoring production chains and Cyber-Physical System (CPS). SCADA systems are subject to technical limitations [3], those legacy systems use proprietary protocols making the integration with modern technology more difficult, maintaining them is expensive and their replacement requires a lot of investment. CPS is designed for specific use cases, making them expensive and complex to upgrade. Meanwhile, IoT devices are off the shelf solutions that are cheap, use well-known protocols, operate either on Internet or Ethernet and are easily replaced. But, those devices lack good security measures as they have constrained resources, computational power, or energy. Even though IoT devices use common protocols that aren't compatible and as a whole IoT based systems lack standardization. And so IoT devices are usually the weakest links in IoT systems and require special precautions when integrating them to an information system.

There are several ways to describe IoT and Industrial Internet of Things (IIoT), the NIST released a publication [4] showing a convergence of the definitions present in the academic literature of the CPS and IoT (Figure 5.1), the former come from the industry and the latter from the general public. The IIoT is at that point of convergence. Serpanos and Wolf [5] argue that IIoT systems need more consideration on safety and continuous operation when designing those systems because a potential failure results in significant financial loss or a life-threatening situation. IoT systems are part of multiple domains of applications. For this chapter, IIoT are IoT systems used in the following domains: Industry 4.0, Logistic, Retail, Smart Grid, and Agriculture. IoT devices are the bridge between the physical and digital world. The major distinction between an IIoT and IoT device is if the device is part of a product used by a consumer. For example, a geolocalized electric scooter is an IoT but an autonomous robot in a warehouse [6] is an IIoT. IIoT

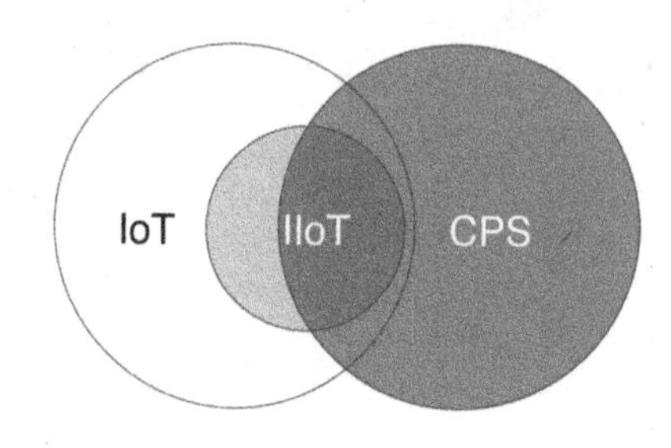

FIGURE 5.1 IoT, IIoT, and CPS in Venn Diagram, Inspired from ref. [7].

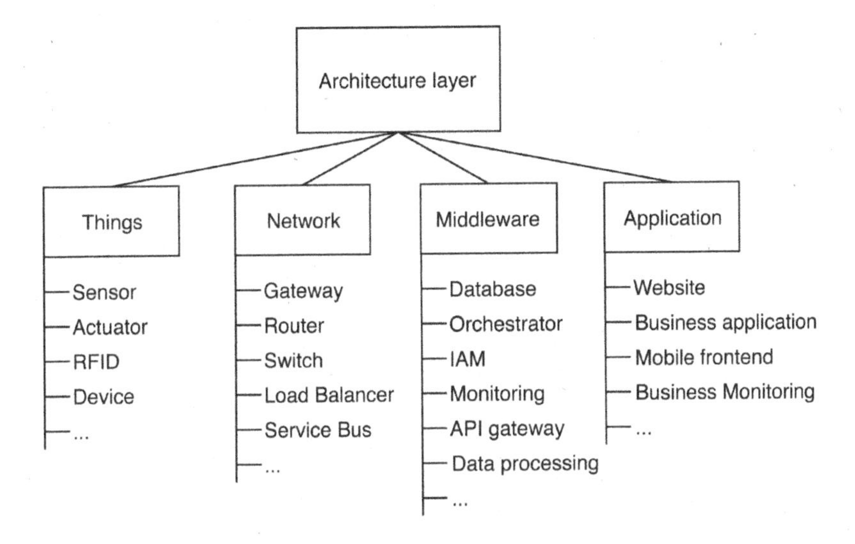

FIGURE 5.2 IoT reference architecture.

devices in an industrial system that enable a ubiquitous monitoring essential for Logistic, Smart Grid, and Agriculture.

Security is a major concern for every system. Spathoulas and Katsikas [8] argue that securing IIoT is a trade-off between security and availability. Indeed, a data leak will not damage a machine while a crucial sensor detecting an anomaly must transmit the data or will result in a significant financial loss or even a life-threatening situation. The heterogeneity of IIoT devices increases the complexity of designing a cohesive system. Integration with already existing equipment is the most problematic, especially if they are highly customized to meet production needs, as they may not be compatible with recent network protocols or lack up-to-date security mechanisms.

This chapter is organized as follows: first, we introduce the layered reference architecture of an IIoT system and the paradigm of platform enterprise. In a second part, we describe the security services into three categories, the information assurance, dependability, and access control, followed by a presentation of the cyber-security threats for the IIoT system for each layer of the reference architecture (Figure 5.2).

5.2 IIoT ARCHITECTURE

IIoT systems are summarized into four layers architecture [9–11].

The **Thing layer** also known as the perception layer contains the physical component of an IIoT system. Here we find the sensors, actuators, SCADA systems, RFID chips.... Depending of the use case and the implementation. The sensors are used to bring a ubiquitous perception, providing data on location, temperature, pressure, motion, humidity, etc. While the actuators collaborate based on self-decision-making or direct control from the upper layers. All those elements are the main components of any IIoT systems.

The **Network layer** is the communication layer between all the other layers. Its role is to transport the data between the layers and the components. The transmissions are to other devices, gateway, or between applications and use tethered connection (Ethernet) short range wireless (Wifi, Zigbee, Bluetooth Low Energy, NFC, RFID) or long range wireless (3/5G, LoRa, Sigfox, WIMAX) depending on the implementation in the IIoT system. This layer is also composed of the traditional network functions (physical or virtualized), router, switch, load-balancer, ESB… used to manage the network.

The **Middleware layer** is the service backbone of the system. It is composed of many application, middleware, database, and management software and is the central layer to each IIoT system. IoT devices generate a large amount of data, the processing needs specialized infrastructure that can adapt to modulation in the workload. This data platform is positioned in this layer. The management of the system like the configuration of network services, especially with SDN, and the configuration of the security services (IAM, LDAP, Key store, VPN…) is found in this layer. In addition, the cloud, on-premise, edge, and fog infrastructures for the system are located in the middleware layer, including, for instance, the servers for hosting applications and services.

The **Application layer**, also called the information layer, is the layer where the users interact with the system. The administration of the Middleware layer, the IoT device, and the application is provided by the Information layer. The business applications are driving the systems, allowing operators to monitor and supervise it. The data can be accessed through web applications, mobile applications, or traditional computer software. These interfaces provide a control interface for operators or administrators to manage an off-site manufacturing plant.

5.3 PLATFORM ENTERPRISE

Nowadays, companies' IT systems can be summarized into a five functions model (Figure 5.3) based on data platform reference architectures [12, 13], composed of the following functions:

- The **Collect** function, the data are collected by the multiple sensors of the systems, also known as the data acquisition phase. The type of data varies from environmental data (temperature, hydrometry…) to message (logs, error message…). The data are transferred from the Things layer through the Network layer.

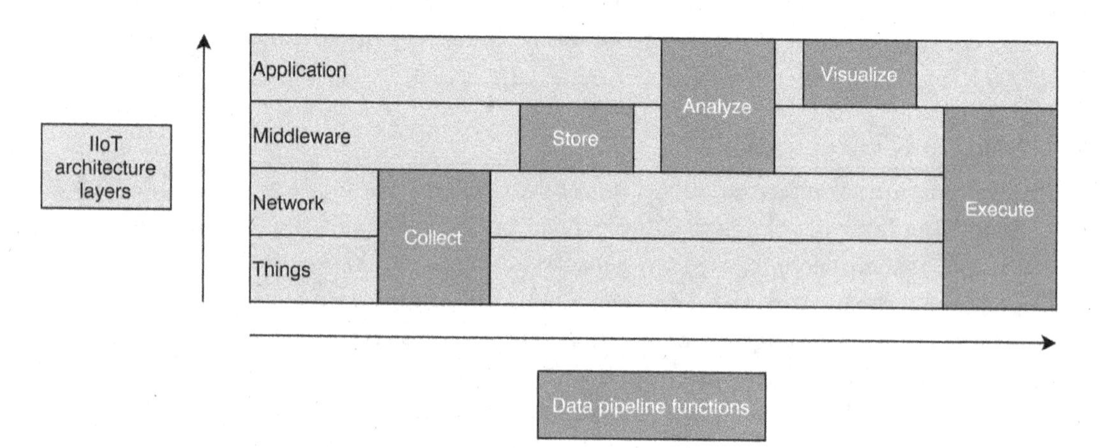

FIGURE 5.3 Mapping of data functions on IIoT architecture.

- The **Store** function, the data are stored inside a database, HDD or HADOOP cluster, pre- and post-processing, it is retrievable for the other layer such and its access are controlled by an Identity and Access Management service. The data can be stored locally close to the machine, centralized in a cloud environment or distributed between the different industrial sites.
- The **Analyze** function, the data are processed through algorithms by compute unit to transform and make them readable by software or humans. The data is utilized to optimize the system, predict behavior or operation, and detect anomalies.
- The **Visualize** function, the data is aggregated to applications to monitor the systems. Decisions can be taken based on metrics or Key Performance Indicator (KPI); this process can be automated by algorithms or AI or by human intervention.
- The **Execute** function, once interpreted, the data are used to give instructions to the system, which can be destined for middleware tools or machines. The instructions can be to optimize the production or certain actions to be carried out remotely.

This data pipeline is fundamental to every IIoT system. The objective is to change the status of the system based on the information collected. This proposed model can be applied to every domain of the industrial sector. It is derived from our previous research [14] and academics literature [15, 16], we concluded that the data and business perspective is often absent from proposed IIoT reference architectures. We present the different industrial domains (Figure 5.4), ergo Industry 4.0, Logistic, Retail, Smart Grid, Agriculture, and Healthcare, where the context needs a ubiquitous view of the environment and remote control. These use cases imply that an important amount of data is collected and processed in data center, edge, fog, or cloud computing, in the so-called middleware layer of a traditional IoT system architecture. As systems become more autonomous, the need for human intervention will lessen and we will see an increase in productivity. Before, IT systems were considered a necessity expense, but nowadays, thanks to innovations in infrastructure, automation, and security, IT systems are providing new use cases and creating value for businesses. Those innovations transform businesses into platforms of exchange, where cooperation between companies will increase productivity, reduce maintenance costs, and raise their observability of the ecosystem. This new paradigm of companies is called platform enterprise [17] supported by the concept of IT as a platform [18, 19]. This concept is embodied in the IoT platform cloud solutions (Google IoT, Azure IoT hub, AWS IoT...) that provide services centralizing all the needs and applications. These platforms at the heart of the middleware layer are supported by management and security tools. Operating an IT system as a platform requires culture and technology. Approaches such as Agile Devops and SRE (Site Reliability Engineering) change the management of IT systems, while innovative companies (Netflix, Uber, AirBnB) are natively agile in culture, they also demonstrate it in their technology approach of innovation. Netflix can afford to run multiple instances of chaos monkeys [20] in their infrastructure, meanwhile an industrial might be reluctant because their IT systems aren't resilient against insider attackers. A machine

	Collect	Store	Analyze	Visualize	Execute
Agriculture	Data about the fields or the cattle : temperature, luminosity, hydrometry, GPS coordinate ...	The data are stored in the production site locally for remote place or in the cloud	Data are used to make prediction and view the state of the fields or cattle ...	Application that allow the producer to review his farm and take decision based of the data. The data can be accessible to other adjacent producer for enabling cooperration	The producer can remotely control the farm, for example feeding the cattle or watering the fields
Logistic	Data about the goods and the delivery fleets : GPS coordinate, fuel consumption, traceability in the supply chain ...	The data can be stored distributed or centralized in a cloud	Data are used to estimate the time to arrival, optimize the supply route, anticipate shortage in a supply chain ...	Companies can visualize their goods in transit. Delivery compagnies manage their fleets	Reroute the supply chain ...
Healthcare	Data about the patients : Body temperature, heart rate, blood pressure ...	The data can be stored locally in the hospital or an certified datacenter	Data can be used to predict patient condition, to find the best treatment, to detect anomaly or health issues ...	Doctors can remotely monitor the patients, adjust the treatment, exchange the data between experts	Doctors can remotely control heath equipment
Industry 4.0	Data about the production : rate of production, maintenance state of the machines, sensors ...	The data can be stored distributed between multiple factory or companies or centralized in a cloud	Data can be used to predict maintenance, to optimize the supply chain, to adapt the production ...	Employe can remotely monitor the the factory plant, adjust the volume of production	Machines adapt their behaviors based on the data or order given by humans
Retail	Data about the product : Temperature in storage area, position of the goods, position ...	The data can be stored locally in each store and recentralized in a datacenter	The data are used to optimize the storage, predict sell ...	Employe have an insight of their store, manager can see their stock and they can follow and detect misplaced product	Temperature in the storage area can be adjusted remotely ...
Smart Grid	Data about the energy grid : The maintenance state of the grid, the consumption of the client, intensity ...	The data can be stored distributed or centralized in a cloud	The data are used to optimize the power flow, predict maintenance ...	Employe see the incident in the power grid, redirect the power flow ...	The smart grid reroute the power based of the data and incident, add new section (plug and play) ...

FIGURE 5.4 Application of our model and examples for each industrial sector.

shutting down on an assembly line will have a larger impact on the physical world rather than a virtual component (e.g., router, server). Industrial need to change their approach to design information system and take into consideration the benefits of IoT complemented with cloud computing.

5.4 CHALLENGES

The interconnectivity of people, devices, and organizations in today's digital world opens up a whole new field of vulnerabilities and access points where cybercriminals can operate. Today the organizations' landscape of risk is a combination of real and potential threats that come from unexpected and unforeseen actors with unpredictable consequences. In this post-economic crisis world, companies are evolving rapidly. New product launches, mergers, acquisitions, market expansion, and the introduction of new technologies are all on the rise: these changes invariably have a complicated impact on the strength and scale of an organization's cyber security and its ability to keep pace. Thus, we will cover the current cybersecurity challenges of industrial information systems.

5.4.1 Physical Limitations

Physical limitations are the physical capabilities of a computer system, they can be characterized by computing power, available memory, battery life, environment, distance latency, etc. An IoT system is built around these constraints, each function will consume these physical capabilities. In addition to addressing use cases, adding security increases the consumption of its resources, needing more computational power that consumes more energy. We can distinguish IIoT devices in two categories: power-supplied devices and battery-powered devices. In the first case, the device has less restriction and allows the constructor to design it with more computational power and capabilities. In the second category, the device is designed to be energy efficient. The lifespan for this kind of device can reach many years in most extreme cases. Battery-powered devices are generally used in places that are difficult to access. The common behavior for a sensor is to wake up, power on, acquire data, send it, and go back into sleep mode. Such processes need to use as little battery life as possible and as stated earlier adding any form of security (encryption, signature…) will decrease the device longevity. The evolution of the communication protocols improves the energy consumption and allows the use of simple cryptographic algorithms. Nowadays, the Elliptic Curves are proven to be more efficient than RSA [21], for the same key size. Data encryption for IoT is improving over the years but the key exchange protocols are still a challenge [22]. IoT devices are constrained by their storage capacity, which requires them to transfer data to a data processing center; the devices must be able to send reliably over the network. This is a problem if they are not designed with sufficient means of communication, or in the case of a wireless network with enough power to reach the receiving antenna. Designing IoT systems according to these technical constraints is a matter of compromising functionality, cost, and security.

5.4.2 Heterogeneity

In the manufacturing sector, there are a lot of highly specific components and machines built by a vast array of companies. Some are standalone and some are part of bundle solutions, in either case, they might be designed with features making them difficult to integrate in an industrial information system. IoT devices aren't exempt from these shortcomings. There are many standards for communication between devices: RFID (e.g., ISO 18000 6c EPC class 1 Gen2), NFC, IEEE 802.11 (WLAN), IEEE 802.15.4 (Zigbee), IEEE 802.15.1 (Bluetooth), Multihop Wireless Sensor/Mesh Networks, IETF Low power Wireless Personal Area Networks (6LoWPAN), Machine to Machine (M2M), IP, IPv6, etc. [9]. This multitude of standard challenge IoT devices designer, adding compatibility to all of them increases the complexity, the cost, and the energy consumption for the IoT. While an Arduino board might include most of these protocols, it is several orders of magnitude more powerful than a simple sensor. Examining further than the network layer, in the application layer, the devices may use proprietary protocols or they are only compatible with a restricted set of appliances or applications, for example, the data collected are formatted according to specific criteria, which makes them easily readable by some data processing software and not by others. A high level of heterogeneity widen the attack surface for the whole system [23], keeping track of all vulnerability of each chosen solution is the challenge hindering the deployment of IIoT devices. Interoperability is the key factor for adoption of IoT by industrials.

5.4.3 Authentication and Identity

Identity management is the combination of assigning identity to each device or user of the systems, defining their permission and validating their identity through authentication process. Identity is essential for managing an important number of components, potentially billions. Authentication is a security service critical protecting from attack such as identity usurpation or unauthorized access to service or data. Identity management becomes challenging when the number of identified entities increases over time. Moreover, the assignment of identities is difficult in environments where actors can come and go unexpectedly. Administrators must ensure that each actor is unique and there is no usurpation. The assignment of device identities needs an Identity Provider (IdP) that will create, maintain, and manage the identity. IdP is a piece of software interconnected to the company information system and the Access Control management each IoT device may have a unique identifier assigned depending on its context of operation. The use of certificates adds a way to provide identities, but its management is complicated for distributed systems, as Won et al. discussed in their research [24], and such we see research and effort on building new methods to identify IoT devices [25, 26]. Identity management remains manageable as long as it remains in a specific environment, such as a factory, farm, or vehicle but in an open environment like a smart city where enthusiast weather stations can freely participate in the systems or a distributed system like a supply chain with multiple actors the administration is a challenge. An attacker who manages to gain access to the Identity and Access Management can compromise all security services of the system, thereby compromising data integrity, breaking confidentiality, and so on.

Identity is nothing without authentication. Means of controlling identities, such as certificates or multifactor authentication, are needed. Authentication is divided into two categories, for users and for devices [27]. Authentication by username/password is the most commonly used method, but it is difficult to implement in some contexts, for example, when directly connecting to a sensor, the latter may not have the technical capabilities (connection to an external authentication server, internal software…) to perform the user's authentication. In addition, in a supply chain case, a device can communicate with several information systems of different companies and must be able to be identified in each. Access control and identity management are the first step to ensure accountability on your system, especially for IIoT systems.

Access control is the control of the permission for each identity. Depending on their identity, a panel of actions or resources is available to them. The definition of permissions is a challenge in itself in information systems, traditionally an administrator needs to keep track of every permission attributed to every user and must ensure that when a user leaves, his or her permissions are removed. The IoT context increases the complexity of this management as they are more dynamic systems, in constant evolution where devices come and go on the network. One of the challenges is also to anticipate new cases, where a device would need authorizations not initially planned, which poses the problem of automatically assigning permissions. Whatever the permission management mechanism used, whether it is an access control dictionary (DAC), access defined by their roles (RBAC), or attributes (ABAC), they operate on the principle of access control list, which requires a prior definition of the permissions [28]. The propagation of this information is a challenge in distributed networks where synchronization may have latency and each component needs to adjust their behavior in accordance to the new rules. Devices might need to function in isolation without access to a central control unit; they require other means of authentication. In addition, the access control center must be particularly protected, once it is compromised, it would open the doors to all types of attacks that would then be undetectable.

5.4.4 Availability and Reliability

In the manufacturing world, some security characteristics are more important than others. Availability and resilience are among the main requirements for industrial sites [29]. Indeed, availability guarantees accessibility to and between services and devices. Resilience is the implementation of countermeasures or redundancy to ensure the continuous operation of the system. However, IoT devices are not known for their durability. By their simplicity they tend to be fragile and easily malfunction. These devices are very sensitive to cyber attacks as well as physical attacks; their low computing capabilities do not give them the cards to defend themselves effectively. They are primarily designed for their functionality (e.g., a sensor must collect the data and transmit it) and security is often overlooked or an afterthought.

The unavailability of a device or data may lead to serious human, material, or financial damage. A message that does not reach its destination may prevent the safety measurement from being activated, for example, a temperature sensor that does not transmit an overheating in a foundry. The origin of an unavailability can be multiple, ranging from a DDoS attack to a physical interference. Detecting the cause of such incident is a real challenge for today's industrial systems. A failure is inevitable, and therefore it is necessary to set up resilience mechanisms. It englobes different concepts as dependability, fault tolerance, and robustness [30], Laprie in [31] defines resilience as "the persistence of dependability when facing changes". The resilience of the components ensures the security and transmission of information even in hostile environments. Although, the implementation of redundancy does not come without financial cost and in addition to complicating the management of the IIoT system. The management of the IoT devices does not come without its own challenges. This starts with the design of the architecture, the choice of management software, deployment patterns, and operational recovery processes. Each of these steps need to be carefully planned and tested which goes against recent trends in development cycle of technology solutions.

5.4.5 Maintainability

The maintenance of IoT devices poses several challenges even though they are inexpensive and can be easily replaced, some are sensitive to their environment and subtle change can render them inoperative. First, physical access: devices are often positioned [32–34] in places that may be difficult to access, geographically dispersed or simply inaccessible. In such case, physical maintenance needs to be as little as possible and software one needs to be done remotely. In addition, updating software in a distributed environment is a challenge. Forcing the system to keep entities at the obsolete security level causing a high security risk, especially when the devices are too simple and cannot be updated. An update of a device can change the behavior of certain features making it incompatible with the rest of the system. An additional issue is the responsibility of an unpatched vulnerability and determining the people accountable for updating. Industrial systems are complex and the digital transformation allows more interactions between manufacturers through exposed API or remote access. Therefore, an infected system in company A could infect the system in company B. In any case, ensuring the right version is a complex task that requires remote access and awareness of the devices connected in your network.

To achieve maintainability, visibility of the devices fleet is crucial yet difficult [35]. Thus, equipment for discovering devices to map the network is needed. The IoT devices must be selected to meet this requirement. The discovery of the devices connected to the system is particularly complex when the components can come and go and due to the liveliness of IoT environment. Future predictions [36] advance that IIoT systems will be self-aware and self-maintained by monitoring their health and detecting anomalies, allowing them to reconfigure their interactions and operations to accommodate a defective machine. The edge computing and artificial intelligence are the enabler technologies or self-maintenance.

5.4.6 Integrity

IoT devices operate in highly hostile environments. However, the information they send can be vital, an interception and modification of a message can cause damage just as serious as the activity of a malicious component. The functioning of an IoT system depends on the integrity of the data as well as the integrity of its components. Aman et al. state two challenges of data integrity [37]; first, at the network level cost a high amount of energy and have many overheads for the calculation. Second, most protection methods assume that the device isn't physically attackable and storing secret key on the device is safe. Messages are transmitted through two types of environments: the Internet or the company network. The difference between these two environments is trust. The Internet is said to be trustless while an enterprise network is trustful. On the Internet, data needs to be encrypted with VPN, IPSec, or TLS, during transit for keeping the confidentiality and IoTs often lack the computing capacity to encrypt using strong cryptographic parameters. To verify the integrity, mechanisms such as MAC, checksum algorithms, PKI, or even blockchain [38] can be used.

Integrity also includes the integrity of devices, software, and operating systems. A hardware can be compromised and then become under the control of a hacker, who will exploit the permission of the device to force harmful behavior, such as during the attack against the DNS Dyn where an army of IP cameras, printers, and other connected consumers IoT devices has performed a DDoS attack impacting the global Internet [39]. No antivirus or antimalware are installed on machines yet they are hacked [40]. The implementation of countermeasures must be a coordinated action between the IoT equipment supplier and the manufacturer. The lifecycle management of IoT devices poses many challenges. First, the supply chain problem, it is necessary to ensure the device is not compromised during its fabrication, transport, and installation [41]. At the end of their life the devices must be destroyed safely, for example, a server not properly erased will leak its data to the new purchaser or a small device will keep its network information or store its credentials that will serve as intelligence for a future attacker.

5.4.7 Observability

Observability [42] extends the notion of monitoring a system in the form of three pillars [43]: traceability, logging, and metrics. Traceability is the ability to know an action or messages' information through metadata, network observation, or direct communication of components. Logging is the ability to save action records in a format processable by big data platforms. Logging enables, coupled with traceability, to perform forensic analysis. Finally, metrics are the data collected by monitoring. The metrics depend on the service provided by the system or devices and must be chosen carefully. Observability allows you to view an information system, react to an attack, prevent an attack, and consolidate security measures. A mature observable system can respond to any request required to verify the status of its components. The major benefit of observability is from a business perspective; it provides information on the capacity of its systems and thus enables decisions such as production adjustments or changes in safety rules to be taken.

Two things are monitored, the health of the systems and the production environment. The first case is the monitoring of connected devices, network status, application traffic, applications, and users in order to detect malfunctions and suspicious activities. In the second case, the monitoring is at a macro level of the system and takes information measured by IoT sensors giving, for example, the production status of a smart factory.

For the industrial sector, the implementation of this type of infrastructure is complicated mainly due to the heterogeneity of IoT and the hostile environment in which IoT equipment operates. There may be discrepancies between the data collected by the multiple sensors, the information collected may be incomplete or corrupt [30] once it reaches the user. Thus, to produce quality data, it must come from multiple sources of information, which implies the production of a large volume of data requiring high computing power to be processed efficiently.

5.5 SECURITY OF IIoT SYSTEMS

5.5.1 Security Services

In this section, we describe the security services used in the proposed taxonomy to characterize attacks on the IIoT system. The initial classification is based on the work of Sterbenz et al. [44], which presents a set of security disciplines for resilient communication network. We choose a subset of the disciplines focusing on security relevant for IIoT systems.

5.5.1.1 Security Services Relating to Information Assurance

Also known as the five pillars of information security, Information assurance [45] (Authenticity, Availability, Confidentiality, Integrity, Nonrepudiability) extend the CIA (Confidentiality, Integrity, Availability) principle (Figure 5.5). The CIA triad is considered obsolete to characterize a system [28]; therefore, the principles of information assurance have been introduced.

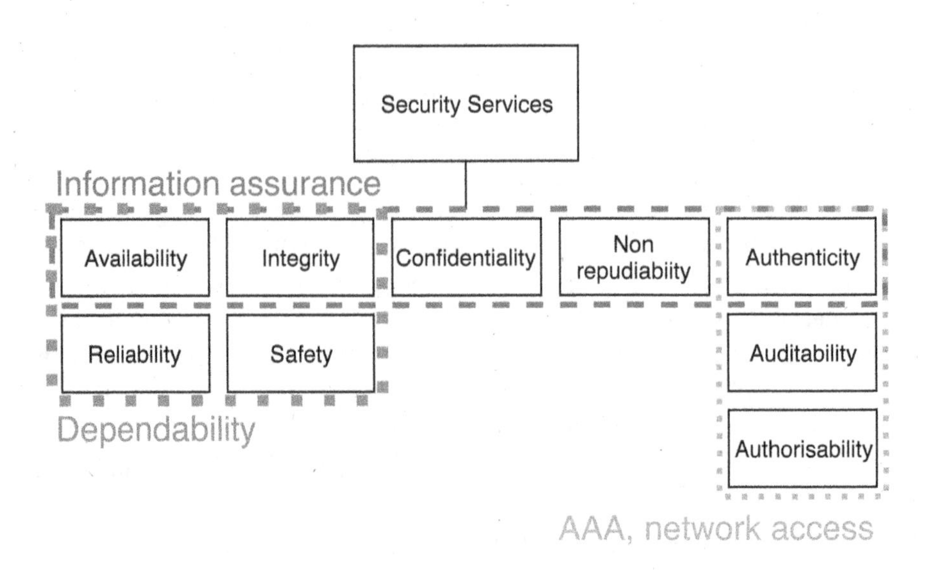

FIGURE 5.5 Taxonomy of security services for industrial Internet of Things systems.

Confidentiality allows users to exchange information securely and ensure the message isn't readable by anyone, making it only readable for authorized users. The goal is to ensure the privacy of the message transmitted; for this purpose, several encryption algorithms should be used. Proper confidentiality brings trust between entities and prevents data leaks. In the IIoT context, confidentiality is essential when sensors collect sensible data (personal data or production critical data).

Integrity is the protection against any intentional or unintentional alteration. Criminals will attempt to modify the data to conceal illegal activities or to send incorrect information to industrial systems that would prevent the plant from operating properly. But natural phenomena can also alter data; electromagnetic interference or packet loss can make transmission incomplete. Simple methods such as checksum and digital signature are used to ensure the integrity of messages without requiring large computations.

Availability is the assurance that information or components are accessible. Data availability is important for industrial systems; in a ubiquitous plant, to maintain optimized production, production data must be accessible for all control devices. In the same way, machines need to communicate without interruption; if a stop message cannot reach a machine, disastrous events can occur. To prevent these scenarios, there are different solutions depending of the architecture. The network topology can be a mesh network or by making every route redundant. Firewall and anti-DDoS technologies can be deployed to protect the most vulnerable devices or applications.

Authenticity is the property validating the identity of actors (devices, users, applications) in an IIoT system. By using an authentication mechanism, entities can safely exchange messages and ensure that only the authorized and authentified can send and read messages. It also uses mechanisms such as digital signatures to guarantee the identity of the sender.

Nonrepudiability is the security property where an entity cannot deny sending or creating a message. This property ensures traceability of data in the system. Proper nonrepudiation mechanism allows the detection of man in the middle attack and track back the source of incorrect messages.

5.5.1.2 Security Service Relating to Access Control

Access control (AAA) are three security principles working conjointly: Authenticity, auditability, and Authorizability. AAA defines access control protocols extended to information systems [46].

Authenticity, known as authentication, is the property that validates the identities of entities. There are multiple processes to verify an identity, each depending on the infrastructure implementation or use cases. Once their identity is accepted by the process, the entities can interact with the system.

Auditability is the principle that every action, identification, or communication is recorded for traceability. Constant monitoring allows real-time security operations to be carried out while keeping logs allows forensic analysis after an incident.

Authorizability is identity control, the device, application, or user has its credential controlled before giving it access to a resource or a service. Depending on the context and the identity, the access can vary; for example, a web developer has only access to the development environment.

5.5.1.3 Security Service Relating to Dependability

Dependabilities are security services that characterize the reliance of a service. In this discipline, there are four services: **Availability** and **Integrity** are the same security service from Information assurance, but they extend to components of a system instead of only data. The availability of a service or device as opposed to the availability of data; for example, two devices that need to cooperate need to reach each other. The integrity property is the guarantee that a system component is uncompromised and will function as intended.

Reliability in an industrial system is the rate of failure caused by incidents, malfunctions, bugs, etc. A 100% reliable system is a chimera, incidents will always occur and machines will break. Thus, building redundant systems by eliminating single point of failure (SPOF), by duplicating network routes or adopting new topology such as mesh networks achieve a more reliable system. Designing the industrial system to be antifragile [47] or fault tolerant will improve the reliability.

Safety is a property characterizing the impact of a failure. It is the security of the peripheral elements of the system, the implementation of mitigation processes to minimize human, material, and financial risks. IIoT devices are an integral part of factory safety; their ubiquity property allows the implementation of contextual rules controlling machines that would, for example, prevent the operation of a mechanical arm as long as a human being is within its area of operation.

5.5.2 Attacks

Every system is vulnerable and industrial systems aren't exempt of weakness. Attackers don't break defense but exploit these weaknesses. Nonetheless, security measures are to be deployed to reduce the potential attack surface. To help architects designing IIoT systems, we propose a taxonomy of attacks on IIoT systems through a layer approach. Industrials have specific requirements that differ from consumer needs [48, 49]. Consumer IoT devices are more sensible to privacy issues because they collect personal data on the end user. Reliability and availability will depend on the applications, but most of the time they are not critical for the consumer. Meanwhile for IIoT systems, the concern will not be on the privacy but on the confidentiality of the data and the reliability and availability are essential for most use cases.

We categorized the attack into four categories based on the affected component level in the reference architecture.

5.5.2.1 Threats on the Thing Layer

Physical access: Data or programs of an IoT device can be compromised if an attacker can physically connect to it. Critical systems should reduce the number of physical I/O to reduce voluntary or involuntary attacks like the Stuxnet case [50]. Furthermore devices can be damaged by malicious people to malfunction, the sensor will then send distorted data and the actuator will not follow instructions correctly.

Compromised hardware: Companies trust their suppliers although they can be the Achilles heel of their security. Even with major investment, the security level of all the supply chains needs to be identical for all the actors [51, 52]. In 2018, Bloomberg published a story about compromised hardware during supply chain [41], although the article is strongly criticized for its veracity, the scenario was plausible enough cause the drop of 40% of the value for Super Micro [53]. This attack causes a breach in the confidentiality of the data collected by the IoT devices. A compromised device can be used to perform simple instruction that will not hinder its normal functioning but when a massive number of devices take a joint action such as a DDoS attack, the impact can be severe for the victim. In the last few years, the amount of DDoS attacks from IoT and the volumetry of the attack are at their all time high [54].

Spoofing: This attack allows the attacker to masquerade its identity by forging messages or falsifying data. The attacker will infiltrate the system by passing for a legitimate IoT device [55] and then carry out a man in the middle attack.

Jamming: By overwhelming the radio waves in a wireless network the attacker can block or delay the communication between the devices and the system by sending noise signals [56]. It can also be used to isolate part of the network in order to blind the system and allow the attacker to act without being monitored because the sensors will be unable to transmit the wrongdoings.

Sleep deprivation: This attack is aimed at devices running on battery. When an IoT device isn't working it will be in a sleep state for energy saving. The attacker will send messages that will wake up the device thus consuming the battery [57–59], usually by pretending to be a node of the network and transmitting mundane network protocol messages that will pass for legitimate communication between two nodes.

DoS attack: As the IoT devices have constrained resources making them vulnerable to most DoS attacks [60]. This kind of attack can affect different components such as the network bandwidth, CPU time, or memory. IoT devices that are accessible through the Internet are particularly sensible to DoS attacks.

5.5.2.2 Threats on the Network Layer

Sybil attack: The attacker will create multiple identities that will send false information to the system [61]. The goal can be to change the network topology like network routes or sending false information to the system; for example, an array of fake sensors can be created to send wrong data about the weather to a smart city or a smart farm. This attack is known to affect mainly distributed and Peer to Peer network where the nodes function as an autonomous organization and nodes are expected to connect and disconnect frequently [62]. By separating a node or a part of the network the attacker can divert the network traffic to a sinkhole, consequently interrupting the communication [63, 64].

Privileged access: The attacker either gets the credential, through malicious means, or exploits weaknesses to gain privileged access to a network component (router, gateway, switch…) allowing him to control them [65]. The configuration of the component can be exploited by an attacker to set up an Advanced Persistent Threat (APT) by collecting metadata and information about the network topology.

DoS attack: Denial of service attacks in the Network layer affects the whole IIoT system. The network is flooded with useless traffic or vital network components (router, switch, gateways, load balancer) have their resources depleted [66]. The goal is to slow down or block the communication on the network.

Man in the middle (Eavesdropping, traffic analysis): The attacker infiltrates the network to intercept, create, and modify the network communication between nodes [67]. End to end encryption is a current challenge for IIoT systems [68]; IIoT devices may not be capable to encrypt data with strong algorithm, allowing the data to be easily captured and decrypted by a packet inspector, allowing an attacker to breach the confidentiality and to analyze the network behavior [69].

5.5.2.3 Threats on the Middleware Layer

Privileged access: The middleware layer is particularly sensible to privileged access. The components of this layer are the backbone of the whole IoT system and rely on centralized IAM system once a hacker manages to infiltrate the IAM systems; the attacker will be able to tamper the data, control the access, and change the security policy [70]. In cloud-based implementation of IIoT platform, the access control is delegated to the cloud services. A poor management of these services and account permissions put the whole system at risk [71], especially in a platform environment where actors come from different companies.

Virtualization-based attack: In this layer, components are either hosted on hardware or virtualized; in the second case, an attacker can exploit a less secure virtual machine to do an escape attack and take control of another virtual machine hosted on the same server [72]. If the Hypervisor is compromised, the system will leak data or the hacker can stop virtual machines or take control and spread virus.

Database corruption: Database can be modified even though the attacker doesn't have credentials. For example, the SQL injection attack, the exploit uses insecure fields in applications to send SQL commands to the database, allowing the attacker to delete, insert, or modify the data [73]. An attacker that manages to poison the database will create wrong behavior for the IIoT systems by manipulating the data or alter the visualization for the Application layer.

5.5.2.4 Threats on the Application Layer

Code execution: Through the front-end of most business applications code can be injected to the back-end in the Middleware layer. The goal is to gain control of the application or the server; in the worst case, the attack can escalate to the whole system. Malware such as ransomware can have a major impact on an Information system and the source of financial and reputation loss.

Man in the middle attack: Web applications can be masqueraded to steal information about the user or their credentials. There are multiple ways to do phishing, by sending an email with a malicious link or by poisoning the DNS server to redirect the traffic to the phishing application [74]. The attacker installs traffic listeners to collect data on the user and the application behavior. The listener can be found in the user devices, in the host of the application. The potential risk is a breach of confidentiality for the business [69].

DoS attack: Denial of service attack on the application layer affects the users. They might lose the vision or control of the system, leaving the attacker free to take any malicious actions.

5.6 CONCLUSION

IIoT system and platform enterprise are complex, while they enable innovation, increase productivity, and give a better monitoring of the system. They bring new potential threats to business. Security architects need to be particularly vigilant to their infrastructure to keep them at the state of the art of system security. The majority of these threats already impact traditional IT systems, countermeasures are well documented in the literature but the constraints of the industrial context bring a complex ecosystem of technologies, regulations, infrastructures, and company. This chapter addresses the security challenges of IIoT systems and presents the similarities between the platform enterprise paradigm and IIoT system. We present an overview of the major threats for such systems through the four layers of an IIoT architecture, showing the vulnerabilities of the devices and the complexity of securing the middleware layer.

REFERENCES

1. Gartner, "Gartner Says 5.8 Billion Enterprise and Automotive IoT Endpoints Will Be in Use in 2020", Gartner, 29-Aug-2019. [Online]. Available: https://www.gartner.com/en/newsroom/press-releases/2019-08-29-gartner-says-5-8-billion-enterprise-and-automoti ve-io. [Accessed: 28-Oct-2019].
2. K. Schwab, "The Fourth Industrial Revolution". Currency, 2017.
3. C.-W. Ten, C.-C. Liu, and G. Manimaran, "Vulnerability Assessment of Cybersecurity for SCADA Systems", IEEE Transactions on Power Systems, vol. 23, no. 4, pp. 1836–1846, 2008.
4. C. Greer, M. Burns, D. Wollman, and E. Griffor, "Cyber-Physical Systems and Internet of Things", National Institute of Standards and Technology, Gaithersburg, MD, NIST SP 1900-202, Mar. 2019.
5. D. Serpanos and M. Wolf, "Industrial Internet of Things", in Internet-of-Things (IoT) Systems: Architectures, Algorithms, Methodologies, D. Serpanos and M. Wolf, Eds. Cham: Springer International Publishing, 2018, pp. 37–54.
6. "Amazon Robotics", Amazon Robotics. [Online]. Available: https://www.amazonrobotics.com/#/. [Accessed: 20-Mar-2019].

7. E. Sisinni, A. Saifullah, S. Han, U. Jennehag, and M. Gidlund, "Industrial Internet of Things: Challenges, Opportunities, and Directions", IEEE Transactions on Industrial Informatics, vol. 14, no. 11, pp. 4724–4734, Nov. 2018, doi: 10.1109/TII.2018.2852491.
8. G. Spathoulas and S. Katsikas, "Towards a Secure Industrial Internet of Things", In Security and Privacy Trends in the Industrial Internet of Things, C. Alcaraz, Ed. Cham: Springer International Publishing, 2019, pp. 29–45.
9. L. Da Xu, W. He, and S. Li, "Internet of Things in Industries: A Survey", IEEE Transactions on Industrial Informatics, vol. 10, no. 4, pp. 2233–2243, 2014.
10. P. P. Ray, "A Survey on Internet of Things Architectures", Journal of King Saud University – Computer and Information Sciences, vol. 30, no. 3, pp. 291–319, Jul. 2018, doi: 10.1016/j.jksuci.2016.10.003.
11. H. Suo, J. Wan, C. Zou, and J. Liu, "Security in the Internet of Things: A Review", In 2012 International Conference on Computer Science and Electronics Engineering (ICCSEE), Hangzhou, China, vol. 3, pp. 648–651, 2012.
12. P. Pääkkönen and D. Pakkala, "Reference Architecture and Classification of Technologies, Products and Services for Big Data Systems", Big Data Research, vol. 2, no. 4, pp. 166–186, Dec. 2015, doi: 10.1016/j.bdr.2015.01.001.
13. R. Kune, P. K. Konugurthi, A. Agarwal, R. R. Chillarige, and R. Buyya, "The Anatomy of Big Data Computing", Software: Practice and Experience, vol. 46, no. 1, pp. 79–105, 2016.
14. V. Vallois, F. Guenane, and A. Mehaoua, "Reference Architectures for Security-by-Design IoT: Comparative Study", In 2019 Fifth Conference on Mobile and Secure Services (MobiSecServ), Miami Beach, FL, 2019, pp. 1–6, doi: 10.1109/MOBISECSERV.2019.8686650.
15. H. P. Breivold, "A Survey and Analysis of Reference Architectures for the Internet-of-Things", In The Twelfth International Conference on Software Engineering Advances, ICSEA, 2017, p. 7.
16. B. Di Martino, M. Rak, M. Ficco, A. Esposito, S. A. Maisto, and S. Nacchia, "Internet of Things Reference Architectures, Security and Interoperability: A Survey", Internet of Things, vol. 1–2, pp. 99–112, Sep. 2018, doi: 10.1016/j.iot.2018.08.008.
17. P. C. Evans and A. Gawer, "The Rise of the Platform Enterprise: A Global Survey", 2016.
18. C. M. Felipe, D. E. Leidner, J. L. Roldán, and A. L. Leal-Rodríguez, "Impact of IS Capabilities on Firm Performance: The Roles of Organizational Agility and Industry Technology Intensity," Decision Sciences, vol. 51, no. 3, pp. 575–619, 2019.
19. V. Sambamurthy, A. Bharadwaj, and V. Grover, "Shaping Agility Through Digital Options: Reconceptualizing the Role of Information Technology in Contemporary Firms", MIS Quarterly, pp. 237–263, 2003.
20. A. Basiri et al., "Chaos Engineering", IEEE Software, vol. 33, no. 3, pp. 35–41, 2016.
21. N. Jansma and B. Arrendondo, "Performance Comparison of Elliptic Curve and RSA Digital Signatures", p. 1–20, 2004.
22. E. Vasilomanolakis, J. Daubert, M. Luthra, V. Gazis, A. Wiesmaier, and P. Kikiras, "On the Security and Privacy of Internet of Things Architectures and Systems", In 2015 International Workshop on Secure Internet of Things (SIoT), Vienna, Austria, 2015, pp. 49–57.
23. S. Sicari, A. Rizzardi, L. A. Grieco, and A. Coen-Porisini, "Security, Privacy and Trust in Internet of Things: The Road Ahead", Computer Networks, Vienna, Austria, vol. 76, pp. 146–164, 2015.
24. J. Won, A. Singla, E. Bertino, and G. Bollella, "Decentralized Public Key Infrastructure for Internet-of-Things", In 2018 IEEE Military Communications Conference (MILCOM), Los Angeles, CA, 2018, pp. 907–913, doi: 10.1109/MILCOM.2018.8599710.
25. M. Eldefrawy, N. Pereira, and M. Gidlund, "Key Distribution Protocol for Industrial Internet of Things without Implicit Certificates," IEEE Internet of Things Journal, vol. 6, no. 1, pp. 906–917, 2018.

26. M. Ma, D. He, N. Kumar, K.-K. R. Choo, and J. Chen, "Certificateless Searchable Public Key Encryption Scheme for Industrial Internet of Things", IEEE Transactions on Industrial Informatics, vol. 14, no. 2, pp. 759–767, 2018.
27. A. K. Das, S. Zeadally, and D. He, "Taxonomy and Analysis of Security Protocols for Internet of Things", Future Generation Computer Systems, vol. 89, pp. 110–125, Dec. 2018, doi: 10.1016/j.future.2018.06.027.
28. C. Maple, "Security and Privacy in the Internet of Things", Journal of Cyber Policy, vol. 2, no. 2, pp. 155–184, 2017, doi: https://doi.org/10.1080/23738871.2017.1366536.
29. A.-R. Sadeghi, C. Wachsmann, and M. Waidner, "Security and Privacy Challenges in Industrial Internet of Things", In Design Automation Conference (DAC) on 52nd ACM/EDAC/IEEE, San Francisco, CA, 2015, pp. 1–6.
30. D. Ratasich, F. Khalid, F. Geissler, R. Grosu, M. Shafique, and E. Bartocci, "A Roadmap Towards Resilient Internet of Things for Cyber-Physical Systems", IEEE Access, vol. 7, pp. 1–1, 2019, doi: 10.1109/ACCESS.2019.2891969.
31. J.-C. Laprie, "From Dependability to Resilience", In 38th IEEE/IFIP International Conference on Dependable Systems and Networks, 2008, pp. G8–G9.
32. B. Chen, J. Wan, L. Shu, P. Li, M. Mukherjee, and B. Yin, "Smart Factory of Industry 4.0: Key Technologies, Application Case, and Challenges," IEEE Access, vol. 6, pp. 6505–6519, 2017.
33. K. A. Kumar and K. Ramudu, "Precision Agriculture Using Internet of Things and Wireless Sensor Networks", Precision Agriculture, vol. 07, no. 03, p. 5, 2019.
34. M. Lee, J. Hwang, and H. Yoe, "Agricultural Production System Based on IoT", In 2013 IEEE 16th International Conference on Computational Science and Engineering, Sydney, 2013, pp. 833–837.
35. C. Adaros Boye, P. Kearney, and M. Josephs, "Cyber-Risks in the Industrial Internet of Things (IIoT): Towards a Method for Continuous Assessment", Information Security, vol. 11060, pp. 502–519, 2018.
36. J. Lee, H.-A. Kao, and S. Yang, "Service Innovation and Smart Analytics for Industry 4.0 and Big Data Environment", Procedia CIRP, vol. 16, pp. 3–8, Jan. 2014, doi: 10.1016/j.procir.2014.02.001.
37. M. N. Aman, B. Sikdar, K. C. Chua, and A. Ali, "Low Power Data Integrity in IoT Systems", IEEE Internet Things Journal, vol. 5, no. 4, pp. 3102–3113, Aug. 2018, doi: 10.1109/JIOT.2018.2833206.
38. B. Liu, X. L. Yu, S. Chen, X. Xu, and L. Zhu, "Blockchain Based Data Integrity Service Framework for IoT Data", In 2017 IEEE International Conference on Web Services (ICWS), Honolulu, HI, 2017, pp. 468–475.
39. C. Kolias, G. Kambourakis, A. Stavrou, and J. Voas, "DDoS in the IoT: Mirai and Other Botnets", Computer, vol. 50, no. 7, pp. 80–84, 2017.
40. R. M. Lee, M. J. Assante, and T. Conway, "German Steel Mill Cyber Attack", Industrial Control Systems, vol. 30, p. 62, 2014.
41. J. Robertson and M. Riley, "China Used a Tiny Chip in a Hack That Infiltrated U.S. Companies", Bloomberg.com, 04-Oct-2018.
42. "Guide Achieving Observability". HoneyComb.io, Jul-2018.
43. C. Sridharan, "Distributed Systems Observability: A Guide to Building Robust Systems". O'Reilly Media, Sebastopol, CA, USA, 2018.
44. J. P. Sterbenz et al., "Resilience and Survivability in Communication Networks: Strategies, Principles, and Survey of Disciplines", Computer Networks, vol. 54, no. 8, pp. 1245–1265, 2010.
45. C. E. Landwehr, "Computer Security", International Journal of Information Security, vol. 1, no. 1, pp. 3–13, Aug. 2001, doi: 10.1007/s102070100003.
46. C. Metz, "AAA Protocols: Authentication, Authorization, and Accounting for the Internet", IEEE Internet Computing, vol. 3, no. 6, pp. 75–79, 1999.

47. A. Abid, M. T. Khemakhem, S. Marzouk, M. B. Jemaa, T. Monteil, and K. Drira, "Toward Antifragile Cloud Computing Infrastructures", Procedia Computer Science, vol. 32, pp. 850–855, 2014, doi: 10.1016/j.procs.2014.05.501.
48. ITU-T, "Series Y: Global Information Infrastructure, Internet Protocol Aspects and Next-Generation Networks: Common Requirements of the Internet of Things". ITU-T, Jun-2014.
49. "CYBER; Cyber Security for Consumer Internet of Things". ETSI, Feb-2019.
50. N. Falliere, L. O. Murchu, and E. Chien, "W32. Stuxnet Dossier", White Paper, Symantec Corp., Security Response, vol. 5, no. 6, p. 29, 2011.
51. B. Turnbull, "Cyber-Resilient Supply Chains: Mission Assurance in the Future Operating Environment", Army, p. 41, 2018.
52. T. Omitola and G. Wills, "Towards Mapping the Security Challenges of the Internet of Things (IoT) Supply Chain", Procedia Computer Science, vol. 126, pp. 441–450, 2018.
53. "The Strangest Technology Story of The Year | Bruceb News", Bruceb News, 09-Jan-2019. [Online]. Available: https://www.brucebnews.com/2019/01/bloomberg-and-chinese-spies-the-strangest-technology-story-of-the-year/. [Accessed: 24-Jun-2019].
54. Akamai, "State of the Internet/Security Q4 2016", 2016. [Online]. Available: https://blogs.akamai.com/2017/02/state-of-the-internet-security-q4-2016.html. [Accessed: 18-Apr-2017].
55. A. S. Sastry, S. Sulthana, and S. Vagdevi, "Security Threats in Wireless Sensor Networks in Each Layer", International Journal of Advanced Networking and Applications, vol. 4, no. 4, p. 1657, 2013.
56. E. Shi and A. Perrig, "Designing Secure Sensor Networks", IEEE Wireless Communications, vol. 11, no. 6, pp. 38–43, 2004.
57. T. Bhattasali, R. Chaki, and S. Sanyal, "Sleep Deprivation Attack Detection in Wireless Sensor Network", International Journal of Computer Applications, vol. 40(15):19–25, 2012.
58. M. Brownfield, Y. Gupta, and N. Davis, "Wireless Sensor Network Denial of Sleep Attack", In Proceedings from the Sixth Annual IEEE SMC Information Assurance Workshop, 2005, pp. 356–364.
59. A. Gallais, T.-H. Hedli, V. Loscri, and N. Mitton, "Denial-of-Sleep Attacks Against IoT Networks", Presented at the CoDIT 2019 - 6th International Conference on Control, Decision and Information Technologies, Paris, 2019, p. 6.
60. D. R. Raymond and S. F. Midkiff, "Denial-of-Service in Wireless Sensor Networks: Attacks and Defenses", IEEE Pervasive Computing, no. 1, pp. 74–81, 2008.
61. J. R. Douceur, "The Sybil Attack", Peer-to-Peer Systems, 2002, pp. 251–260, doi: 10.1007/3-540-45748-8_24.
62. J. Newsome, E. Shi, D. Song, and A. Perrig, "The Sybil Attack in Sensor Networks: Analysis & Defenses", In Third International Symposium on Information Processing in Sensor Networks, Berkeley, CA, 2004, pp. 259–268.
63. C. Cervantes, D. Poplade, M. Nogueira, and A. Santos, "Detection of Sinkhole Attacks for Supporting Secure Routing on 6LoWPAN for Internet of Things", In 2015 IFIP/IEEE International Symposium on Integrated Network Management (IM), Ottawa, ON, 2015, pp. 606–611.
64. Y.-C. Hu, A. Perrig, and D. B. Johnson, "Packet Leashes: A Defense Against Wormhole Attacks in Wireless Ad Hoc Networks", In Proceedings of INFOCOM, San Francisco, CA, 2003, vol. 2003.
65. M. J. Covington and R. Carskadden, "Threat Implications of the Internet of Things", In 2013 5th International Conference on Cyber Conflict (CYCON 2013), Tallinn, Estonia, 2013, pp. 1–12.
66. J. Mirkovic and P. Reiher, "A Taxonomy of DDoS Attack and DDoS Defense Mechanisms", ACM SIGCOMM Computer Communication Review, vol. 34, no. 2, pp. 39–53, 2004.
67. M. Conti, N. Dragoni, and V. Lesyk, "A Survey of Man in the Middle Attacks", IEEE Communications Surveys Tutorials, vol. 18, no. 3, pp. 2027–2051, third quarter 2016, doi: 10.1109/COMST.2016.2548426.

68. I. Yaqoob et al., "Internet of Things Architecture: Recent Advances, Taxonomy, Requirements, and Open Challenges", IEEE Wireless Communications, vol. 24, no. 3, pp. 10–16, 2017.
69. Q. Xu, P. Ren, H. Song, and Q. Du, "Security Enhancement for IoT Communications Exposed to Eavesdroppers with Uncertain Locations", IEEE Access, vol. 4, pp. 2840–2853, 2016.
70. A. Arsenault and S. Farrell, "Securely Available Credentials-Requirements", IETF, RFC 3157, 2001.
71. K. S. Tep, B. Martini, R. Hunt, and K.-K. R. Choo, "A Taxonomy of Cloud Attack Consequences and Mitigation Strategies: The Role of Access Control and Privileged Access Management", In 2015 IEEE Trustcom/BigDataSE/ISPA, Helsinki, Finland, 2015, vol. 1, pp. 1073–1080.
72. K. Hashizume, D. G. Rosado, E. Fernández-Medina, and E. B. Fernandez, "An Analysis of Security Issues for Cloud Computing", Journal of Internet Services and Applications, vol. 4, no. 1, p. 5, Feb. 2013, doi: 10.1186/1869-0238-4-5.
73. W. G. Halfond, J. Viegas, and A. Orso, "A Classification of SQL-Injection Attacks and Countermeasures", In Proceedings of the IEEE International Symposium on Secure Software Engineering, 2006, vol. 1, pp. 13–15.
74. S. Gupta, A. Singhal, and A. Kapoor, "A Literature Survey on Social Engineering Attacks: Phishing Attack", In 2016 International Conference on Computing, Communication and Automation (ICCCA), Noida, 2016, pp. 537–540.

CHAPTER 6

Network Intrusion Detection with XGBoost

Arnaldo Gouveia and Miguel Correia

CONTENTS

6.1 INTRODUCTION

The detection of security-related events using *Machine Learning* (ML) has been extensively investigated [1–9]. *Intrusion Detection Systems* (IDSs) are security tools used to detect malicious activity. *Network Intrusion Detection Systems* (NIDSs) are one of the best-known contexts of machine learning application in the security field [4]. IDSs can be classified using several criteria [1]. One of these criteria is the detection approach, in terms of which IDSs (and NIDSs) can be signature-based or anomaly-based. The former class detects attacks by comparing the data flow under analysis to patterns stored in a signature database of known attacks. The later detects anomalies using a model of normal behavior of the monitored system and flagging behavior lying outside of the model as anomalous or suspicious. Signature-based IDSs can detect well-known attacks with high accuracy but fail to detect or find unknown attacks [5, 9], whereas anomaly-based IDSs have that capacity. In this chapter, we focus on using ML classifiers for detection, which is a form of signature-based detection.

A sophisticated attacker can bypass these techniques, so the need for more intelligent intrusion detection is increasing by the day. Researchers are attempting to apply ML techniques to this area of cybersecurity. Supervised machine learning algorithms, when applied to historical alert data, can significantly improve classification accuracy and decrease research time for analysts. It can supplement analysts with additional data and insights to make better judgment calls. Though prediction models based on historical data can improve analyst productivity, they will never replace security analysts altogether.

NIDSs serve as one of the critical components for a defense-in-depth solution. NIDS appliances allow for active, inline protection for known and unknown threats passing across a network segment at all layers of the OSI model. The employment, tuning, and upkeep of machine learning models on a NIDS may lead to a negative impact on production traffic if not properly maintained. This document serves as guidance to help shape the development of an NIDS machine learning model management approach. Through proper maintenance, placement, and tuning of models, an unwanted impact to network traffic can be kept to a minimum while also achieving an optimal balance of security and network performance.

XGBoost (EXtreme Gradient Boosting) [10] is a recent decision-tree-based ensemble machine learning algorithm. XGBoost won Kaggle's Higgs Machine Learning Challenge in 2014, that consisted in searching for Higgs bosons [11] in a large dataset provided by CERN[1] [12, 13]. The dataset and the subject of the challenge correspond to a particular physical phenomenon, the Higgs boson to *Tau* particles decay ($H \rightarrow \tau^{+}\tau^{-}$) [14]. The Higgs boson has many different processes (called channels by physicists) through which it can decay, that is, produce other particles. The standard model of particle physics has been being developed for decades, and the existence of almost all subatomic particles has been confirmed experimentally in the past century. The Higgs boson was the exception, as it was detected experimentally only in 2012, after experiments in the Large Hadron Collider (LHC) [14, 15]. XGBoost found again Higgs bosons on CERN data only 2 years later.

XGBoost is gaining popularity due to its performance and scalability [10, 13, 16]. It has been proved that XGBoost is faster than other popular algorithms in a single machine and to scale to billions of examples in distributed or memory-limited settings, as shown empirically in Kaggle competitions [16]. Kaggle is an online community focused on machine learning, popular for the competitions it organizes.

[1] Conseil Européen pour la Recherche Nucléaire.

The goal of a NIDS is to generate alerts when adversaries try to penetrate or attack the network. Consider a *flow* to be a sequence of Internet Protocol (IP) packets with similar features (e.g., same destination IP address and port). Typically an intrusion involves a few flows hidden among many legitimate flows. In statistical terms, the problem of detecting a few flows in a large set of flows is similar to the problem of detecting Higgs bosons. In this scope, we aim at answering the following questions:

1. How effectively can XGBoost be used in the context of NIDSs?
2. How can machine learning model management be optimized in NIDS?
3. What insights can we devise in this approach?

We use XGBoost with two datasets carefully designed for evaluating NIDSs machine learning algorithms: UNB NSL KDD [8] and UNSW NB15 [6, 7]. XGBoost has been barely used for intrusion detection, as explained in the next section.

The contributions of the this work are: (1) a study of the use of XGBoost in the context of network intrusion detection; (2) an optimization method for XGBoost for achieving performance in that context; and (3) a set of XGBoost parameters with strong impact/correlation on performance. We used the H2O XGBoost implementation, written in R, in the experiments. H2O is an open source software package machine learning.

6.2 RELATED WORK

There are a few works related to ours in the sense that they use XGBoost for intrusion detection. However, they are very limited, for example, because they consider only the detection of DDoS attacks or focus on the specific case of Software-Defined Networks (SDNs).

Chen et al. 17] focused on applying XGBoost to learn to identify a DDoS attack in the controller of an SDN cloud. They used the old 1999 ACM KDD Cup dataset[2]. The authors have shown that XGboost can detect DDoS attacks by analyzing attack traffic patterns. The XGBoost algorithm has shown higher accuracy and lower false-positive rate than other algorithms tested (Gradient Boosting Decision Tree, Random Forest, and Support Vector Machines), according to the authors. In addition, XGBoost proved to be quicker than the other algorithms tested.

Bansal and Kaur [18] used the CICIDS 2017 dataset again to detect DDoS attacks, not other classes [19]. XGBoost proved again to provide better accuracy values than KNN, AdaBoost, MLP, and Naïve-Bayes.

In a short paper, Amaral et al. 20] used XGBoost to classify malicious traffic in SDNs. The authors present a simple architecture deployed in an enterprise network that gathers traffic data using the OpenFlow protocol and shows how several machine learning techniques can be applied for traffic classification, including XGBoost. Traffic of a number of applications (e.g., BitTorrent, YouTube) was classified and the comparative results of

[2] https://www.kdd.org/kdd-cup/view/kdd-cup-1999

XGboost in face of other classifiers (Random Forests and Stochastic Gradient Boosting) proved comparable or better.

Dhaliwal et al. also used XGBoost for detection [2]. However, they present essentially an exercise of running XGBoost on a single dataset that does not cover the second and third questions mentioned earlier.

Mitchell et al. [21] presented a CUDA-based implementation of a decision tree construction algorithm by using XGBoost. The algorithm has been executed entirely on the GPU and has shown high performance with a variety of datasets and settings, including sparse input matrices.

6.3 XGBOOST AND BOOSTING TREES

XGBoost is based on a set of machine learning concepts:

- *Weak learners:* XGBoost uses *decision trees* as *weak* learners;
- *Boosting:* combines the contributions of many weak learners to produce a strong learner;
- *Gradient boosting:* the notion of doing boosting by minimizing errors by introducing a gradient term.

Three main forms of gradient boosting are supported:

- Gradient Boosting algorithm also called gradient boosting machine including the learning rate;
- Stochastic Gradient Boosting with subsampling at the row, column, and column per split levels;
- Regularized Gradient Boosting with both L1 and L2 regularization.

6.3.1 XGBoost General Formalism and Tuning Parameters

The model concept in supervised learning (and in machine learning in general) refers to the mathematical processing by which the prediction y_i are made from the input x_i.

The model parameters are the undetermined factors that to learn from data in order to have a model that can make good predictions. The model optimization is executed by means of a function to be maximized or minimized in order to find out such parameters—the objective function.

Modeling XGBoost. An important characteristic of objective functions is that they consist of two parts, training loss and the regularization term:

$$\text{obj}(\theta) = L(\theta) + \Omega(\theta) \tag{6.1}$$

where L is the training loss function and Ω is the regularization term. The training loss measures the capacity of the model to predict data that is also in the training dataset. A common choice of L is the mean squared error, which is given by $L(\theta)=\Sigma_i(y_i-\hat{y}_i)^2$.

Another commonly used loss function is logistic loss, to be used for logistic regression:

$$L(\theta)=\sum_i [y_i \ln(1+e^{-\hat{y}_i})+(1-y_i)\ln(1+e^{\hat{y}_i})] \tag{6.2}$$

Mathematically, we can write the XGboost model in the form

$$\hat{y}_i=\sum_{k=1}^{K} f_k(x_i), f_k \in \mathrm{F} \tag{6.3}$$

where K is the number of trees, f is a function in the functional space $s\mathrm{F}$ set of all possible classification and regression trees (CARTs). The objective function to be optimized is given by

$$\mathrm{obj}(\theta)=\sum_i^{n} l(y_i,\hat{y}_i)+\sum_{k=1}^{K}\Omega(f_k) \tag{6.4}$$

The tree parameters to be used are those that can be determined by functions f_i each containing the structure of the tree and the leaf scores. It is considered intractable to learn all the trees at once. Instead, an additive strategy is used: iterate adding one new tree at a time. The prediction value at step t as $\hat{y}_i^{(t)}$. Then we have:

$$\begin{aligned}
&\hat{y}_i^{(0)}=0\\
&\hat{y}_i^{(1)}=f_1(x_i)=\hat{y}_i^{(0)}+f_1(x_i)\\
&\hat{y}_i^{(2)}=f_1(x_i)+f_2(x_i)=\hat{y}_i^{(1)}+f_2(x_i)\\
&\ldots\\
&\hat{y}_i^{(t)}=\sum_{k=1}^{t} f_k(x_i)=\hat{y}_i^{(t-1)}+f_t(x_i)
\end{aligned} \tag{6.5}$$

In each step, a tree is added that optimizes the objective function.

$$\begin{aligned}
\mathrm{obj}^{(t)}&=\sum_{i=1}^{n} l(y_i,\hat{y}_i^{(t)})+\sum_{i=1}^{t}\Omega(f_i)\\
&=\sum_{i=1}^{n} l(y_i,\hat{y}_i^{(t-1)}+f_t(x_i))+\Omega(f_t)+\text{constant}
\end{aligned} \tag{6.6}$$

If we consider using mean squared error (MSE), as in the Ridge approach, as our loss function, the objective becomes

$$\begin{aligned}\text{obj}^{(t)} &= \sum_{i=1}^{n}(y_i - (\hat{y}_i^{(t-1)} + f_t(x_i)))^2 + \sum_{i=1}^{t}\Omega(f_i) \\ &= \sum_{i=1}^{n}[2(\hat{y}_i^{(t-1)} - y_i)f_t(x_i) + f_t(x_i)^2] + \Omega(f_t) + \text{constant}\end{aligned} \tag{6.7}$$

With the Taylor expansion of the loss function up to the second order we obtain:

$$\text{obj}^{(t)} = \sum_{i=1}^{n}[l(y_i, \hat{y}_i^{(t-1)}) + g_i f_t(x_i) + \frac{1}{2}h_i f_t^2(x_i)] + \Omega(f_t) + \text{constant} \tag{6.8}$$

where the g_i and h_i are defined as

$$\begin{aligned}g_i &= \partial_{\hat{y}_i^{(t-1)}} l(y_i, \hat{y}_i^{(t-1)}) \\ h_i &= \partial^2_{\hat{y}_i^{(t-1)}} l(y_i, \hat{y}_i^{(t-1)})\end{aligned} \tag{6.9}$$

After we remove all the constants, the specific objective at step t becomes

$$\sum_{i=1}^{n}[g_i f_t(x_i) + \frac{1}{2}h_i f_t^2(x_i)] + \Omega(f_t) \tag{6.10}$$

This becomes our optimization goal for the new tree. One important advantage of this definition is that the value of the objective function depends only on the terms g_i and h_i. This is the reason why XGBoost allows using different custom loss functions. We can optimize every loss function, including logistic regression and pairwise ranking, using exactly the same solver that takes g_i and h_i as input.

Model complexity. To determine the complexity of the tree $\Omega(f)$ the definition of the tree $f(x)$ can be defined as:

$$f_t(x) = w_{q(x)}, w \in R^T, q: R^d \rightarrow \{1, 2, \cdots, T\} \tag{6.11}$$

where T is the number of leaves and w is the vector of scores on leaves, q is a function assigning each data point to the corresponding leaf.

A commonly accepted definition of complexity in XGboost is given by:[3]

$$\Omega(f) = \gamma T + \frac{1}{2}\lambda\sum_{j=1}^{T}w_j^2 \tag{6.12}$$

[3] https://xgboost.readthedocs.io/en/latest/tutorials/model.html

The structure score. After reformulating the tree model, we can write the objective value with the tth tree as:

$$\begin{aligned} \text{obj}^{(t)} &\approx \sum_{i=1}^{n}[g_i w_{q(x_i)} + \frac{1}{2}h_i w_{q(x_i)}^2] + \gamma T + \frac{1}{2}\lambda\sum_{j=1}^{T} w_j^2 \\ &= \sum_{j=1}^{T}[(\sum_{i\in I_j} g_i)w_j + \frac{1}{2}(\sum_{i\in I_j} h_i + \lambda)w_j^2] + \gamma T \end{aligned} \tag{6.13}$$

The term $I_j = \{i \mid q(x_i) = j\}$ refers to the set of indices of data points assigned to the jth leaf. In the second line the index of the summation changed because all the data points on the same leaf get the same score. Defining $G_j = \sum_{i\in I_j} g_i$ and $H_j = \sum_{i\in I_j} h_i$ the expression becomes:

$$H_j = \sum_{i\in I_j} h_i \tag{6.14}$$

The terms w_j are independent with respect to each other, the form $G_j w_j + \frac{1}{2}(H_j + \lambda)w_j^2$ is quadratic and the best w_j for a given tree structure $q(x)$ and the best objective reduction we can get is:

$$\begin{aligned} w_j^* &= -\frac{G_j}{H_j + \lambda} \\ \text{obj}^* &= -\frac{1}{2}\sum_{j=1}^{T}\frac{G_j^2}{H_j + \lambda} + \gamma T \end{aligned} \tag{6.15}$$

$$Gain = \frac{1}{2}\left[\frac{G_L^2}{H_L + \lambda} + \frac{G_R^2}{H_R + \lambda} - \frac{(G_L + G_R)^2}{H_L + H_R + \lambda}\right] - \gamma \tag{6.16}$$

Further insight into regularization approaches. To address overfitting, that is, to avoid modeling the training data too tightly, XGBoost in the H2O R Package [22] uses L1 and L2 *regularization*. The difference between these two penalty terms is mainly:

- *Lasso regression (L1):* adds the modulus of the coefficient vector as penalty term to the loss function.
- *Ridge regression (L2):* adds a squared modulus term of the coefficient vector as penalty term to the loss function.

The key difference between these techniques is that Lasso shrinks the less important features' coefficients to zero, thus effectively removing some features, something that Ridge does not manage to do. Feature reduction is, therefore, a more defined consequence of using Lasso. XGBoost's regularization strategy involves also configuring other parameters to be effective, for example, *Maximum Depth*, *Minimum Child Weight*, and *Gamma*. Without proper regularization, the tree may split until it can predict the training set perfectly, that is, may overfit. Therefore, it will end up losing generality and will not perform as good on new data (i.e., on test data or on production data).

The trees used play the role of weak learners, providing decisions only slightly better than a random guess. XGBoost used an ensemble approach, that is, it combines a large number of decision trees to produce a final model consisting of a forest of decision trees. An individual decision tree is grown by ordering all features and checking each possible split for every feature. The split that results in the best score for the chosen objective function becomes the rule for that node.

For a given data set with n examples and m features $D=\{(x_i, y_i)\}$ $(|D|=n, x_i \in R^m, y_i \in R)$, a tree ensemble model uses K additive functions to predict the output [10]:

$$\hat{y}_i = \phi(x_i) = \sum_{k=1}^{K} f_k(x_i), \quad f_k \in F, \tag{6.17}$$

where $F=\{f(x)=w_{q(x)}\}(q: R^m \rightarrow T, w \in R^T)$ is the space of regression trees. In the equation, q represents the structure of each tree that maps an example to the corresponding leaf index, T is the number of leaves in the tree, each f_k corresponds to an independent tree structure q, and w is leaf weights.

In the general case as stated before an objective functions has two terms: training loss and regularization [10]: $L(\theta)=L(\theta)+\Omega(\theta)$. A common choice of $L(\theta)$ is the mean squared error. With proper choices for y_i we can express regression, classification, and ranking. Training the model amounts to finding the parameters θ that best fit the training data x_i labels y_i. In order to train the model, we need to define the objective function to measure how well the model fits the training data. To learn the set of functions used in the model, we minimize the following regularized objective:

$$L(\phi) = \sum_i l(\hat{y}_i, y_i) + \sum_k \Omega(f_k) \tag{6.18}$$

where l is a differentiable convex loss function that measures the difference between the prediction $\hat{y}_i$ and the target y_i.

The second term Ω penalizes the complexity of the model (i.e., the regression tree functions). Depending on the type of regularization chosen, $\Omega(f)$ is usually defined as:

$$\Omega(f) = \gamma T + \frac{1}{2}\lambda \| w \| \, (L1 \text{ or } Lasso) \text{ or} \tag{6.19}$$

$$\Omega(f) = \gamma T + \frac{1}{2}\lambda \| w \|^2 \, (L2 \text{ or } Ridge) \tag{6.20}$$

where w is the vector of scores on leaves, T is the number of leaves, and γ and λ are adjustable gain factors. This regularization term helps to avoid overfitting.

TABLE 6.1 Dataset Comparison. The UNB NSL KDD Files Used Were the *KDDTest+.arff* and the *KDDTrain+.arff* Dataset Files

Year	Dataset	Size	Types of Attack
2016	NSL-KDD	Train = 18.7 MB Test = 3.4 MB	Probing, DoS, R2L, U2R
2017	UNSW NB15	Train = 14.7 MB Test = 30.8 MB	Fuzzers, Shellcode, Analysis, Backdoors, DoS, Exploits, Generic, Reconnaissance, Shellcode, Worms

6.4 THE DATASETS

An important component of an effective NIDS evaluation is a good benchmark dataset. We use two recent datasets that have been designed specifically and carefully to evaluate NIDSs. A comparative summary of the two can be found in Table 6.1.

6.4.1 The UNB NSL KDD Dataset

The University of New Brunswick (UNB) NSL KDD intrusion detection evaluation dataset was created using a principled approach, that involved defining features that maximize three costs [23]. The data are sequences of entries labeled as *normal* or *attack*. Each entry contains a set of characteristics of a *flow*, that is, of a sequence of IP packets starting at a time instant and ending at another, between which data flows between two IP addresses using a transport protocol (TCP, UDP) and an application-layer protocol (HTTP, SMTP, SSH, IMAP, POP3, or FTP). The dataset is fairly balanced with prior class probabilities of 0.465736 for the *normal* class and 0.534264 for the *anomaly* class. The attacks represented in dataset fall in four classes.

The UNB NSL KDD dataset is composed of two subdatasets: a *train dataset*, used for training a NIDS, and a *test dataset*, used for testing. Both have the same structure and contain all four types of attacks. However, the test dataset has more attacks as shown in Table 6.2, to allow evaluating the ability of algorithms to generalize. Each record of the dataset is characterized by features that fall into three categories: basic, content related, and traffic related. These features are represented in Table 6.3.

TABLE 6.2 Attacks in the UNB NSL KDD Train and Test Datasets (All Attacks from the First Exist Also in the Second)

Class	Train Dataset Attacks	Test Dataset Only Attacks
Probing	portsweep, ipsweep, satan, guesspasswd, spy, nmap	snmpguess, saint, mscan, xsnoop
Denial of Service (DoS)	back, smurf, neptune, land, pod, teardrop, buffer overflow, warezclient, warezmaster	apache2, worm, udpstorm, xterm
Remote to Login (R2L)	imap, phf, multihop	snmpget, httptunnel, xlock, sendmail, ps
User to Root (U2R)	loadmodule, ftp write, rootkit	sqlattack, mailbomb, processtable, perl

TABLE 6.3 Features Used to Characterize Each Flow in the UNB NSL KDD Dataset: Basic (Top), Content (Middle), and Traffic (Bottom)

Feature	Description
duration	length of the flow in seconds
protocol-type	type of protocol, e.g., TCP, UDP, ICMP
service	network service, e.g., HTTP, Telnet
src-bytes	num. of data bytes from source to destination
dst-bytes	num. of data bytes from destination to source
flag	status of the flow, normal or error
lang	1 if flow is for the same host/port; 0 otherwise
wrong-fragment	num. of erroneous fragments
urgent	num. of urgent packets
hot	num. of hot indicators
num-failed-logins	num. of failed login attempts
logged-in	1 if successfully logged in; 0 otherwise
num-compromised	num. of compromised conditions
root-shell	1 if root shell is obtained; 0 otherwise
su-attempted	1 if su root command attempted; 0 otherwise
num-root	num. of root accesses
num-file-creations	num. of file creation operations
num-shells	num. of shell prompt
num-access-files	num. of operations on access control files
num-outbound-cmds	num. of outbound commands in a ftp session
is-host-login	1 if the login belongs to the hot list; 0 otherwise
is-guest-login	1 if the login is a guest login; 0 otherwise
count	num. of connections to the same host as current
serror-rate	% of connections that have SYN errors
rerror-rate	% of connections that have REJ errors
same-srv-rate	% of connections to the same service
diff-srv-rate	% of connections to different services
srv-count	num. of connections to the same service as current
srv-serror-rate	% of connections that have SYN errors
srv-rerror-rate	% of connections that have REJ errors
srv-diff-host-rate	% of connections to different hosts
dst-host-count	num. of connections to the same destination host
dst-host-srv-count	num. of connections to the same service as current
dst-host-same-srv-rate	% of connections to the same service
dst-host-diff-srv-rate	% of connections to different services
dst-host-same-src-port-rate	% of connections from same source and port
dst-host-srv-diff-host-rate	% of connections to different services
dst-host-serror-rate	% of connections that have SYN errors
dst-host-srv-serror-rate	% of connections that have SYN errors per service
dst-host-rerror-rate	% of connections that have REJ errors
dst-host-srv-rerror-rate	% of connections that have REJ errors per service

6.4.2 The UNSW NB15 Dataset

The second dataset comes from the Australian Centre for Cyber Security (ACCS) [6, 7]. The IXIA PerfectStorm tool[4], an application traffic simulator, has been used in the ACCS Cyber Range Lab to create a dataset with normal traffic and abnormal synthetic network traffic. The abnormal traffic is created with the IXIA tool, which contains information about new attacks updated using MITRE's Common Vulnerabilities and Exposures (CVE) site. The number of records of the training set is 175,341; the testing set has 82,332 records from the different types, attack and normal. The UNSW NB15 data set includes nine categories of attacks:

- *Fuzzers*: in this type of attack, randomly generated data is feed into a suspended program or network;
- *Reconnaissance*: the attacker gathers information about a system for later attacking it;
- *Shellcode*: code is used as the payload in an attack packet;
- *Analysis*: includes port scan, spam, and HTML files penetrations;
- *Backdoors*: access to a system is gained by bypassing any secured layers;
- *Denial of Service*: the perpetrator intents resource unavailability by temporarily or indefinitely disrupting services;
- *Exploits*: the attacker exploits vulnerabilities of the system through known loopholes of the system;
- *Generic*: the attack is implemented without knowing how the cryptographic primitive is implemented and works for all block ciphers;
- *Worms*: the attack mechanism replicates itself through the network.

The UNSW NB15 dataset includes 9 different moderns attack types (compared to 21 older attack types in UNB NSL KDD dataset) and wide varieties of real normal activities as well as 49 features inclusive of the class label consisting total of 2,540,044 records. These features are categorized into six groups of features: flow related, Basic Features, content related, time related, additional generated, and labeled features. These features include a variety of packet-based features and flow-related features. The packet-based features assist the examination of the payload beside the headers of the packets. On the contrary, for the flow-based features and maintaining low computational analysis instead of observing all the packets going through a network link, only connected packets of the network traffic are considered. Moreover, the flow-related features are based on traffic direction, interarrival time, and interpacket length. The matched features are categorized into several groups (Table 6.4).

---

[4] https://www.ixiacom.com/products/perfectstorm

TABLE 6.4 Features of the UNSW NB15 Dataset: Flow-Related (Top), Basic (Next), Content-Related (Next), Time-Related (Next), and Connection (Bottom)

#	Name	Type	Description
1	srcip	N	Source IP address.
2	sport	I	Source port number.
3	dstip	N	Destination IP address.
4	dsport	I	Destination port number.
5	proto	N	Transaction protocol.
6	state	N	The state and its dependent protocol, e.g., ACC, CLO, else (-).
7	dur	F	Record total duration.
8	sbytes	I	Source to destination bytes.
9	dbytes	I	Destination to source bytes.
10	sttl	I	Source to destination time to live.
11	dttl	I	Destination to source time to live.
12	sloss	I	Source packets re-transmitted or dropped.
13	dloss	I	Destination packets re-transmitted or dropped.
14	service	N	http, ftp, ssh, dns .., else (-).
15	sload	F	Source bits per second.
16	dload	F	Destination bits per second.
17	spkts	I	Source to destination packet count.
18	dpkts	I	Destination to source packet count.
19	swin	I	Source TCP window advertisement.
20	dwin	I	Destination TCP window advertisement.
21	stcpb	I	Source TCP sequence number.
22	dtcpb	I	Destination TCP sequence number.
23	smeansz	I	Mean of the flow packet size transmitted by the src.
24	dmeansz	I	Mean of the flow packet size transmitted by the dst.
25	trans depth	I	The depth into the connection of http request/response transaction.
26	res bdy len	I	The content size of the data transferred from the server's http service.
27	sjit	F	Source jitter (mSec).
28	djit	F	Destination jitter (mSec).
29	stime	T	Record start time.
30	ltime	T	Record last time.
31	sintpkt	F	Source interpacket arrival time (mSec).
32	dintpkt	F	Destination interpacket arrival time (mSec).
33	tcprtt	F	The sum of "synack" and "ackdat" of the TCP.
34	synack	F	The time elapsed between the SYN and the SYN ACK packets of the TCP.
35	ackdat	F	The time between the SYN ACK and the ACK packets of the TCP.
36	is sm ips ports	B	If source (1) equals to destination (3) IP addresses and port numbers (2) (4) are equal, this variable takes value 1 else 0
37	ct state ttl	I	No. for each state (6) according to specific range of values for source/ destination time to live (10) (11).
38	ct flw http mthd	I	No. of flows that has methods such as Get and Post in http service.
39	is ftp login	B	If the ftp session is accessed by user and password then 1 else 0.
40	ct ftp cmd	I	No. of flows that has a command in ftp session.
41	ct srv src	I	No. of connections that contain the same service (14) and source address (1) in 100 connections according to the last time (26).

(*Continued*)

TABLE 6.4 *(Continued)* Features of the UNSW NB15 Dataset: Flow-Related (Top), Basic (Next), Content-Related (Next), Time-Related (Next), and Connection (Bottom)

#	Name	Type	Description
42	ct srv dst	I	No. of connections that contain the same service (14) and destination address (3) in 100 connections according to the last time (26).
43	ct dst ltm	I	No. of connections of the same destination address (3) in 100 connections according to the last time (26).
44	ct src ltm	I	No. of connections of the same source address (1) in 100 connections according to the last time (26).
45	ct src dport ltm	I	No. of connections of the same source address (1) and the destination port (4) in 100 connections according to the last time (26).
46	ct dst sport ltm	I	No. of connections of the same destination address (3) and the source port (2) in 100 connections according to the last time (26).
47	ct dst src ltm	I	No. of connections of the same source (1) and the destination (3) address in 100 connections according to the last time (26).

N: Name, I: Integer, F: Float, T: Time and B: Binary

6.5 PERFORMANCE METRICS

We have chosen *logloss* as the primary criteria for assessing the model's performance, and used the *Area Under Curve* (AUC) to obtain additional insights. Moreover, we show that there are a number of other metrics that are closely correlated to those, so optimizing one will likely provide comparable results for others (Section 6.7.2).

logloss (logarithmic loss) is a metric used to evaluate the performance of both binomial and multinomial classifiers [24]. logloss evaluates how close the values predicted by a model (uncalibrated probability estimates) are to the actual target value. In other words, this metric captures the extent to which predicted probabilities diverge from class labels: it incentives well-calibrated probabilities. The metric is shown in Equation 6.21. In the equation, N is the total number of rows (observations) of the dataset; w is user-defined weight per-row (defaults is 1); p is the predicted value (uncalibrated probability) assigned to a given row by the algorithm; and y is the actual target value.

$$H_p(q) = -\frac{1}{N}\sum_{i=1}^{N} w_i(y_i \ln(p_i) + (1-y_i)\ln(1-p_i)) \tag{6.21}$$

Receiver operating characteristic (ROC) graphs (or curves) are useful for visualizing the performance of classifiers [25]. ROC graphs have, in recent years, been increasingly used in machine learning and data mining research. The AUC, that is, the area under a ROC curve, represents the probability that a given sample classification is (correctly) rated or ranked with greater suspicion than a randomly chosen classification [26, 27].

The ROC curve is a useful tool for a few reasons: The curves of different models can be compared directly in general or for different thresholds. Also the AUC can be used as a summary of the model skill. The shape of the curve is informative in terms of asserting the balance between the classification for the classes. Smaller values on the x-axis of the

TABLE 6.5 Feature Set Used in the Automatic Training Phase (Discrete Values) Totaling $2^7 = 128$ Different Combinations

XGBoost Model Parameter	Values
ntrees	5/10
max_depth	4/9
eta	0.1/0.5
sample_rate	0.5/0.9
gamma	0/0.001
reg_lambda	0/1.0
reg_alpha	0/1.0

plot indicate lower false positives and higher true negatives. Larger values on the y-axis of the plot indicate higher true positives and lower false negatives. The correlation among the models metrics was calculated having as input data the specific figures for these metrics for each of the models produced. A model has been obtained for each point of the grid search space (explained next).

6.6 APPROACH AND PARAMETERS

Our approach can be summarized as follows:

- To define the values of the model's parameters for the two datasets, we used a full grid search approach, that is, we did a search over a hyperparameter space subset using H2O's grid search functionality for a full search. The range of discrete values used for the parameters in the search is in Table 6.5.
- The models are ranked via the validation logloss minimization criteria;
- The best model is set to be the model with the minimum validation logloss value;
- A correlation calculation of the models parameters versus logloss has been performed in order to seek which types of metrics can be used as proxy to estimate others;
- The two correlations were validated with a significance test.

XGBoost, configured for acting as a tree booster, has been configured with the following parameters,[5] explicitly defined in the training phase with the ranges listed in Table 6.5:

- *ntrees*: number of trees to build;
- *max-depth* (default = 6): maximum depth of a tree. Used to control over fitting as higher depth will allow model to learn relations very specific to a particular sample;
- *eta* (default = 0.3): learning rate by which to shrink the feature weights. Shrinking feature weights after each boosting step makes the boosting process more conservative and prevents overfitting;

[5] http://docs.h2o.ai/h2o/latest-stable/h2o-docs/data-science/xgboost.html#defining-an-xgboost-model

- *gamma* (default = 0): a node is split only when the resulting split gives a positive reduction in the loss function. Gamma specifies the minimum loss reduction required to make a split. Makes the algorithm conservative. The values can vary depending on the loss function and should be tuned;
- *subsample* (alias sample-rate) (default = 1): row sampling ratio of the training instance;
- *reg lambda* (default = 1): L2 regularization term on weights (Ridge regression);
- *reg alpha* (default = 0): L1 regularization term on weight (Lasso regression).

The values used in the grid search are listed in Table 6.5. All other parameters took the default values.

6.7 EVALUATION

The objective of the experimental evaluation was to answer the three questions posed in the introduction. A number of performance metrics have been applied as per Table 6.6.

The parameters for the best models obtained for both datasets are depicted in Table 6.7. The main performance numbers are depicted in Figures 6.1(a), 6.1(b), 6.2(a), and 6.2(b). Complementarily, Figures 6.4(a), 6.5(a), and 6.6(a) show performance values for the UNB NSL KDD dataset. Also Figures 6.4(b), 6.5(b), and 6.6(b) show the same for the UNSW NB15 dataset.

6.7.1 Logloss and AUC Results

Regarding logloss, we can see in Figures 6.1(b) and 6.2(b) the fast convergence obtained with the training dataset as seen by the number of trees in the x axis (top [a] and bottom [b] lines).

TABLE 6.6 Performance Values for the Best Models Obtained Per Dataset

Dataset	KDD-NSL	UNSW NB15
Validation AUC	0.9187	0.9861
Validation Error	0.1136	0.0670
Validation Accuracy	0.8864	0.9334
Training Logloss	0.0577	0.0598
Validation Logloss	0.5816	0.2583

TABLE 6.7 Relevant Experimental Best Model Features for the Two Datasets. It Is Relevant to Remark the Similarity Between the Two Sets of Values

Parameter	NSL KDD	UNSW NB15
gamma	0.001	0
eta	0.5	0.5
sample_rate	0.9	0.9
max_depth	9	9
ntrees	10	10
reg_alpha	0	0
reg_lambda	0	0

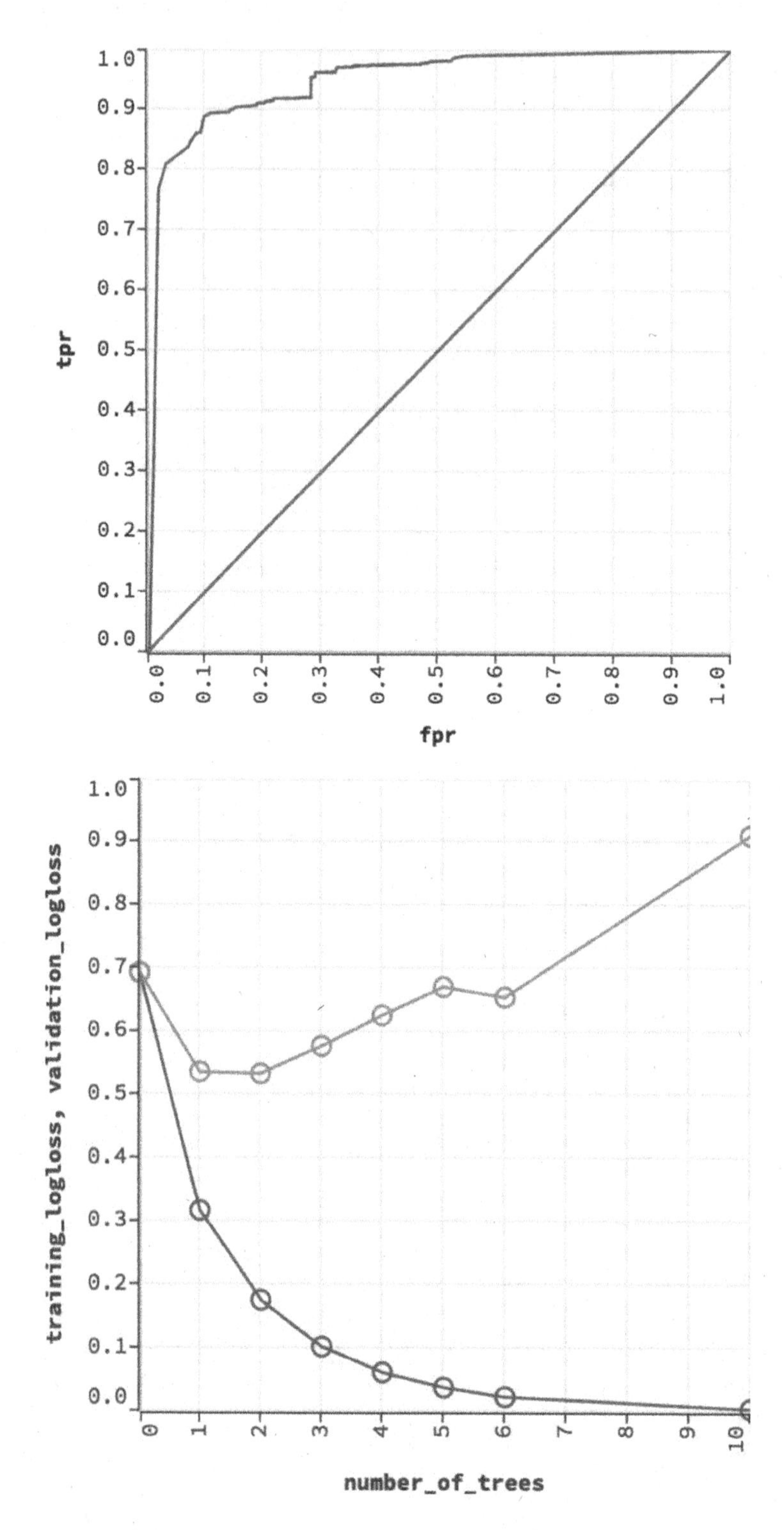

FIGURE 6.1 UNB NSL KDD ROC Curves and logloss. It is relevant to note the divergence in (b) regarding the validation Logloss that diverges denouncing a training dataset that is statistically dissimilar when compared with the testing dataset. (a) [ROC curve for best model validation. AUC = 0.9187.] (b) [Logloss by number of trees (1–10); training (**black**) and validation datasets (**grey**).]

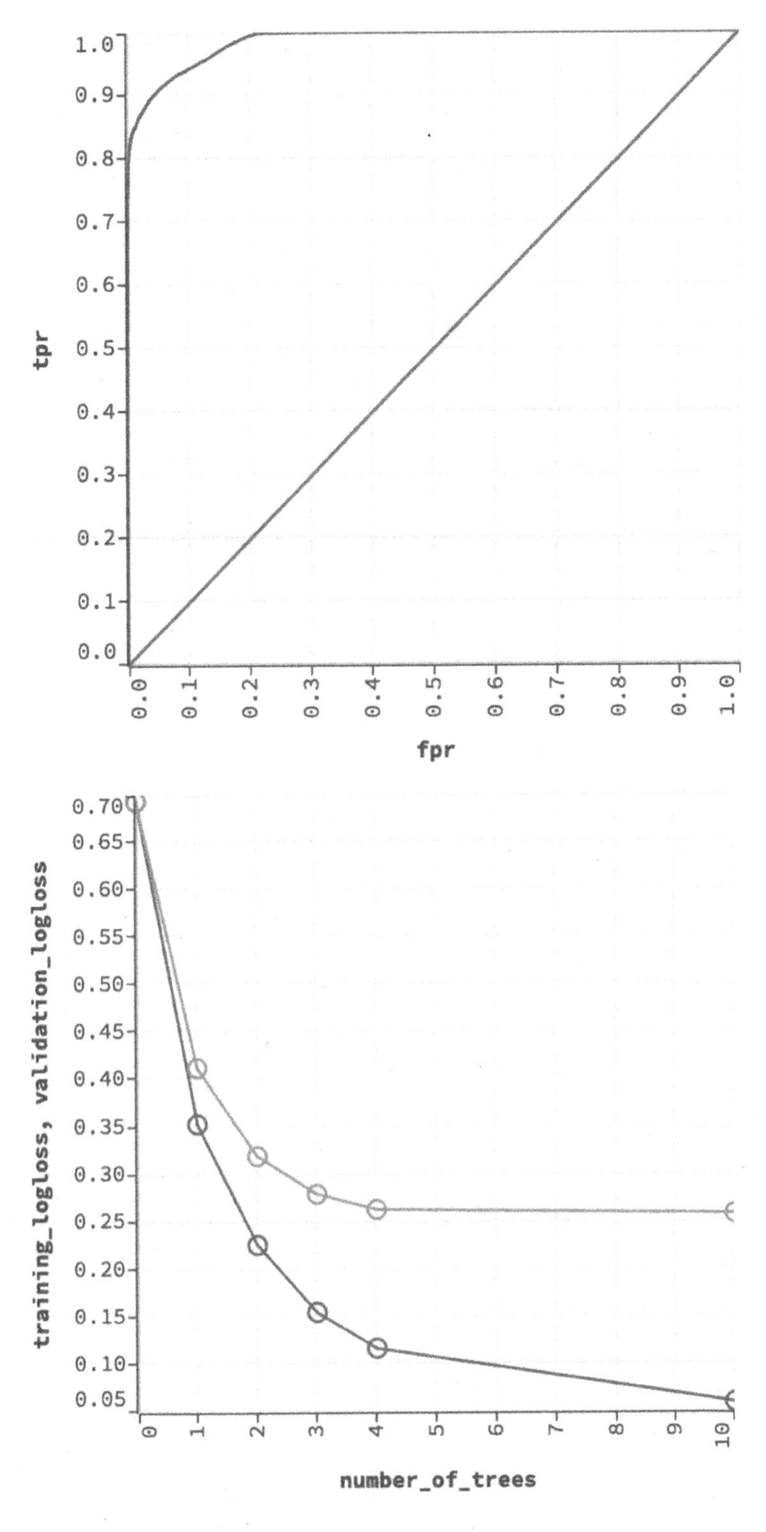

FIGURE 6.2 UNSW NB15 ROC curves and Logloss. It is relevant to note a much lesser divergence in (b) regarding the validation Logloss that diverges denouncing a training dataset that is statistically more resembling when compared with the testing dataset. (a) [ROC curve for best model validation. Cross Validation AUC = 0.9861.] (b) [Cross Validation Logloss by number of trees (1 –10); training (**black**) and validation datasets (**grey**).]

Focusing on the validation dataset, for the KDD dataset the logloss figures did not converge to a minimum value (Figure 6.1(b), bottom line). The values obtained for AUC were high (see Table 6.6). As can be observed, the curve obtained is relatively well-balanced (Figure 6.1(a)). The predictor scored better with the UNSW NB15 dataset, in line with previous results by Moustafa [28].

6.7.2 Interpreting the Models

Figures 6.3(a) and 6.3(b) show that there are clusters of metrics highly correlated between each other (one central cluster with strong negative correlation and two triangle vertices clusters with strong positive correlation). The figures represent values of the Pearson correlation coefficient, probably the most widely used measure of the extent to which two variables are linearly related: $X,Y : \rho(X,Y) = \frac{\mathrm{cov}(X,Y)}{\sigma_x \sigma_y}$. The metrics considered in this

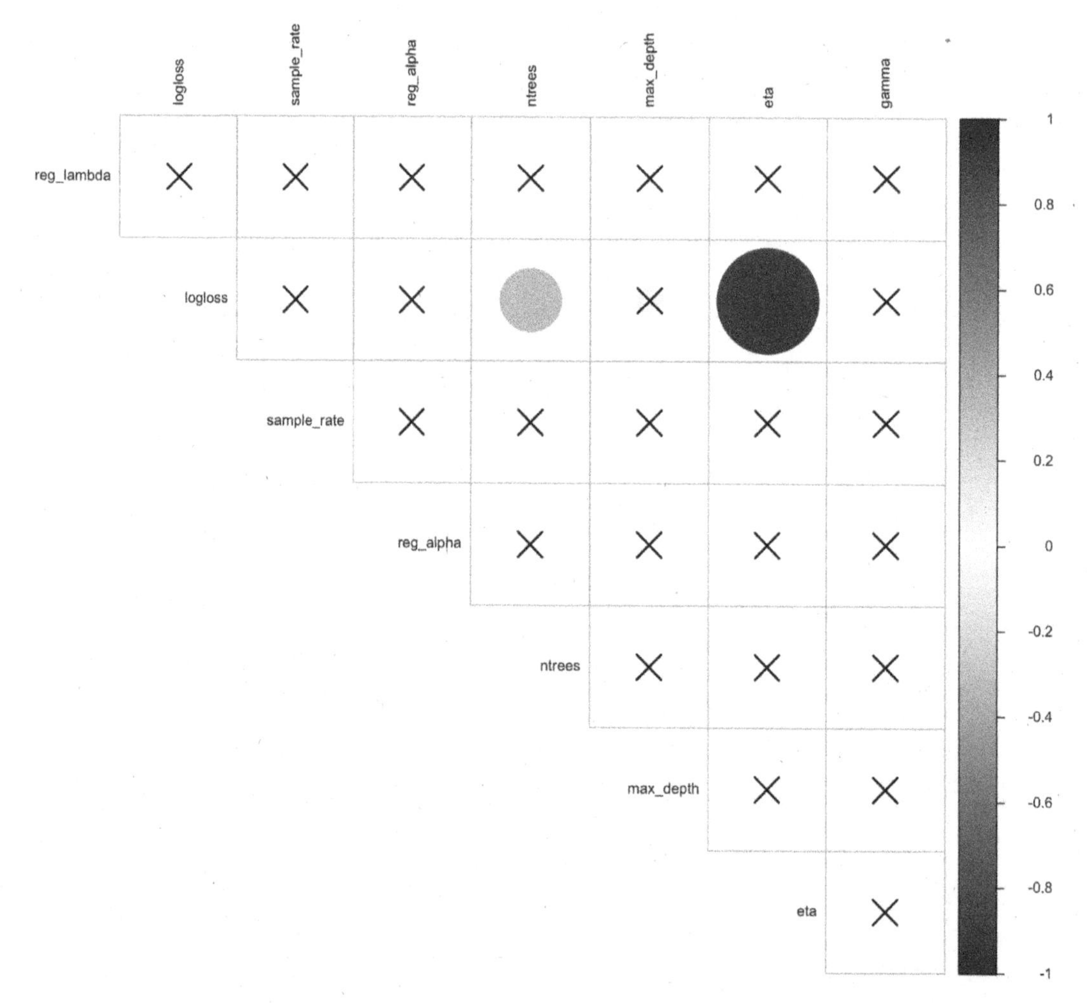

FIGURE 6.3 Pearson correlation for the model variables. Correlation values found to be not significantare crossed. The magnitude of the correlation links to the size of the circles. **Darker tones relate to stronger correlation either negative or positive.** Values processed for all the iterated models with a significance level forthe p-test = 0.001. (a) [UNB NSL KDD.] (b) [UNSW NB15.] *(Continued)*

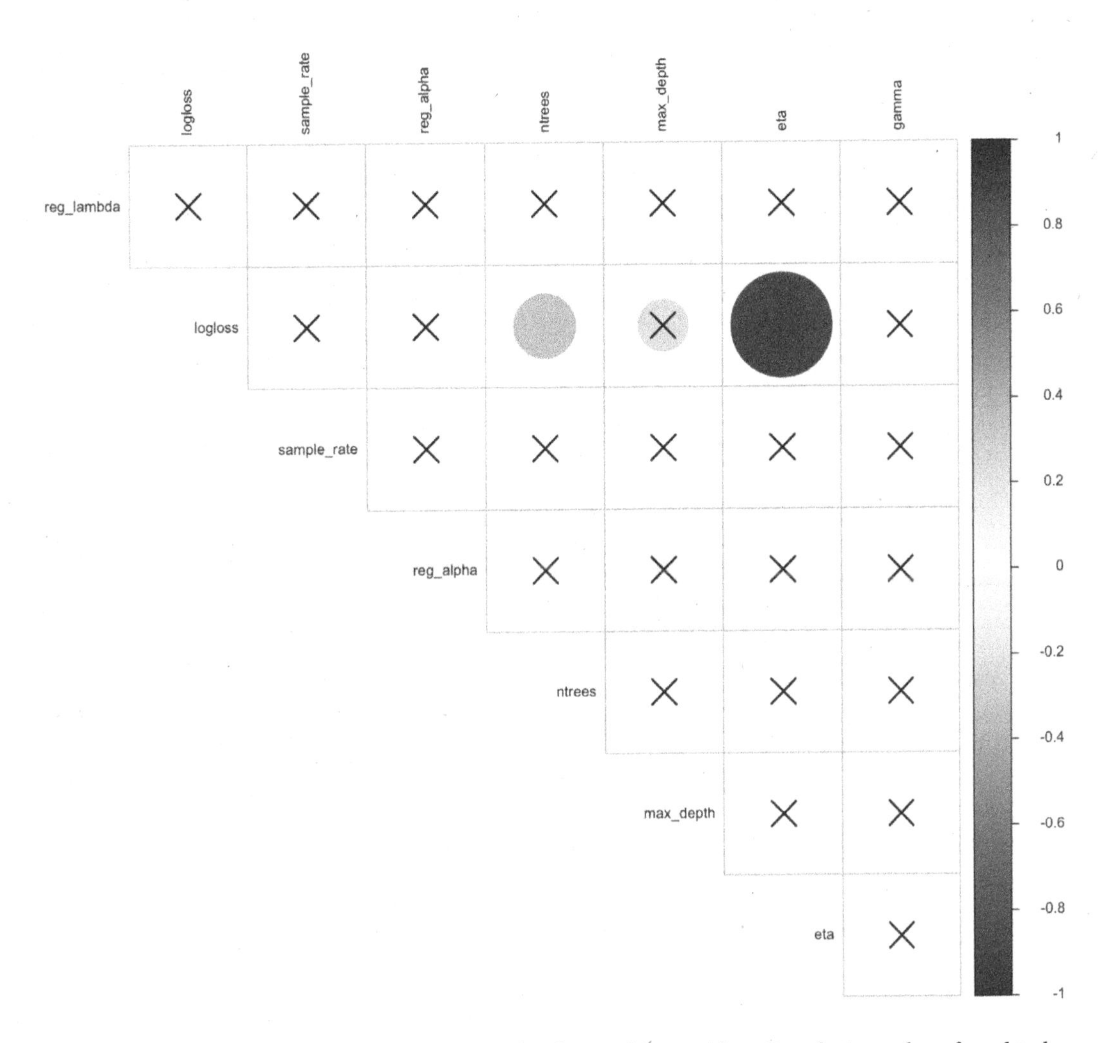

FIGURE 6.3 (*Continued*) Pearson correlation for the model variables. Correlation values found to be not significant are crossed. The magnitude of the correlation links to the size of the circles. **Darker tones relate to stronger correlation either negative or positive**. Values processed for all the iterated models with a significance level for the p-test = 0.001. (a) [UNB NSL KDD.] (b) [UNSW NB15.]

correlation test are[6]: *Validation R2, Validation Classification Error, Training Logloss, Cross Validation Logloss, Training RMSE, Cross Validation RMSE, Training Classification Error, Cross Validation Classification Error, Validation AUC, Validation Lift, Validation RMSE, Validation Logloss, Training Lift, Cross Validation Lift, Training R2, Cross Validation R2,* and *Training AUC*. The values of the Pearson correlation are shown with a p test at a significance level of 0.001, so many correlations were not found significant and are shown crossed.

Two main clusters are evident for both the UNB NSL KDD (Figure 6.3(a)) and UNSW NB15 cases (Figure 6.3(b)), each with a high positive correlation and another with a null

[6] http://docs.h2o.ai/h2o/latest-stable/h2o-docs/performance-and-prediction.html

correlation among parameters. In Figures 6.7 and 6.8, the training features for both models show correlations with the training metric used, logloss. It is noticeable that for both training models the single most relevant features to affect the logloss results are the *eta* and *ntrees* features. Some of the metric correlations also did not pass the significance test. These values are shown as crossed.

The divergent testing logloss in Figure 6.1 is caused by the fact that the testing dataset for KDD-NSL has considerable different statistical properties when compared with the train dataset. This situation can be identified by a learning curve for training loss that shows improvement and similarly a learning curve for validation loss that shows improvement, but a large gap remains between both curves.

6.7.3 Correlation Validation

The p test is used to check statistical significance of the correlation values obtained. Small p values occur when the null hypothesis is false.

The validity of the test depends on the p value used. Apparently small values of p like 0.01 still open a relatively large probability for the null hypotheses. Expanding on earlier work, especially Edwards et al. [29], it is shown that actual and relevant evidence against a null can differ by an order of magnitude from the p value. The choice here is to choose a p value equal or smaller than 0.001 as per Table 6.8.

6.7.4 How Much Information Is Hidden in the Performance Metrics?

The graphs in Figures 6.4(a), 6.5(a), and 6.6(a) can be understood as a way of empirically assessing the amount of information XGBoost manipulate in the datasets. This information amount can be assessed as an entropy value by means of the parallel with the entropy concept (Equation 6.22): given the graphs are density functions, the entropy magnitude can be inferred as proportional to the graphs area, as per Equation 6.22.

$$H(X) = -\int_x p(x)\log p(x)dx \tag{6.22}$$

It is straightforward to observe that the bigger this area is (entropy-like magnitude) for the training metrics the narrower and focused the values for the validation metrics are. The inverse is also observable. The relevance of this topic seems pertinent from the observations and deserves further study.

TABLE 6.8 Calibration Table as a Function of the p-values Tested for Significance

p	0.1	0.5	0.01	0.005	0.001
α(p)	0.385	0.289	0.111	0.067	0.0184

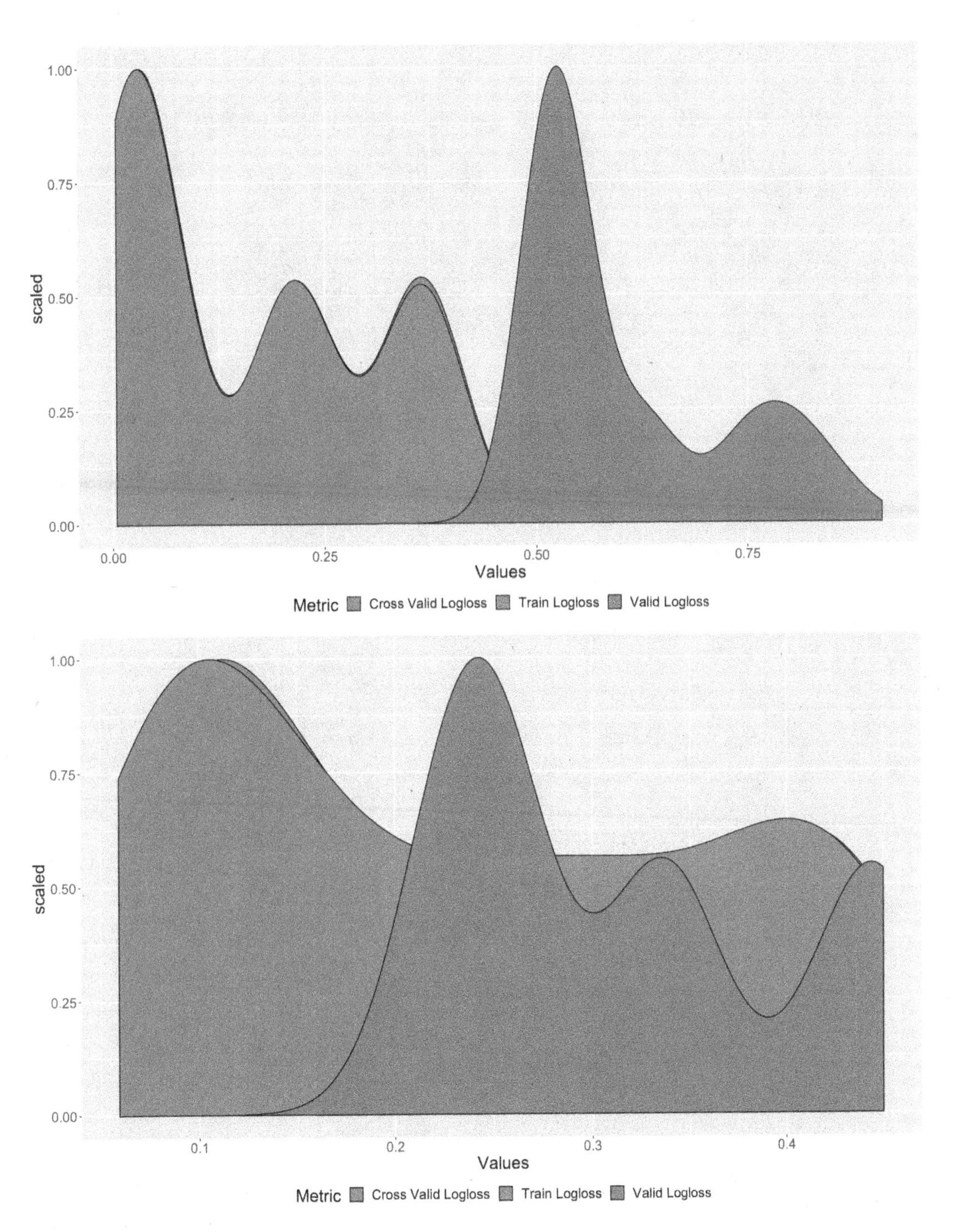

FIGURE 6.4 Logloss values dispersion. Logloss has a relatively large spread. Cross Validation logloss mostly superimposes with Train logloss for the NB15 case denoting more similitude in terms of statistical characteristics between the train and test datasets. (a) [UNB NSL KDD.] (b) [UNSW NB15.]

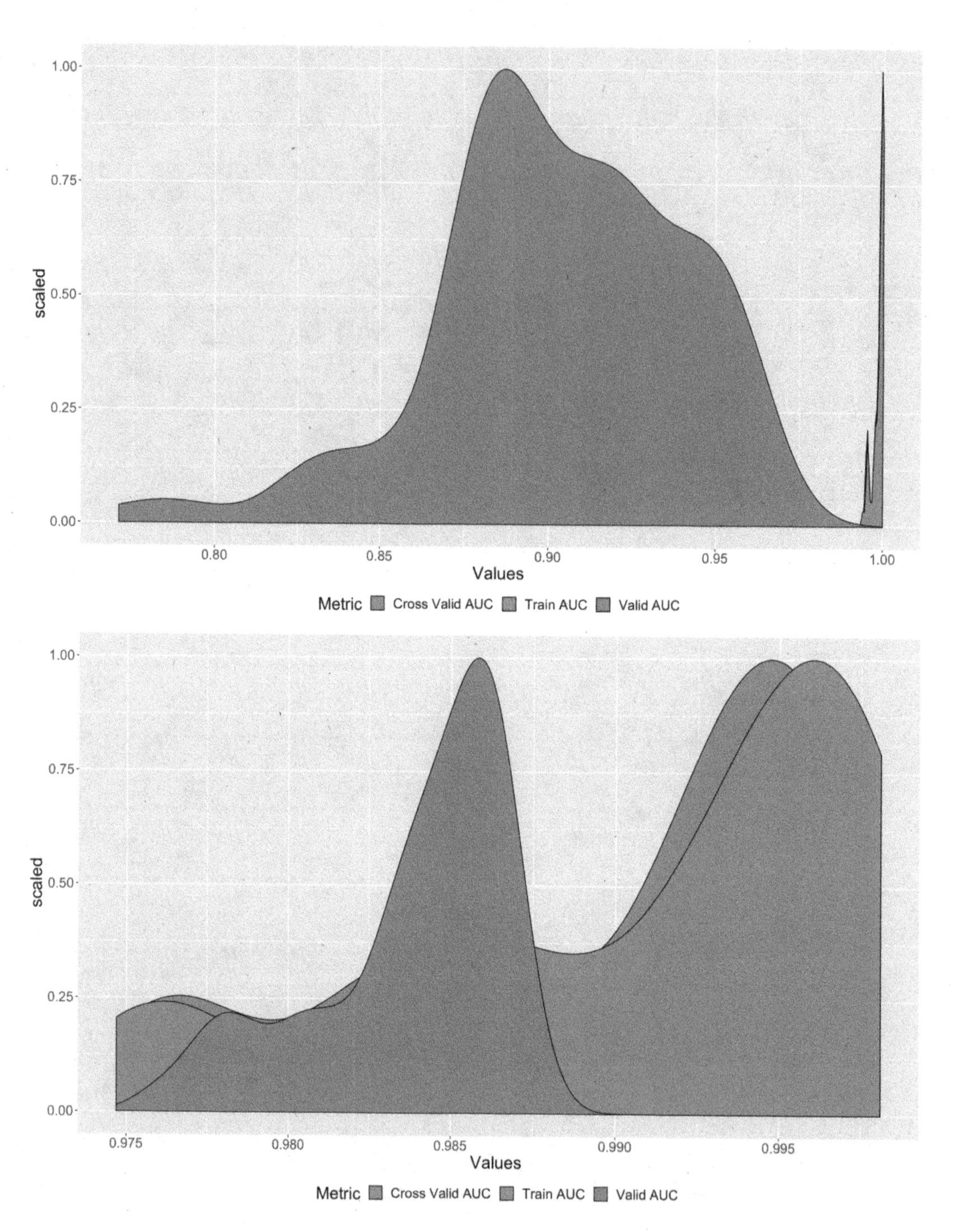

FIGURE 6.5 AUC values dispersion. Validation AUC has a large spread for NSL KDD. On the opposite Cross Validation AUC and Training AUC have low spread. Again this different behavior reveals differences between the test and train datasets for KDD, which are not so marked for the NB15 case. (a) [UNB NSL KDD.] (b) [UNSW NB15.]

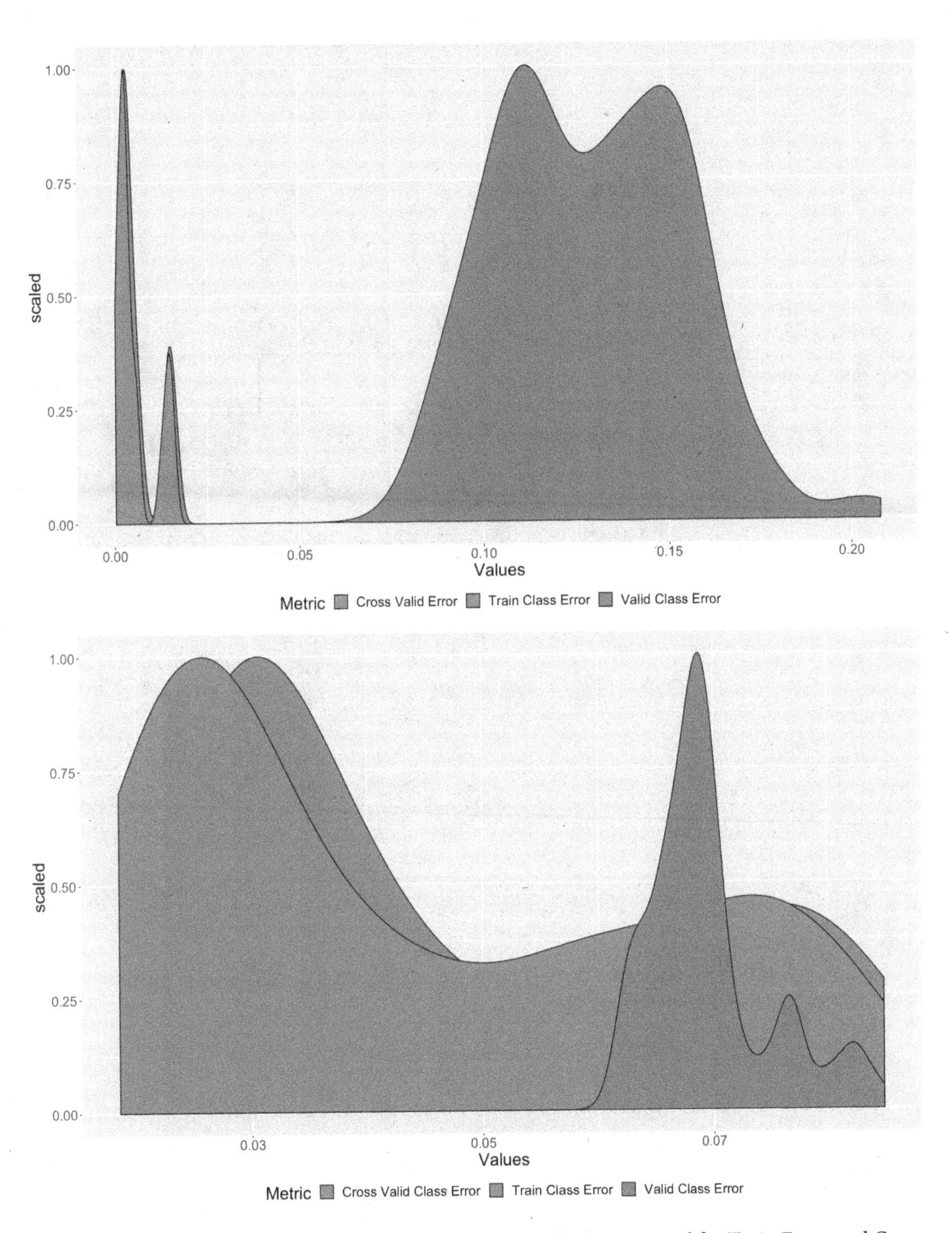

FIGURE 6.6 Classification error values dispersion. Note the large spread for Train Error and Cross Validation error in the NB15 case. On the contrary the Validation error has a much smaller dispersion. This suggests a different distribution subjacent to the dataset features. (a) [UNB NSL KDD.] (b) [UNSW NB15.]

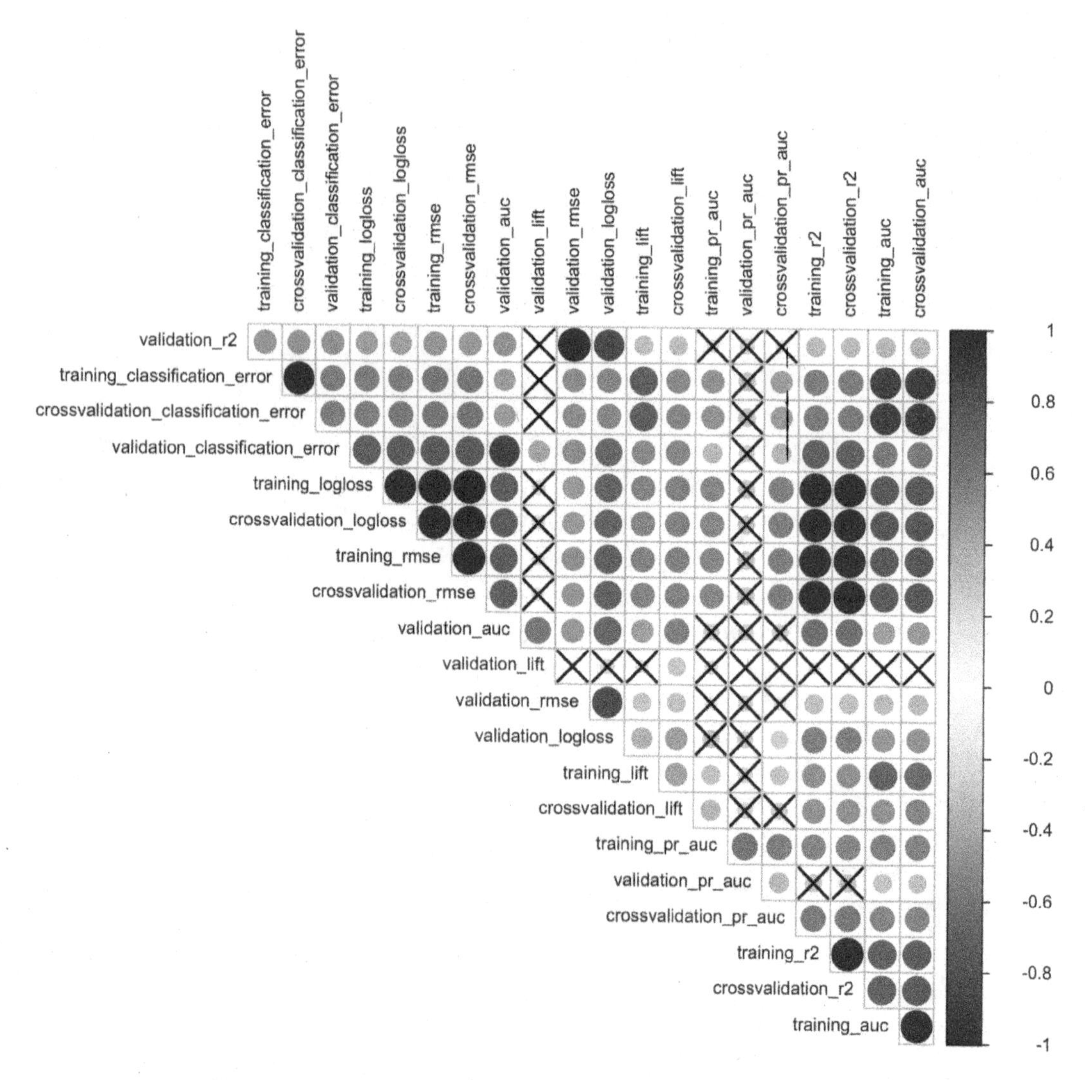

FIGURE 6.7 UNB NSL KDD Pearson correlation tested against significance at 0.001 level. Nonsignificant correlation values are crossed.

6.7.5 Discussion

Here we present a short explanation on how we answer the three questions posed in the introduction:

1. How effectively can XGBoost be used in the context of NIDSs? Our experiments show that XGBoost is a good approach for network intrusion detection, as the metrics we obtained can be considered very good, for example, AUC circa 0.990 for the NB15 dataset case (Table 6.6).

2. How to optimize XGBoost automatic training for this application? We proposed a full grid search approach (Section 6.6). We used it to obtain the parameters for the

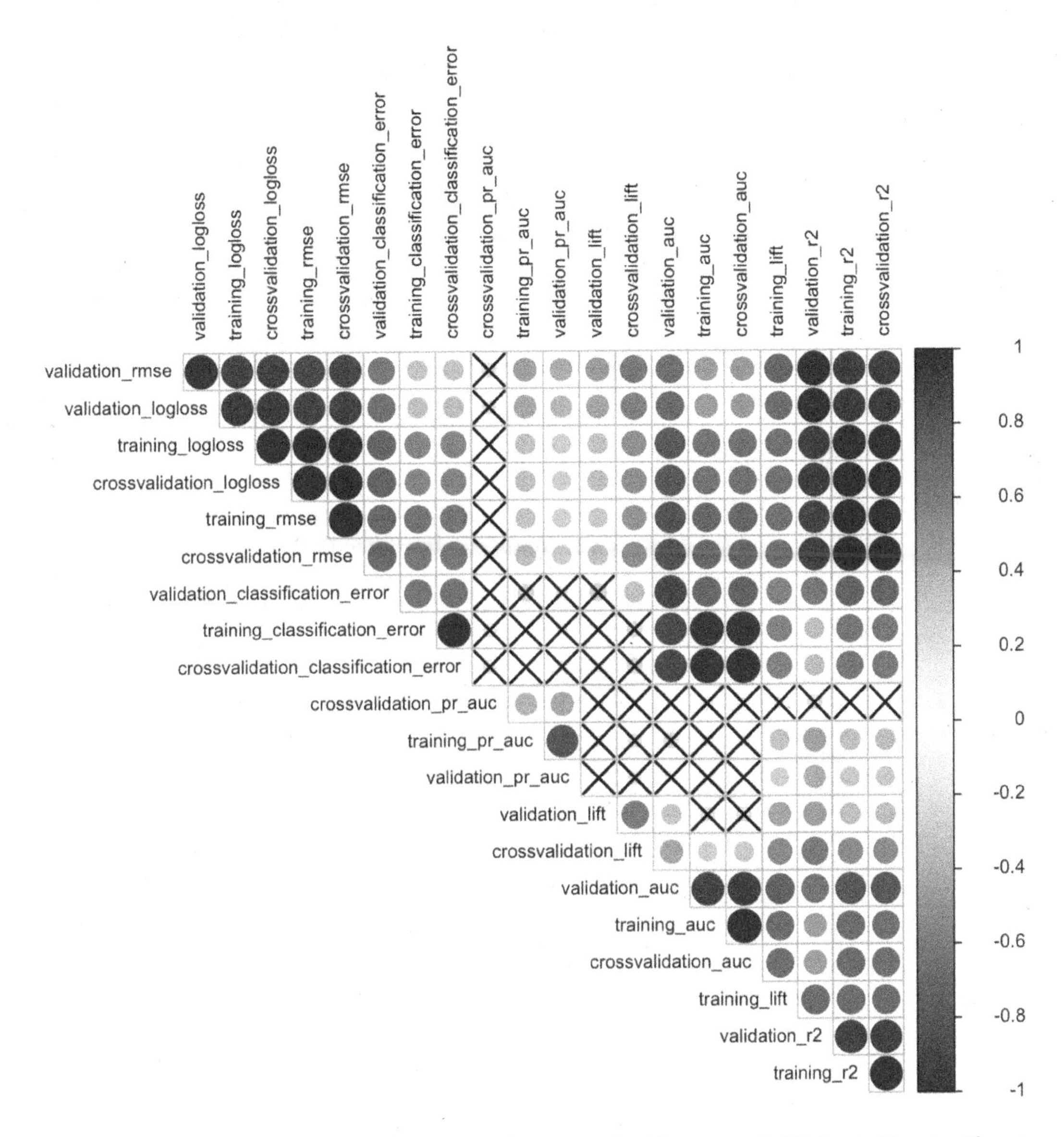

FIGURE 6.8 NB15 Pearson correlation tested against significance at 0.001 level. Nonsignificant correlation values are crossed.

detector for the two datasets (Table 6.7). The values for both datasets were similar, a fact that suggests that XGBoost is indeed a good approach as parameters can be reasonably stable for different datasets.

3. Among the parameters configured for XGBoost model training, can we find correlations with performance predictors, for example, area under curve or logloss? The answer is positive, as shown in Figures 6.3(a) and 6.3(b).

6.8 CONCLUSIONS

We describe our approach to NIDS using XGBoost. The results show that the techniques proposed are effective in fulfilling the objectives:

- We have shown and illustrated an effective method of automatic training for XGBoost models relying on only a limited fraction of the complete parameter space, attaining good performance metrics.
- The method allowed the most relevant model parameters for attaining such performance metrics to be identified, as illustrated in Figures 6.3(a) and 6.3(b).
- In Figures 6.7 and 6.8 the method proves its consistency by means of illustrating the consistent correlations among different performance metrics.
- The gain in model complexity is linear and it is remarkable when observing the most relevant features for the models performance as illustrated in Figure 6.9.

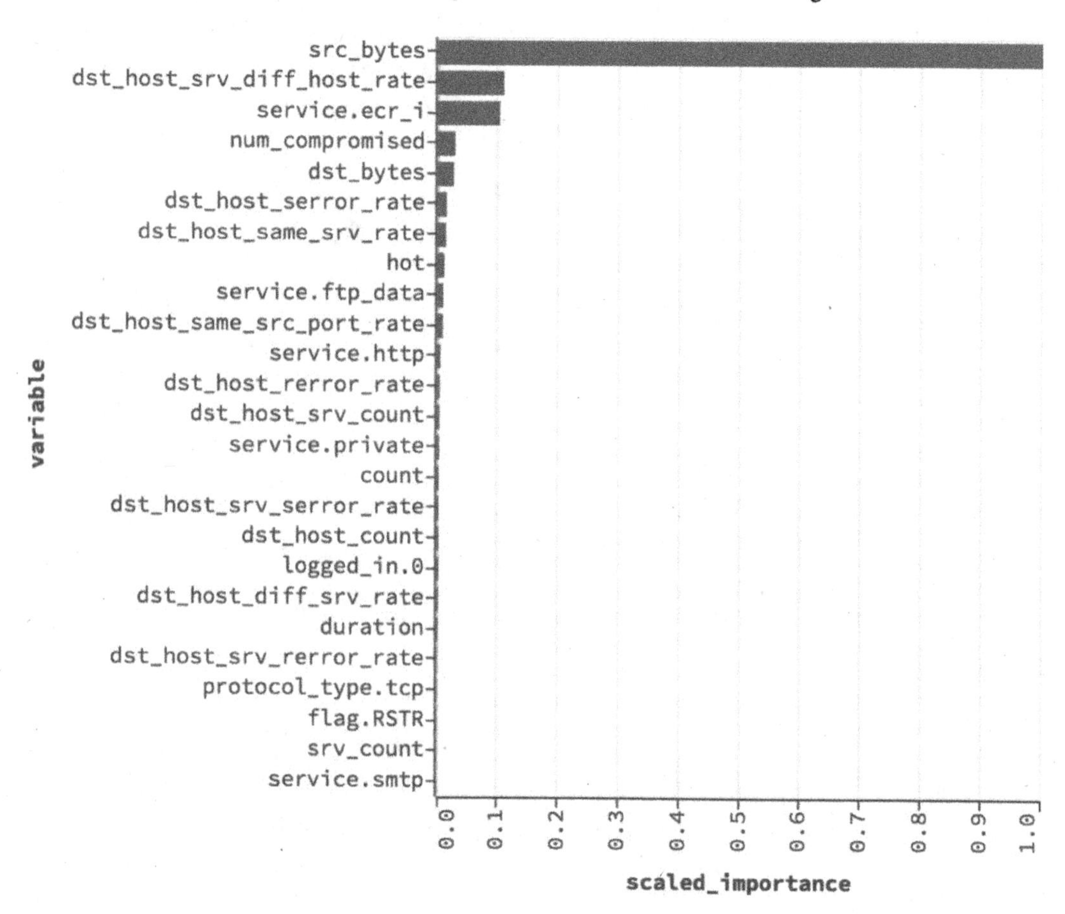

FIGURE 6.9 Variable importance for the best model relative to NSL-KDD and NB15 datasets. The most commonly used threshold to discard irrelevant features is 0.1 in the scaled importance axis. For both cases the expected gains in complexity model if using a reduced set of features is remarkable. *(Continued)*

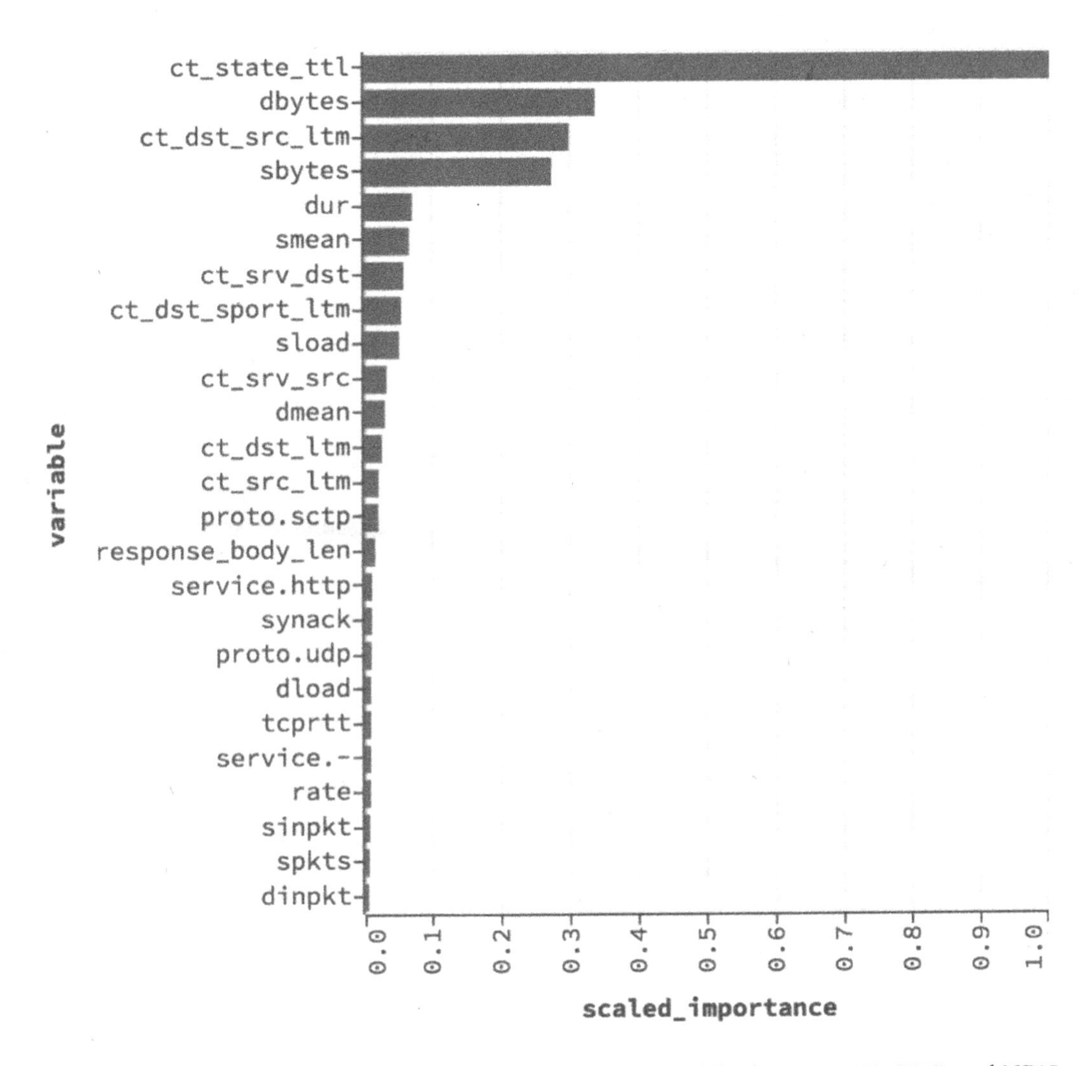

FIGURE 6.9 (*Continued*) Variable importance for the best model relative to NSL-KDD and NB15 datasets. The most commonly used threshold to discard irrelevant features is 0.1 in the scaled importance axis. For both cases the expected gains in complexity model if using a reduced set of features is remarkable.

An evaluation with two well-known NIDS datasets has been performed using the XGBoost R implementation. Performances compatible with the best values obtained with other experiments have been confirmed. We proved the potential XGBoost has to be used in the context of network intrusion detection, in applying XGBoost to two representative datasets often used in machine learning research. A method for automatic best model parameter set identification is described. The method is based on a grid search approach, with the objective of logloss minimization as a criteria for model selection. By minimizing logloss and imposing closed grid search results have been obtained balancing computational time and model results.

Elements that contribute to the model interpretability were added, specifically elements linked to entropy of the resultant metrics (Section 6.7.4). Clear correlations were found among many important performance metrics. The correlation calculation executed with the performance metrics allowed to identify which metrics are correlated with others, and above all, identify which metrics do not correlate at all among each other.

ACKNOWLEDGMENTS

This research was supported by the European Commission under grant agreement number 830892 (SPARTA) and by national funds through Fundação para a Ciência e a Tecnologia (FCT) with reference UIDB/50021/2020 (INESC-ID).

REFERENCES

1. Debar, H., Dacier, M., Wespi, A.: A revised taxonomy of intrusion detection systems. Annales des Télécommunications **55**(7), 361–378 (2000).
2. Dhaliwal, S., Nahid, A.A., Abbas, R.: Effective intrusion detection system using XGBoost. Information **9**(7), 1–24 (2018).
3. Dias, L.F., Correia, M.: Big data analytics for intrusion detection: An overview. In: Handbook of Research on Machine and Deep Learning Applications for Cyber Security. pp. 292–316. IGI Global (2020).
4. García-Teodoro, P., Díaz-Verdejo, J., Maciá-Fernández, G., Vázquez, E.: Anomaly-based network intrusion detection: Techniques, systems and challenges. Computers & Security **28**(1-2), 18–28 (Feb 2009).
5. Mitchell, R., Chen, I.R.: A survey of intrusion detection techniques for cyber-physical systems. ACM Computing Surveys **46**(4), 55:1–55:29 (Mar 2014).
6. Moustafa, N., Slay, J.: UNSW-NB15: A comprehensive data set for network intrusion detection systems. In: 2015 Military Communications and Information Systems Conference. pp. 1–6 (Nov 2015).
7. Moustafa, N., Slay, J.: The evaluation of network anomaly detection systems: Statistical analysis of the UNSW-NB15 data set and the comparison with the KDD99 data set. Information Security Journal: A Global Perspective **25**(1-3), 18–31 (2016).
8. Tavallaee, M., Bagheri, E., Lu, W., Ghorbani, A.A.: A detailed analysis of the KDD CUP 99 data set. In: Proceedings of the 2nd IEEE International Conference on Computational Intelligence for Security and Defense Applications. pp. 53–58 (2009).
9. Wheeler, P., Fulp, E.: A taxonomy of parallel techniques for intrusion detection. In: Proceedings of the 45th Annual Southeast Regional Conference. pp. 278–282 (2007).
10. Chen, T., Guestrin, C.: XGBoost: A scalable tree boosting system. In: Proceedings of the 22nd ACM SIGKDD International Conference on Knowledge Discovery and Data Mining. pp. 785–794 (2016).
11. Higgs, P.W.: Broken symmetries and the masses of Gauge Bosons. Physical Review Letters **13**, 508–509 (1964).
12. Adam-Bourdarios, C., Cowan, G., Germain, C., Guyon, I., Kégl, B., Rousseau, D.: How machine learning won the Higgs boson challenge. In: European Symposium on Artificial Neural Networks, Computational Intelligence and Machine Learning (Apr 2016).
13. Adam-Bourdarios, C., Cowan, G., Germain, C., Guyon, I., Kégl, B., Rousseau, D.: The Higgs boson machine learning challenge. In: NIPS 2014 Workshop on High-energy Physics and Machine Learning. pp. 19–55 (2015).
14. The CMS Collaboration: Observation of the Higgs boson decay to a pair of τ leptons with the CMS detector. Physics Letters **B779**, 283–316 (2018).
15. The CMS Collaboration: Evidence for the direct decay of the 125 GeV Higgs boson to fermions. Nature Physics. 10, pp. 557–560 (2014).

16. Nielsen, D.: Tree Boosting with XGBoost—Why does XGBoost win "every" machine learning competition? Master's thesis, NTNU (2016).
17. Chen, Z., Jiang, F., Cheng, Y., Gu, X., Liu, W., Peng, J.: XGBoost classifier for DDoS attack detection and analysis in sdn-based cloud. In: BigComp. pp. 251–256. IEEE Computer Society (2018).
18. Bansal, A., Kaur, S.: Extreme gradient boosting based tuning for classification in intrusion detection systems. In: Advances in Computing and Data Sciences. pp. 372–380. Springer (2018).
19. Sharafaldin., I., Lashkari., A.H., Ghorbani., A.A.: Toward generating a new intrusion detection dataset and intrusion traffic characterization. In: Proceedings of the 4th International Conference on Information Systems Security and Privacy. pp. 108–116. INSTICC (2018).
20. Amaral, P., Dinis, J., Pinto, P., Bernardo, L., Tavares, J., Mamede, H.S.: Machine learning in software defined networks: Data collection and traffic classification. In: IEEE 24th International Conference on Network Protocols (ICNP). pp. 1–5 (Nov 2016).
21. Mitchell, R., Frank, E.: Accelerating the xgboost algorithm using gpu computing. PeerJ Computer Science **3**, e127 (2017).
22. Aiello, S., Kraljevic, T., Maj, P.: h2o: R Interface for H2O (2015), https://CRAN.R-project.org/package=h2o, r package version 3.6.0.8.
23. Shiravi, A., Shiravi, H., Tavallaee, M., Ghorbani, A.A.: Toward developing a systematic approach to generate benchmark datasets for intrusion detection. Computers & Security **31**(3), 357–374 (May 2012).
24. Masnadi-Shirazi, H., Vasconcelos, N.: On the design of loss functions for classification: Theory, robustness to outliers, and savageboost. In: Proceedings of the 21st International Conference on Neural Information Processing Systems. pp. 1049–1056 (2008).
25. Fawcett, T.: An introduction to ROC analysis. Pattern Recognition Letters **27**(8), 861–874 (2006).
26. Bradley, A.P.: The use of the area under the ROC curve in the evaluation of machine learning algorithms. Pattern Recognition **30**(7), 1145–1159 (Jul 1997).
27. Hanley, J., Mcneil, B.: The meaning and use of the area under a receiver operating characteristic (ROC) curve. Radiology **143**, 29–36 (1982).
28. Moustafa, N.: Designing an online and reliable statistical anomaly detection framework for dealing with large high-speed network traffic. Ph.D. thesis, University of New South Wales, Canberra, Australia (2017).
29. Edwards, W., Lindman, H., J. Savage, L.: Bayesian statistical inference in psychological research. Psychological Review **70**, 193–242 (1963).

CHAPTER 7

Anomaly Detection on Encrypted and High-Performance Data Networks by Means of Machine Learning Techniques

Lorenzo Fernandez Maimo, Alberto Huertas Celdrán and Félix J. García Clemente

CONTENTS

In recent years, the rapid expansion of the Internet of Things along with the recent 5G wireless communications technology has provided an excellent breeding ground for the proliferation of botnets. Similarly, the current ease with which new malware variations are created and the high effectiveness of the phishing campaigns, have carried out an alarming increase in the number of ransomware attacks against companies and institutions. In the 5G context, the high transfer rate and the huge volume of traffic prevent every packet from being examined. Additionally, the extensive use of network traffic encryption makes it impossible to analyze the packet payloads. Due to the difficulty of deep packet inspection in these scenarios, we present in this chapter an anomaly detection approach based on the aggregation of network flows. However, the use of flows makes the anomalous traffic patterns difficult to detect. To deal with that, our proposal utilizes machine learning detection methods which have proved to be effective in the classification of complex patterns. Our solution uses Software Defined Network to transparently capture the flows and isolate compromised devices. It also leverages Network Function Virtualization in order to allow self-adaptation to the traffic fluctuation of 5G, or replace a virtualized device infected. All this is done seamlessly, dynamically and in real time. Our results demonstrate the suitability of our proposal in two scenarios: 5G wireless networks and integrated clinical environments.

7.1 INTRODUCTION

Over the years, professionals and researchers in the field of cybersecurity have designed a variety of systems to protect organizations from malicious attackers. These systems face threats such as viruses, Trojans, worms, and botnets among others. Existing solutions based on Intrusion Detection Systems (IDSs) include proactive approaches to anticipate vulnerabilities in computer systems as well as to be able to execute mitigation actions. However, at the same time, the number of threats has increased enormously, mainly due to the emergence of malware development environments. These environments allow almost automatic generation of different versions of the same virus, allowing any amateur to produce their own variation. Additionally, this proliferation of malware makes the rule databases used by IDS increasingly larger; thus, also increasing the computational time required for detection.

In particular, these detection systems currently have three weaknesses that are worth highlighting: (a) the necessary balance between the effectiveness of threat detection and the speed at which data collected on modern data networks can be examined; (b) the need of detecting such threats even when the data travels encrypted; and (c) the difficulty of new malware detection even with variations of well-known malware families.

The first weakness is caused by the transfer rates that communications networks are reaching (above 10 Gbps) in combination with the increasing volumes of data exchange. In addition, the new fifth-generation (5G) mobile technology provides unprecedented latency and transfer speed to wireless networks, allowing an extraordinary expansion of the Internet of Things (IoT). Due to this, it will be really difficult to capture and analyze every packet transmitted across the network. Therefore, the current detection procedures may become obsolete if we are not able to adapt them properly. Precisely because of the ubiquity of Internet access, the modern wireless technologies and the enormous growth of IoT, there are millions of devices with vulnerabilities that can be used to form botnets.

A botnet is a set of devices connected to the Internet previously infected with software that allows an attacker to manage them remotely to perform all kinds of actions such as distributed denial of service attacks, theft of sensitive or critical information such as bank accounts and personal data, or even theft of CPU cycles to mine cryptocoins. Botnets have become one of the biggest current and future security problems in data networks; suffice it to say that, according to the recent Nokia Threat Intelligence Report, in 2018, botnets based on IoT accounted for 78% of detected malware attacks and 16% of infected IoT devices [1].

Therefore, in the context of 5G communications and IoT, malware detection becomes a challenge. The increasing diversity of malware also increases the number of detection rules that must be applied to each packet. Moreover, the high transfer rates along with the large volume of data, leave little time to examine each packet traveling over the network.

With respect to the second weakness, an increasing number of malware encrypts their communications [2], making unfeasible the deep packet inspection and the usual detection tools ineffective. Added to this is the progressive increase in the amount of encrypted traffic in everyday communications, and the obligation to use encryption in environments where privacy is critical, such as medical environments. These environments are incorporating in their projects for the evolution of the hospital of the future, a growing number of interoperable medical devices in order to implement closed-loop processes (monitoring, analysis, decision making, and reaction or implementation of a treatment). In parallel, they have been suffering in recent years an increasing number of successful malware attacks, demonstrating that current detection systems are ineffective. A particularly problematic malware is ransomware, which infects one of the devices on a network thanks to a vulnerability or human failure. Its next move consists in reaching the rest of the devices by means of horizontal propagation, usually based on system vulnerabilities. When a new device is infected, all the data contained in its hard disks and shared folders is encrypted. Subsequently, the malware demands a certain amount of money to provide the decryption key. To appreciate the potential danger posed by ransomware, it is sufficient to mention the attacks suffered by UK health service hospitals in 2017. They had to close entire services, send patients to other hospitals and even postpone surgery [3].

Finally, the third weakness of our interest is motivated by the proliferation of new malware, usually derived from existing versions in which some characteristic has been modified (i.e., recompiled to run on another architecture or use other encryption algorithm) [4]. IDSs have difficulty identifying these variations as they normally work by examining traffic by means of rules, which makes early detection impossible. In the case of botnets and ransomware, both easily allow the generation of new versions. As an example, a Heimdal Security report from 2016 [5] concludes that nine of the ten most dangerous data filtering botnets are variations of the Zeus botnet. Similarly, in ransomware we also find great diversity of members of the same family, for example, from ransomware Petya-derived NotPetya, ExPetr, or PetrWrap among others.

Botnets and ransomware have in common that they generate network traffic following characteristic patterns. In the case of botnets, they usually have a command and control mechanism (C&C) whereby each infected device communicates with the attacker's computer periodically to receive orders. In the case of ransomware, its traffic patterns come from its eagerness to spread horizontally to maximize the damage (thereby increasing the probability of paying the ransom), its communication with a central server to obtain the encryption keys, and the traffic generated during the encryption of the network shared folders. These patterns can be interpreted as anomalies in normal network traffic.

To overcome the previous challenges, this chapter proposes an architecture that combines SDN and NFV to detect and mitigate cyberthreats in heterogeneous 5G network by using AI applied to network flows. The proposed architecture has been deployed in two scenarios with different purposes.

In the first scenario, a 5G network, the architecture allows self-adaptation to the traffic fluctuation by reassigning virtualized resources on-demand and in real time. Given the high transfer rates of 5G networks, we have also performed experimental evaluation of the execution time of the deep-learning model with a selection of the most well-known deep-learning frameworks. These experimental results provide useful data to establish management policies which control automatically and in real time the network infrastructure.

In the second scenario, a hospital of the future, the architecture provides an automatic and intelligent system to detect, classify and mitigate ransomware attacks affecting hospital rooms of the future equipped with Integrated Clinical Environments (ICE) [6]. A set of experiments has demonstrated the effectiveness of our architecture detecting some of the most recent and dangerous malware (i.e., WannaCry, Petya, BadRabbit, and PowerGhost) as well as its viability in terms of time. An additional contribution is a publicly available labeled dataset with both the captured traffic and the network flows from a realistic integrated clinical environment (ICE) configuration.

The rest of this chapter is organized as follows. First, Section 7.2 presents the necessary background with references to the main relative work. Section 7.3 describes the methodology followed in the design of the experiments. In Section 7.4, the experiments performed to demonstrate the viability of our proposal are detailed. Finally, conclusions are drawn and future work is outlined in Section 7.5.

7.2 BACKGROUND AND RELATED WORK

This section provides the background and state of the art required to understand the key aspects and main contributions of related works.

7.2.1 Anomaly Detection in Networks

An anomaly can be defined as a pattern that does not conform to expected or normal behavior, implying this definition that it appears rarely. Precisely because it is based on the concept of normality, which in itself is not easy to define, the problem is far from simple, with three main categories of anomalies [7]:

Point anomalies. It is the simplest form of anomaly, and concentrates the focus of the majority of the research in this field. A sample that lies outside the region of normal data is considered anomalous with respect to the rest.

Contextual anomalies. A data instance might be anomalous in a given context, but not in another. For example, a temperature of 5 degrees is considered an anomaly or not depending on the season. In this example, the temperature is a behavioral attribute, and the season is a contextual attribute.

Collective anomalies. A collection of data instances can be considered an anomaly with respect to a set of data if, although each of the instances is not considered an anomaly, the appearance of all of them as a collection is. A slight elongation of part of the electrocardiogram waveform can be an example of this type of anomaly. The values belong to the normal range, but the sequence of values itself constitutes the anomaly.

Anomaly detection techniques based on machine learning act as classifiers capable of distinguishing anomalous and normal instances, with approaches based on supervised, semi-supervised and unsupervised learning. Within these categories if a properly labeled data set exists, it is to be expected that the supervised approach will yield the best results by having more complete information. However, among the main challenges of this approach are the difficulty of obtaining a representative dataset and the fact that anomalies tend to be orders of magnitude less numerous than normal cases. This imbalance of the dataset complicates the task of classification.

Network traffic anomalies can belong to any of the aforementioned three types. An example of a point anomaly might be a packet routing to a suspect port. Similarly, an example of a contextual anomaly, is a large packet if the network traffic exchanged by two specific devices contains only small packets. It is contextual because such a packet would not be considered an anomaly if the devices involved were other devices on the same network using other packet size distribution. Finally, in a context where packets are issued in bursts of a known duration, a change in that duration can be considered a collective anomaly. In this research, aggregation of flows over time is used to insert some contextual and collective information into every single feature vector, therefore making point anomaly detection methods suitable.

7.2.2 Use of Network Flows in Anomaly Detection

At the transmission rates that modern network are achieving, a deep inspection of every transmitted packet becomes unfeasible. When we evaluate the volume of packets that the current deep packet inspection tools can manage, we find that the well-known Snort supports wired networks of up to 1 Gbps. The rate at which it begins to discard packets due to overload is 1.5 Gbps [8]. This fact has led to the emergence of hardware solutions based on programmable gate arrays (FPGA) [9] or application-specific integrated circuits (ASIC), which managed to deal with rates of up to 7.2 Gbps [10]. Even so, these speeds are far from those that await us in the near future. Partly because of this, IDS-based detection solutions have had to evolve from analyzing network packets to analyzing network traffic flows using innovative techniques based on artificial intelligence [11]. For example, a block-based neural network model used in an IDS based on flow anomalies was able to handle 22 Gbps traffic using FPGAs [12]. A complete review of solutions for quickly classifying network flows and detecting attacks or malicious code can be found at [13].

In a context of encrypted traffic, deep inspection-based solutions are not applicable, and most IDS that handle encrypted traffic are based on identifying certain basic patterns such as port scans, or brute force attacks [14]. As an alternative, there are proposals based on machine learning that use data calculated from a network flow [15], and even offer imaginative approaches, like this that uses convolutional networks taking the flow as if it were an image [16]. However, if we cannot access the payload because it is encrypted, a single flow does not provide sufficient information to achieve accurate detection.

Using network flows to detect such anomalies has many advantages, including the following: no need to access the payload of the packet, so it is applicable to encrypted traffic; reduction by orders of magnitude of the volume of data to be analyzed, so it can be applied to environments with high-speed networks; and finally, respect for users' privacy. Privacy is crucial when it comes to obtaining permission from organizations to capture the huge amount of traffic that some of the recent machine learning algorithms require for their training. Working with network flows makes it easier to guarantee user anonymity and privacy, as administrators know that only packet headers will be required. All these reasons make them attractive for this research.

Unfortunately, working with flows also has some drawbacks. The main one is the loss of information due to the fact that the payload is not being considered and, additionally, we are using only aggregate information from the headers. However, this chapter aims to show that although a single flow may not be sufficient to extract complex information about the traffic pattern, a properly treated sequence of flows may contain sufficient information for a machine learning algorithm to distinguish anomalous from normal patterns [17, 18]. In this way we would meet the constraints of the proposed scenarios by performing the detection without needing to examine a high volume of data, or worry about the fact that the data travels encrypted.

7.2.3 Machine Learning for Anomaly Detection

In recent years machine learning algorithms, and particularly those based on deep learning, have achieved state-of-the-art in a wide range of difficult domains. These algorithms have proved to be suitable for complex pattern identification, as well as able to deal with unseen samples, thanks to their generalization capability. In the context of anomaly detection based on network flows, we have to tackle the loss of information due to the absence of payload. Moreover, in order to incorporate contextual information in the feature vector, a number of features need to be computed by aggregating information from a network context made up of a sequence of network flows. The use of aggregations blurs the effect of every single flow, making the identification of anomalous samples a hard problem. Therefore, it is expected that machine learning algorithms can detect satisfactorily the nuances associated to anomalous behaviors even after aggregation.

Machine-learning anomaly detection techniques operate in one of the following three modes:

Supervised. This mode requires a dataset with traffic labeled as normal or anomalous. The labels are used to train a model capable of separating both labels, commonly by adjusting a boundary surrounding each traffic class.

Semisupervised. This operation mode is preferable when only one class of traffic is easily labeled; in this context, the normal class. Therefore, the model is trained with the normal traffic, being a sample anomalous if it is sufficiently distant from a normal region according to a given distance metric.

Unsupervised. No label is available. These methods generally try to discover clusters of similar samples in the dataset, or to estimate the density of the data in the input space, or to obtain a projection of the dataset onto a lower-dimensional space.

In supervised operation, the main issue is how to build a sufficiently comprehensive training dataset containing a wide variety of anomalous traffic properly labeled. Such a labeled dataset is not usually available and it can be difficult to collect and maintain [19, 20] due to the effort and expertise required, or even to user's privacy issues. As an illustration, we can take a botnet detection context where traffic is obtained from synthetic environments by using honeypots, or by executing botnet binaries in a controlled framework [21]. On the other hand, normal traffic is commonly gathered from a public or institutional network, and such network captures might contain unnoticed anomalous traffic, for example, a previously active botnet. The two traces are then combined; however, there is not a unique way of doing it, and the resulting combination might not match with any real traffic pattern. Moreover, if each anomalous traffic capture corresponds to one single botnet, the traffic might not be representative of the real traffic generated by concurrent infections on a host.

Regarding clustering unsupervised methods applied to anomaly detection, an issue that arises is the correspondence between the identified classes and the actual anomalous and normal classes. Usually this correspondence has no reason to be correct, so we cannot trust blindly the results. Nevertheless, when used for dimensionality reduction, these algorithms can extract discriminative features, improving the classification performance of other methods.

Finally, semisupervised approaches try to estimate a boundary around the regions occupied by normal traffic in the sample space. The difference with respect to the supervised method is that there is no information about the shape of the anomalous regions. A new sample is classified as anomalous if it exceeds a certain threshold distance.

In computer networks, anomaly detection has been a subject of study for decades, and many approaches have been explored [7, 22] while the machine learning perspective has received special attention in recent years [23]. There exist a wide variety of machine learning method applied to anomaly detection in the literature. For example, One-Class SVM is a semisupervised algorithm that has been extensively used, specially as a final layer in conjunction with a neural network, or a Stacked AutoEncoder (SAE) working as a previous unsupervised stage for feature extraction [24, 25]. Neural networks and SAEs have proved to be effective in learning discriminative features from complex and high-dimensional datasets.

However, machine learning algorithms performance is tightly related to the existence of a suitable dataset, preferably labeled, and containing a sufficient variety of samples of each class. In the context of computer networks, there exist some well-known datasets; however, not all of them provide traffic captured from an actual network, and most of them do not include they packet payload due to privacy legal issues. ISCX 2012 IDS dataset provides feature vectors based on Netflow format, and has been generated in a physical testbed implementation using real devices that generate real traffic that mimics usersâ€™ behavior (e.g., SSH, HTTP, and SMTP). Similarly, CTU is another publicly available dataset suitable to be used in this context. It tries to accomplish all the good requirements to be considered a good dataset: it has actual botnets attacks and not simulations, unknown traffic from a large network, ground-truth labels, as well as different types of botnets. The CTU dataset comprises thirteen scenarios with different numbers of infected computers and seven botnet families [26].

7.2.4 Infrastructure for Anomaly Detection in Networks

With regard to how to integrate the proposals of this research into the communications infrastructure, mechanisms are needed to facilitate tasks such as model update, adaptation to the circumstances of the environment, and even automatic and intelligent mitigation after the detection of threats. The combination of NFV/SDN has been used in our laboratory with different purposes [27–31]. One of the most relevant usages is the seamless integration of the process of network flow data acquisition, and the optimization of resources during the mitigation of attacks. This optimization can be done both in real time and on demand. For example, in a 5G scenario, resources could be reallocated to adapt to a greater volume of traffic to achieve the needed performance. Similarly, in an integrated clinical scenario, NFV/SDN offers the flexibility to isolate or replace a compromised virtual device in real time.

In the literature there exist a variety of solutions capable of detecting and mitigating attacks in computer networks by considering NFV/SDN. For example, [32] proposed to use them to detect and mitigate botnet attacks in 5G networks. This proposal involves the tight combination of both NFV and SDN technologies to provide effective detection and mitigation of cyber-attacks in 5G networks. Another architecture that combines NFV and SDN features to create sophisticated network resilience strategies was proposed in [33]. This architecture uses a control-loop to monitor and analyze the network state, indicating whether specific parts can be reconfigured or replaced to improve the detection capabilities.

Additionally, in the literature there are also proposals that use some of the possibilities offered by NFV/SDN for the detection and mitigation of ransomware [15, 34, 35]. However, to the best of our knowledge, only [17] integrated the combination SDN/NFV with the ICE standard to be able to replace infected elements in a few seconds, addressing the detection of the ransomware spreading in ICE with encrypted traffic.

7.3 METHODOLOGY

The main objective of this work is to propose solutions that use machine learning methods to the detection of anomalies in the data networks of the two aforementioned environments. In the 5G environment, a restriction is imposed by the impossibility of analyzing the payload of all the packets due to the volume of traffic. Similarly, in the ICE environment the main restrictions are both the use of encrypted traffic and the short time available for detection and mitigation.

After analyzing the existing flow-based proposals in the literature to classify network traffic, our hypothesis was that a single flow, without access to the packet payload, does not provide sufficient information for anomaly detection. Due to this hypothesis, we proposed to study whether a context for a given flow, formed by flows previously received over a period of time, would allow a more accurate detection of anomalies in complex traffic patterns. Machine learning methods, both classic and deep, are especially well suited in this case of complex patterns detection. Additionally, our hypothesis also includes the assumptions that traffic evaluation can be done at the rate required by the 5G networks, and that a short enough detection time will mitigate a ransomware attack before it spreads; all this dynamically, intelligently, in real time, and integrated into a suitable architecture for each environment. To achieve this goal, a number of specific actions were taken, as detailed below:

1. Study the concept of anomaly and the main machine learning techniques proposed in the literature for anomaly detection, along with their computational and storage requirements. In particular this includes the study of the state of the art of network traffic anomaly detection systems, as well as their need to inspect the payload of packets.
2. Determine at least two representative environments where we can check the performance of the proposals, and that allow us to measure execution and deployment times.
3. Design a mechanism for generating feature vectors from flows extracted from captured traffic, containing information about the behavior of that traffic over a period of time. This mechanism should allow us to determine or delimit the cause of the anomalous behavior without violating the user's privacy.
4. Integrate this mechanism in an NFV/SDN-based architecture specific for each environment, with a life cycle that allows both a dynamic adaptation of the models to the new circumstances of the network, and facilitate the mitigation of threats.
5. Analyze existing public data sets with actual network traffic containing normal traffic and traffic generated by both botnets and ransomware for possible use in this work. Also evaluate the possibility of generating a new set of data.
6. Evaluate the proposals in each of the selected environments by means of the datasets obtained, in terms of detection capacity, as well as detection and mitigation times.

To achieve the previous objectives, this work has followed a scientific methodology based on the study of the state of the art of anomaly detection systems based on machine learning methods in the context of data networks. Following the first objective, we started studying the concept of anomaly, especially in the context of data networks, and examining a selection of the principal machine learning techniques used for its detection, specifically those payload independent. During the research carried out, and following the second action listed, two scenarios were identified where the detection of anomalies was a challenge due to the particular restrictions imposed by each one. The first was the 5G mobile communications environment, where high transfer rates and large volumes of data make it extremely difficult to analyze every packet of circulating traffic, which can delay detection. The second scenario was integrated clinical environments (ICE), where the majority of network traffic travel encrypted.

Once analyzed the state of the art and identified the open challenges and requirements of the chosen environments and aligned with the third objective, it was designed a mechanism to generate relevant feature vectors to detect anomalies in the previous scenarios. Subsequently, and in accordance with the fourth proposed objective, we considered relevant to propose an architecture in the context of network traffic anomaly detection in 5G environments based on NFV/SDN, as well as providing the necessary mechanisms to dynamically adjust the resources assigned to the continuous anomaly detection subsystem. Together with the model and the feature vector, and as part of the fifth objective, we analyzed the available public datasets to determine at least one meeting our requirements for each environment. Once a suitable dataset was determined, and with the goal of achieving the sixth objective, a subsequent evaluation of the datasets in the selected environments was performed.

7.4 RESULTS

This section shows the results of the proposed methodology.

7.4.1 Architecture Design

In this section we show the design details of our 5G-oriented architecture that combines NFV, SDN, and AI techniques to detect and mitigate heterogeneous cyberattacks affecting different scenarios where 5G networks play a key role.

We decided to use network function virtualization (NFV) because it provides flexibility and dynamism in the infrastructure by separating the hardware layer from the software layer, in conjunction with software defined networks (SDN) that gives us control of communications in real time and on demand. 5G networks have been conceived as extremely flexible infrastructures based on an architecture organized by different functional planes. We propose a high-level design of the management and orchestration plane, following the ETSI NFV architecture [36], which has also been used as the basis for other proposals [37, 38].

The proposed architecture is depicted in Figure 7.1. It consists of four groups of components: Virtualized Infrastructure (VI); Virtualized Network Functions (VNF);

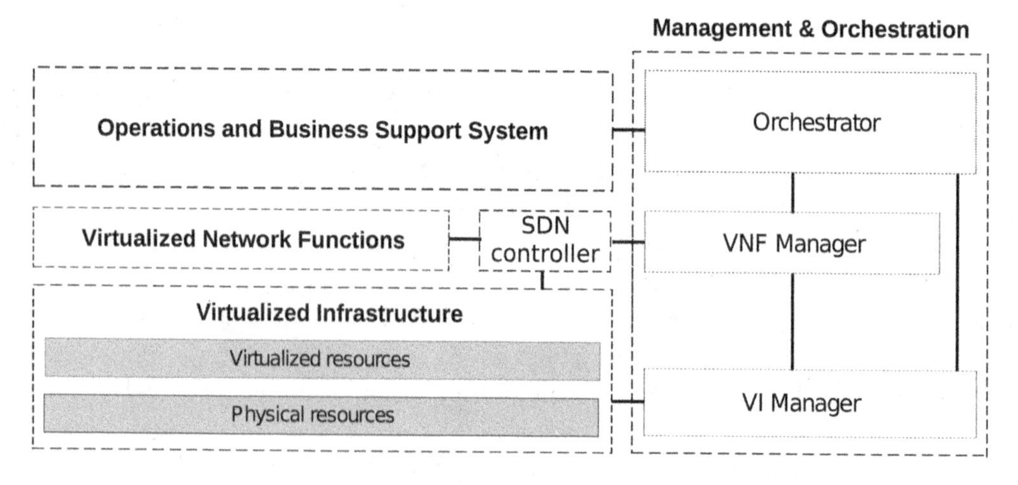

FIGURE 7.1 High-level management and orchestration planes of a ETSI-NFV architecture.

Management and Orchestration (MANO); and Operations and Business Support Systems (OSS/BSS). The VI group virtualizes the physical resources (computing, storage, and network) and exposes them for consumption by VNFs. The MANO group manages the combination of VNFs implementing network services, the full life cycle of VNFs, the deployment of VNFs within virtualized resources, and the network slicing for supporting multitenancy. Therefore, MANO controls the general behavior of the infrastructure by considering the policy set defined in the OSS/BSS. A more comprehensive description of two similar architectures in the context of anomaly detection can be found in [18, 39].

To provide anomaly detection capabilities, we chose a flow-based approach, since it was compatible with the restrictions imposed: the use of network flows reduces by orders of magnitude the amount of data to be examined, allowing large volumes of network traffic and high transfer rates; and network flows do not need access to the packet payload, therefore they are not affected by the fact that the network traffic is encrypted.

Due to the decision of working with network flows and the fact that a single flow does not carry enough information, we decided to generate each feature vector from a contiguous sequence of received network flows for a certain configurable time period. We also decided not to use the usual network flows, but their representation in bidirectional Netflow format. Our feature vectors contain aggregated data from all the flows received in that time period together with other values calculated only from the last flow received. This last flow determines the class of the resulting feature vector (normal/anomalous). A drawback of working with flows is that the process of detecting anomalies becomes even more complex because the differences between normal and anomalous patterns are much more subtle. Figure 7.2 illustrates the feature vector computation with a 10-second time window, whereas Table 7.1 shows a noncomprehensive list of the aggregated features computed.

Feature vectors are computed in the edge of the infrastructure from the network flows collected and used to feed a local anomaly detector. This detector sends the anomalies

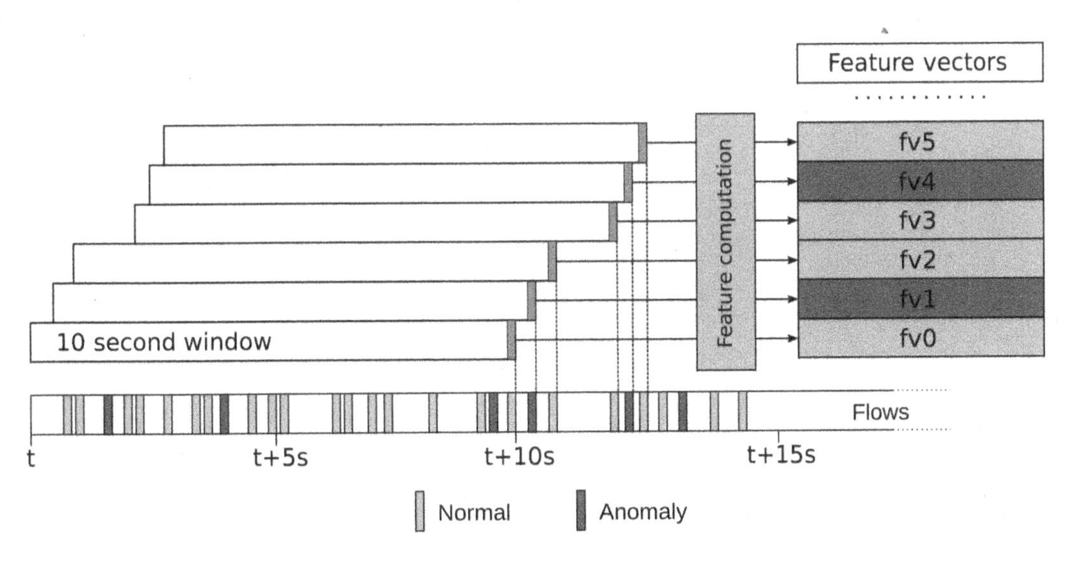

FIGURE 7.2 Illustration of feature vectors computed as an aggregation of the flows received in the previous 10 seconds (10-second time window). The label corresponds to the last flow received.

TABLE 7.1 Noncomprehensive List of TCP/UDP Aggregated Features Computed from the Network Flows

Feature Description
– # of total flows, incoming flows (in-flows) and outgoing flows (out-flows)
– % of incoming/outgoing flows over total.
– % of symmetric/asymmetric incoming flows over total.
– Sum, max., min., mean, and var. of IP packets per in/out/total flows.
– Sum, max., min., mean, and var. of bytes per in/out/total flows.
– Sum, max., min., mean, and var. of source bytes per in/out/total flows.
– # of different source IPs for in flows and destination IPs for out flows.
– # of different source and destination ports for in and out flows.
– Entropy of source IPs for in-flows and destination IPs for out-flows.
– Entropy of source and destination ports for in/out-flows.
– % of source and destination ports >1024 for in/out-flows.
– % of source and destination ports ≤1024 for in/out-flows.

detected to an optional global anomaly detector (Figure 7.3). Both detectors are implemented as machine learning models properly trained. This part of the architecture is implemented as virtualized network functions, allowing both the dynamic deployment of different models and model updates.

7.4.2 Deployment in a 5G Scenario

5G networks expect great fluctuations in traffic, for example, the traffic in the Radio Access Network (RAN) near a stadium increases greatly when there is an event. In this section, we show how our architecture is able to adapt the network computational resources and inspection elements to optimize and ensure the detection of cyberattacks. This adaptation can be done seamlessly thanks to the use of virtualized network functions.

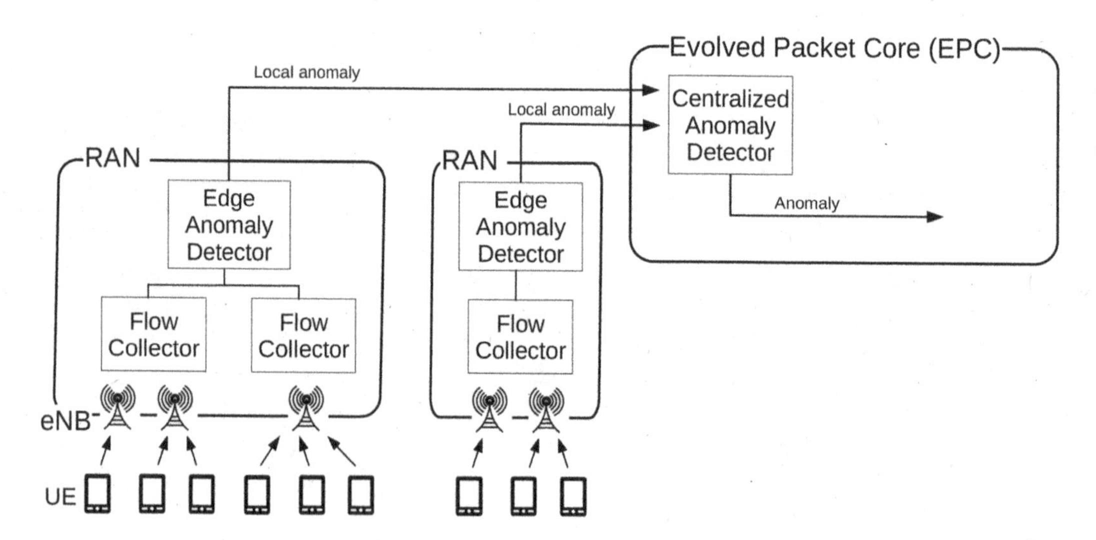

FIGURE 7.3 Flow collectors in each RAN feeding a generic two-level network anomaly detector architecture. In a generic architecture, both local detector and global detector are optional.

The deployment of our architecture in a 5G network scenario combines the *Mobile Edge Computing* and NFV/SDN, allowing a variety of actions to ensure effective and autonomous detection of anomalies in real time. This can be achieved thanks to the use of three types of policies and an orchestrator that integrates the anomaly detection mechanism in charge of the actions associated with these policies.

In order to demonstrate the possibilities of the proposal, we have created an anomaly detection method based on a two-level deep learning model. The lower level runs at the edge of the network. This lower level is a detector of anomaly symptoms, that is, local anomalies at the RAN that can be part of a more global anomaly affecting a large number of end devices. The upper level, on the other hand, is located in the cloud or a central server, it collects the properly labeled symptoms coming from the lower levels executed in the different RANs (optionally, information about the flows responsible for the symptom is also sent in case subsequent processing is required). These symptoms are then sorted by time stamp in order to analyze the sequence to determine the existence of a more global anomaly involving several RANs.

Some of the most popular machine learning models (One-class SVM, Isolation Forest and Local Outlier Factor as semisupervised models, and Dense Deep Neural Network, SVM, Naive Bayes and Random Forest as supervised models) were analyzed in the context of anomaly detection to determine the scalability to large volumes of data, the stability in the prediction time of each model, and the use of parallelism provided by specific purpose hardware, such as GPUs, among other factors. It was determined that the best candidate for the first level of our system was a dense deep neural network because it is a model essentially based on matrix products, which allows a great performance when running on a GPU while maintaining a prediction time independent of the training process. The second level receives symptoms and, therefore, it does not have such critical evaluation time constraints. Due to the fact that this second level must deal with sequences of symptoms from different RAN, interpreting them as a sequence of global symptoms of the system, we proposed using a network LSTM (*Long Short-Term Memory network*). Figure 7.4 illustrates the relationship between the two levels.

In this context, different policies have been developed to illustrate the dynamic adaptation process of the proposed system; for instance, allowing RANs to allocate resources

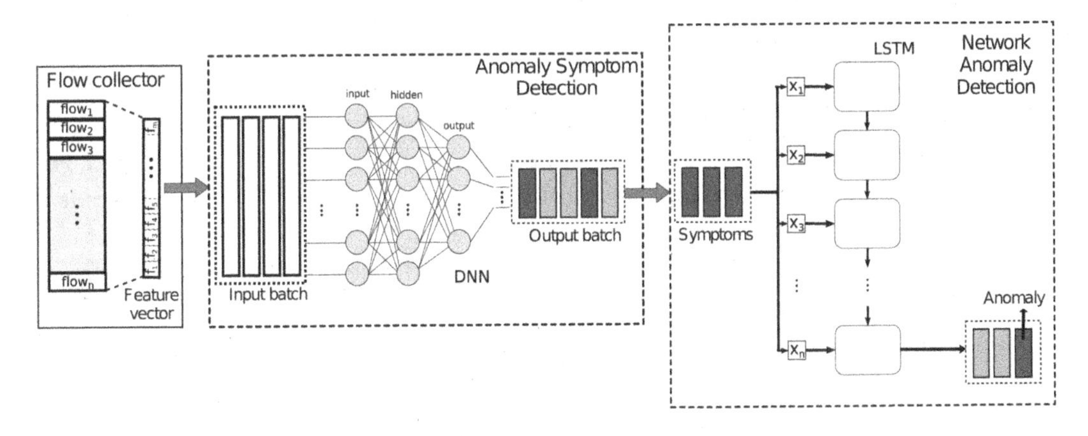

FIGURE 7.4 Detail of the modules for low-level anomaly symptom detection and high-level anomaly detection.

flexibly and dynamically for the detection of anomaly symptoms under traffic fluctuation. This is achieved thanks to the possibilities offered by NFV for both configuration hot swapping and the adjustment of virtualized hardware resources. For example, when network traffic is light, the model can be evaluated on a single CPU, but when traffic is increasing, additional available virtualized hardware, such as a GPU, can be assigned to run the model if the virtual machine allows it. Furthermore, when the administrators receive a new improved machine learning model they can automatically deploy it.

7.4.2.1 Experiments

In our experiments, the symptom detector running on the RANs, implemented as a dense deep network, was trained and validated with the aforementioned CTU dataset used for botnet detection. In this case, our aim was to detect the anomalies produced by the C&C traffic generated by botnets. The tuning of the hyperparameters during the training stage was done with the goal of maximizing the recall, relying on the second level to refine the detection performance.

Due to the usual imbalance between anomalous and normal classes in anomaly detection datasets, accuracy is not considered the most adequate metric for detection performance. Instead, precision and recall are preferred alternatives:

Precision: indicates the percent of correct positive predictions.

$$Precision = \frac{TruePositive}{TruePositive + FalsePositive}$$

Recall: percent of correctly classified positive samples.

$$Recall = \frac{TruePositive}{TruePositive + FalseNegative}$$

When our first level detector was applied to known botnets, a very satisfactory recall value was obtained (0.9934), while the precision achieved was only 0.8126. In the case of unknown botnet detection (none of these botnets were used in the training stage), the precision of this first level was in the range 0.40–0.95 (average of 0.686 and standard deviation of 0.1968) and the recall was in the range 0.38–0.95 (average of 0.71 and standard deviation of 0.2421).

In these first tests, besides proving that the model detected adequately, we wanted to demonstrate the viability of the execution time of the model proposed in the first level. For this purpose, an exhaustive comparison of the performance of the different development libraries of deep learning models when predicting with dense neural networks was carried out. A wide variety of configurations were tested using different number of hidden layers and feature vector sizes, as well as different hardware support (GPU vs. CPU). The results showed that there was no single winning candidate. The implementation selected would be determined by aspects such as the hardware available in a RAN or the reception rate of network flows. Figure 7.5 plots the performance results obtained in evaluation by the GPU version of the selected model for the first level when implemented using six well-known deep learning frameworks. An extended description of these results can be found in [18].

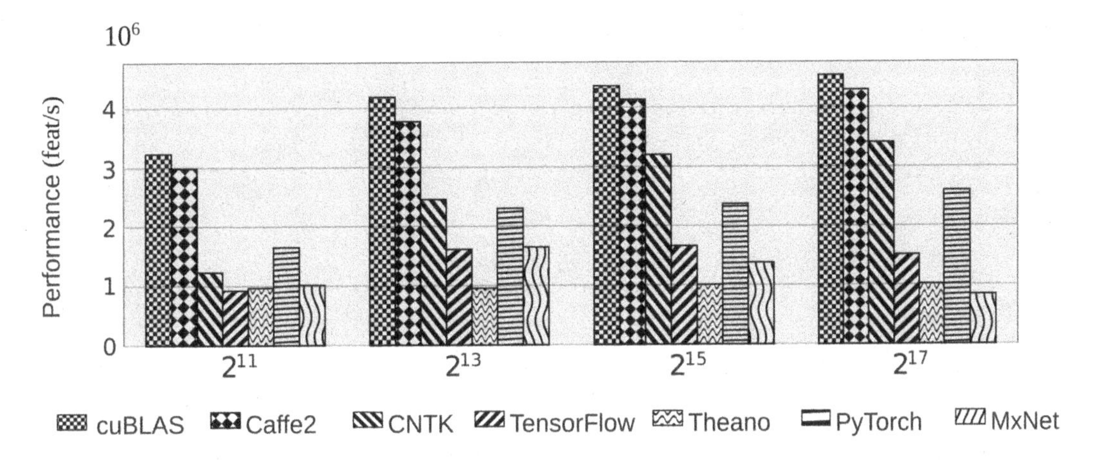

FIGURE 7.5 Evaluation performance of the selected model implemented in six well-known deep learning frameworks (Caffe2, CNTK, TensorFlow, Theano, PyTorch, and MxNet) and an ad-hoc performance upper-bound implementation in cuBLAS.

Additionally, we retrained the model employing a wider exploration of the hyperparameter space in an attempt to maximize precision, achieving a precision of 0.9537 and a recall of 0.9954 on known botnets. With respect to unknown botnets, an average precision of 0.8693 with standard deviation of 0.1928 and an average recall of 0.4803 with standard deviation of 0.2353 were obtained. Since botnets are a threat that remains in the system for long periods of time, even with such a low recall, it is only a matter of time to detect the anomaly. On the contrary, a low precision would flood the second level with false symptoms. As a next step, the model execution times estimated in our fist approach were then used to obtain the expression of an upper boundary of the execution time needed to detect an anomaly. This was done bearing in mind that the model evaluation must use batches of vectors for performance reasons. This was implemented in an example showing the evolution of network traffic over time in a hypothetical situation. The results showed how the system would adapt, even anticipating the critical point in order to perform a progressive adaptation. Figure 7.6 illustrates the operation of this adaptation mechanism. As the intensity of the network traffic increases, its estimated derivative provides a way of anticipating the necessity of more resources. With this information, the orchestrator can reassign hardware resources, or even deploy a new anomaly detection model before system saturation.

7.4.3 Deployment in ICE

Our second scenario of interest is the detection of anomalies caused by ransomware in ICE. ICE is a framework enabling the open coordination of heterogeneous medical devices and applications. Among the proposed components of the ICE framework, the most relevant are the *ICE Equipment Interfaces*, which are attached to medical devices to enable their networking capabilities; the *ICE Supervisor*, focused on hosting medical applications that receive and control the patients' vital signs; the *ICE Network Controller*, in charge of enabling the communications between the supervisor and the ICE equipment interfaces, as well as handling and maintaining the discovery of medical devices and their information;

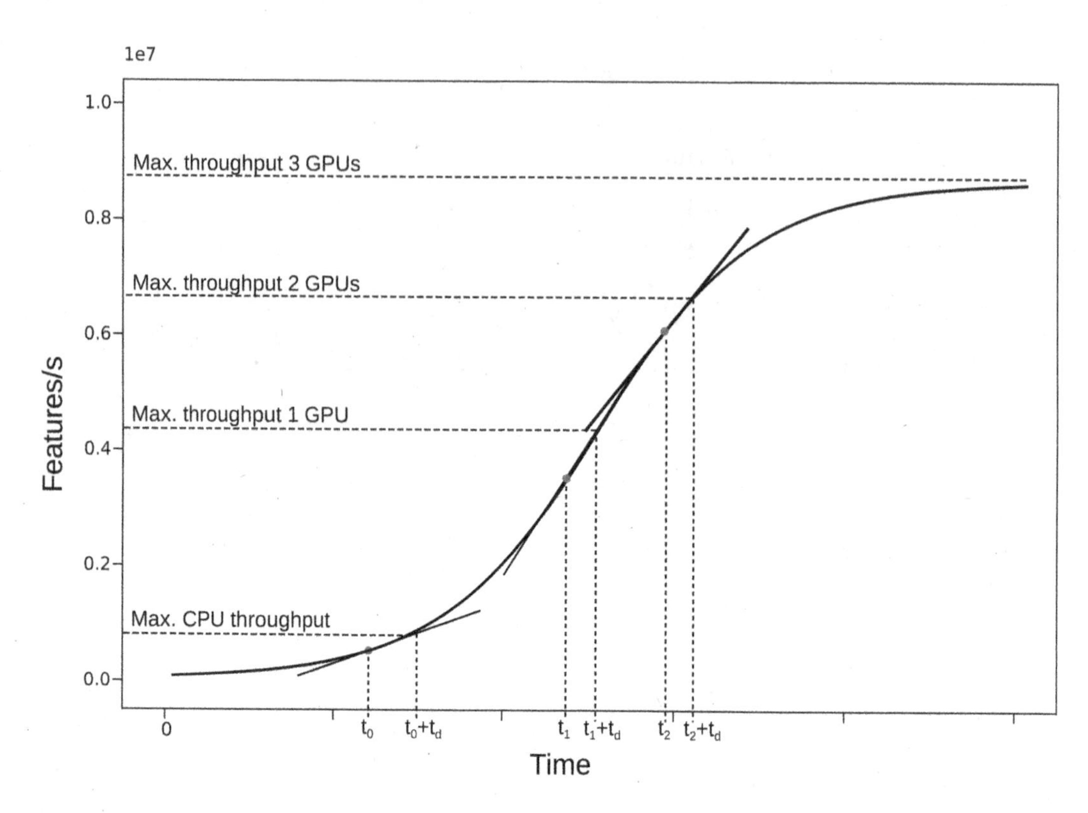

FIGURE 7.6 Resource adaptation to a hypothetical continuous sigmoidal increment of the network traffic due to the dynamic of the attendance at an event. The load at $t + t_d$ is estimated by means of the derivative, and indicates the time at which the load would exceed the maximum throughput supported by the virtualized hardware. t_d is the deployment time of the new model with additional resources.

the *Data Logger*, focused on troubleshooting and forensic analysis; and *External Interface*, which enables the communication with external hospital resources such as Electronic Health Records (EHR).

The contribution of our architecture to ICE is a real-time, automatic, and intelligent system capable of detecting, classifying and mitigating ransomware attacks in the hospital rooms of the future. Our deployment is integrated into the ICE++ [40] architecture and employs machine learning techniques to detect and classify the propagation stage of ransomware attacks affecting ICE. Another relevant contribution is the proposed mitigation mechanism, which uses the NFV/SDN paradigms to stop propagation by isolating and seamlessly replacing infected medical devices.

The proposal presented is also based on the creation of feature vectors by aggregating flow sequences in Netflow format. These feature vectors feed an analysis module containing two classifiers: one semisupervised, used to detect more generic anomalies, and another supervised for ransomware classification. This supervised classifier is probabilistic, that is, their multiple outputs can be interpreted as a probability of belonging to each class. A decision and reaction module, fed by a set of rules that act both on control and data planes, is responsible for taking the output of the two classifiers and reacting to a detection in an intelligent way. For example, after an anomaly detection, a mitigation could be forced, deploying by means of NFV a clean virtual device if there is sufficient consensus between the two classifiers, or alerting the administrator in those cases where there is no consensus. All this happens online and in real time, with event information being logged.

Additionally, these feature vectors are used in an offline phase in which human experts are responsible for creating, labeling, and partitioning the dataset needed for model training. This is followed by a model-dependent feature selection stage, and finally, a training phase, including hyperparameter tuning by cross validation to determine the best semisupervised and supervised models from a preselected list of methods. The trained models are then integrated into the online stage described before (Figure 7.7).

7.4.3.1 Experiments

To test the effectiveness of our proposal, a series of experiments were conducted using some of the most dangerous and recent malware: WannaCry, Petya, BadRabbit, and PowerGhost.

The first step was to recreate, using OpenICE and virtual machines, a realistic clinical environment. We decided to use OpenICE because it is a well-known open-source implementation of the ICE standard. OpenICE simulates interoperable clinical devices and medical controllers capable of deploying closed-loop scenarios. Several hours of normal network traffic were acquired in this virtual environment. For each of the selected ransomware, we created a clean environment as starting point. Then, one of the machines was infected with the corresponding ransomware, and the resulting combined traffic was captured until the ransomware propagated to another machine. From this traffic, the flows and feature vectors used in the experimentation were generated and labeled by us. This dataset was one of the results of this research and it remains available to the scientific community.

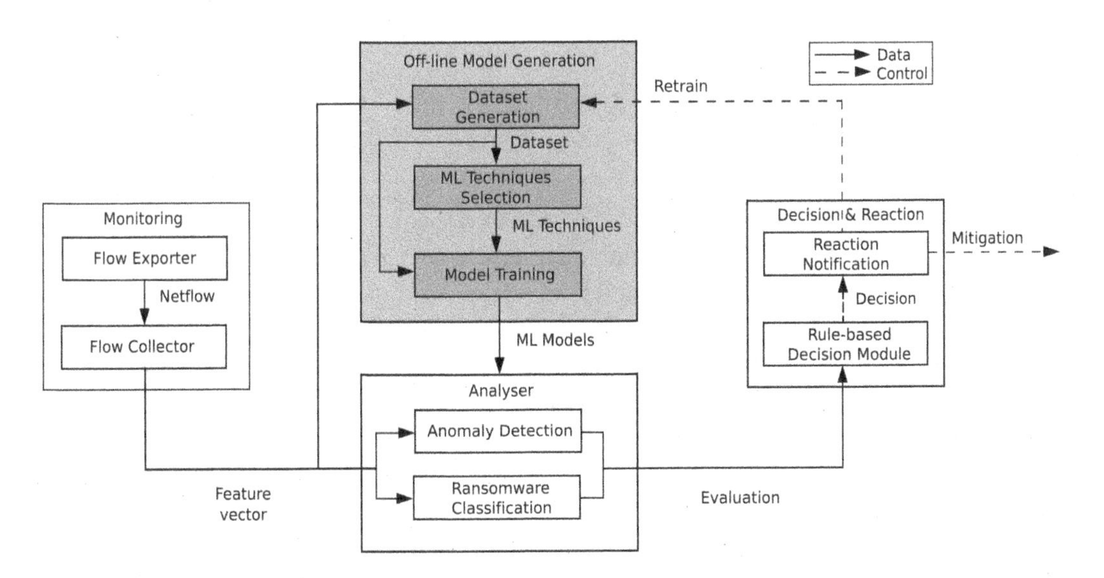

FIGURE 7.7 Offline (gray) and online (white) stages in our system. During the offline phase, a dataset is created from captured network flows; subsequently the machine learning models (anomaly detector and ransomware classifier) are trained and validated by experts. These models are deployed and used in the online phase to detect and mitigate anomalies in real time.

TABLE 7.2 Hyperparameters Tuned in Semisupervised Methods

Method	Hyperparameter	List/Range of Values
OC-SVM	Nu	[0.0005, 0.001, 0.005, 0.01]
	Gamma	[0.0005, 0.001, 0.005, 0.01]
LOF	Neighbors	[10, 20, 30, 50]
	Leaf size	[10, 20, 30, 40]
	Contamination	[0.1, 0.05, 0.01, 0.005]
IF	Estimators	[100, 200, 300]

TABLE 7.3 Hyperparameters Tuned in the Ransomware Classifier Selection Process

Method	Hyperparameter	List/Range of Values
Flow Collector	Window (seconds)	[5, 10, 20, 30, 60]
Neural Network	First layer size	[8, 10, 16]
	Second layer size	[0, 4, 6]
Gaussian Naive Bayes	No parameters	
Random Forest	Estimators	[100, 200, 300]

With regard to the results in classification with this dataset, several algorithms were tested, as listed in Table 7.2 (semisupervised) and Table 7.3 (supervised). The semisupervised method that obtained the best performance in the experiments was OC-SVM (one-class support vector machine), trained with normal traffic, which achieved a precision of 0.9232, a recall of 0.9997 and a false positive rate of 0.046 when classifying joint traffic (normal and anomalous). On the other hand, regarding the supervised methods, all algorithms obtained an acceptable performance for a certain combination of hyperparameters. Among these parameters, the size of the time window used to generate the feature vector, which ended up being set at 10 seconds, must be highlighted for its importance. Naive Bayes was selected for being the simplest of the three algorithms evaluated whereas it offered a similar performance for a window of 10 seconds, having especially good performance in detecting unknown ransomware. The results show a precision and sensitivity of 99.99%. It is important to remember that in the rules that activate mitigation both predictors (OC-SVM and Naive Bayes) are taken into account to make the decision.

Finally, an analysis of the time needed to detect and mitigate the spread of ransomware in our network was carried out. The results demonstrated that, by using containers, our system could recover in a time ranging from 23.5 seconds in a Raspberry Pi 3 to 10.5 seconds in a personal computer. These times are more than sufficient since our lowest measured ransomware propagation time was 63 seconds. The detailed development of these results can be found in [17].

7.5 CONCLUSIONS AND FUTURE WORK

Botnet and ransomware attacks account for a high percentage of the most dangerous cybersecurity threats, causing incalculable losses each year. The networks that the near future brings us present a challenge when it comes to applying methods of detecting these threats in network traffic based on deep packet inspection. This challenge is mainly due to the restrictions imposed by aspects such as high transmission speeds, huge volumes of data, the continuous expansion of the IoT and the increasing use of encrypted network traffic. In addition, nowadays the generation of new modified versions of malware to avoid detection is a common practice.

This chapter has studied three solutions based on NFV/SDN and machine learning for the detection of anomalies in network traffic in three scenarios representative of the near future:

An adaptive system for the detection of network anomalies in the context of 5G mobile communications that allows computational resource optimization, ensuring a suitable execution time for the detection process.

An extension of the previous system to derive an architecture focused on Mobile Edge Computing. This extended architecture incorporates policies that allow to deploy actions in order to guarantee an effective detection in real time of the anomalies.

A real-time, automatic, and intelligent system capable of detecting, classifying and mitigating ransomware attacks in the integrated clinical environments of the hospital rooms of the future.

The connection between these three solutions has been the use of a novel approach to detect anomalies in data network traffic. This approach is based on the idea that a sequence of network flows can be interpreted as a context for the last flow. A vector of aggregated features is generated from a sequence of flows, with the aim of condensing contextual information into a single vector to which specific anomaly detection techniques can be applied. This way of obtaining a vector that condenses contextual information from network traffic for the detection of anomalies, has been integrated in the three solutions proposed, and it has been validated in the three aforementioned scenarios.

Among the advantages of using flows are the following: it does not depend on access to the packet payload, so it allows encrypted traffic and reduces in orders of magnitude the amount of data to analyze. Additionally, it facilitates the acquisition of training datasets because it preserves users' privacy. However, and also due to the fact that the packet payload is not used, the patterns to be identified in the resulting vectors are more complex and, thus, detection rules are not easily applicable. Moreover, the difficulty of the detection is increased by the necessity of protection against unknown malware variations. The results support the hypothesis that machine learning methods, with sufficient data, can identify such patterns and generalize them to new versions of malware, becoming a valuable tool in the near future.

7.5.1 Scenarios

In the 5G scenario and thanks to the flexibility offered by 5G standard, an architecture based on NFV/SDN has been proposed containing an anomaly detection mechanism integrated at two levels. The lower level, running on each RAN, performs a local anomaly symptom detection. The sequence of anomaly symptoms from all the detectors in the different RAN, adequately labeled and accompanied by their context is sent to the upper centralized level. Those symptoms are considered a temporal sequence of local symptoms that is analyzed for global anomalies. The time constraints imposed by the high transfer rates and the volume of data expected in 5G led to the decision to use deep neural networks for the detection of symptoms. This machine learning model, has a stable execution time in prediction, independent of the training procedure and an excellent performance when running on GPUs. A study of the different deep learning environments was carried out and it concluded that the available hardware resources (e.g., if GPU should be used instead of CPU) determine the implementation of the learning model used for detection. The first level of this model, located in the 5G infrastructure RANs, was validated using the CTU public data set. This dataset includes both clean and anomalous traffic from various botnet families, captured in a real network. Our study concluded that the model chosen, in conjunction with the use of flow aggregations, was a combination that allowed obtaining an adequate classification performance (assuming that the second level will be in charge of the final decision) and, more importantly, it also allowed the evaluation of the model in a sufficiently short time period.

In the Mobile Edge Computing scenario, the previous solution was substantially extended to include policies and an orchestration process in charge of the actions associated to those policies. This extension provides a dynamic and efficient management system of the computational resources used in the real-time anomaly detection process. A study was included, in the context of 5G networks, of both the execution time needed for the detection of anomalies and the dynamic behavior of the system in a use case of adaptation to the evolution of the network traffic.

With respect to the performance study of the two-level model as a whole, negotiations were held with a telecommunication company in order to conduct a data acquisition. Unfortunately no agreement was reached, and the capture of a set of real and labeled data from several RANs to validate the second level of our proposed system remains as future work.

In the last scenario, that is, the integrated clinical environments of the future, one contribution is the incorporation to the ICE++ architecture [40] based on NFV/SDN of a mechanism capable of detecting the anomalies caused in network traffic by the propagation of ransomware. Precisely the use of NFV/SDN allows another of the contributions of this proposal, which is the intelligent and dynamic detection and mitigation of ransomware. For detection purposes we proposed the combination of two detection algorithms: one of them semisupervised, and the other, supervised. Subsequently, their predictions are unified by means of a rule-based mechanism, which provides an automatic and intelligent management of both the mitigation procedure and the alerts raised to the administrator. The modification of the ICE++ architecture included both the off-line operation and the online operation. The off-line operation is responsible for the acquisition, cleaning and labeling of data sets, as well as the selection of feature and training of the models. In turn, the online operation is responsible for generating the feature vectors in real time and classify them. Specifically, the semisupervised algorithm classifies each vector as anomalous or normal, whereas the supervised algorithm is a probabilistic classifier providing for every feature vector a probability of belonging to each class of ransomware. Due to the absence of a suitable publicly available dataset, a realistic virtualized clinical scenario was created with OpenICE. This virtualized scenario was used to capture both clean traffic and the anomalous traffic caused by the propagation of some samples of ransomware. This dataset remains available to the community as an additional contribution of this research. The dataset was used to validate the detection capacity of the models and to carry out a study of the reaction speed of the detection/mitigation architecture on different hardware. The system proved capable of detecting and mitigating with great effectiveness both known and new ransomware in a time substantially shorter than that necessary for the ransomware to spread.

7.5.2 Future Work

One of the future lines of this work consists in the application of more complex deep learning models such as convolutional networks and the recently proposed *transformers* based on attention mechanisms. These methods have demonstrated their good performance in tasks of natural language recognition, among others, substantially improving the results achieved with recurrent networks in the identification of complex patterns. Due to the use of flow sequences, it is expected that these techniques achieve state of the art.

Another crucial aspect to be studied is the existence of datasets of realistic and sufficiently varied data in order to be able to apply the techniques of deep learning like the aforementioned. Currently, the research group I belong to is still trying to reach an agreement with a mobile communications company so as to capture actual anonymized traffic.

Finally, another future line of work consists of applying the techniques developed here to other environments, such as industrial networks, which have traditionally been isolated and, precisely due to this isolation, have vulnerabilities that are being exposed with the gradual opening of these systems to the Internet. The use of the network for industrial process control opens the door to all types of sabotage by malware. This malware can alter the network traffic, which in these environments usually have very stable patterns of operation. The existence of stable patterns makes the detection of anomalies a promising option. We have already contacted local industrial companies which have seriously considered the possibility of offering their facilities to capture traffic and validate the solutions proposed.

REFERENCES

1. Nokia. Nokia threat intelligence report. https://pages.nokia.com/T003B6-Threat-Intelligence-Report-2019.html, 2019. Accessed on 19th November 2019.
2. Cisco. Encrypted traffic analytics (white paper). https://www.cisco.com/c/dam/en/us/solutions/collateral/enterprise-networks/enterprise-network-security/nb-09-encrytd-traf-anlytcs-wp-cte-en.pdf, 2019. Accessed on 19th November 2019.
3. G. Mohney. Hospitals remain key targets as ransomware attacks expected to increase. https://abcnews.go.com/Health/hospitals-remain-key-targets-ransomware-attacks-expected-increase/story?id=47416989, 2017. Accessed on 19th March 2019.
4. Palo Alto Networks Threat Intelligence Team (Unit 42). Unit 42 Finds New Mirai and Gafgyt IoT/Linux Botnet Campaigns. https://unit42.paloaltonetworks.com/unit42-finds-new-mirai-gafgyt-iotlinux-botnet-campaigns/, 2018. Accessed on 19th November 2019.
5. Heimdal Security. The Top Ten: Most Dangerous Malware That Can Empty Your Bank Account. https://heimdalsecurity.com/blog/top-financial-malware/, 2016. Accessed on 19th November 2019.

6. Medical Devices and Medical Systems—Essential safety requirements for equipment comprising the patient-centric integrated clinical environment (ICE)—Part 1: General requirements and conceptual model. Standard, ASTM International, West Conshohocken, PA, 2013.
7. V. Chandola, A. Banerjee, and V. Kumar. Anomaly detection: A survey. *ACM Computing Surveys (CSUR)*, 41(3):15, 2009.
8. V. Richariya, U.P. Singh, and R. Mishra. Distributed approach of intrusion detection system: Survey. *International Journal of Advanced Computer Research*, 2(6):358–363, 2012.
9. J. Yu, B. Yang, R. Sun, and Y. Chen. FPGA-based parallel pattern matching algorithm for network intrusion detection system. In *2009 International Conference on Multimedia Information Networking and Security*, pages 458–461, November 2009.
10. Y.M. Hsiao, M.J. Chen, Y.S. Chu, and C.H. Huang. High-throughput intrusion detection system with parallel pattern matching. *IEICE Electronics Express*, 9(18):1467–1472, September 2012.
11. A.L. Buczak and E. Guven. A survey of data mining and machine learning methods for cyber security intrusion detection. *IEEE Communications Surveys & Tutorials*, 18(2):1153–1176, October 2016.
12. Q.A. Tran, F. Jiang, and J. Hu. A real-time NetFlow-based intrusion detection system with improved BBNN and high-frequency field programmable gate arrays. In *2012 IEEE 11th International Conference on Trust, Security and Privacy in Computing and Communications*, pages 201–208, June 2012.
13. J. Gardiner and S. Nagaraja. On the security of machine learning in malware C&C detection: A survey. *ACM Computing Surveys*, 49(3):59:1–59:39, December 2016.
14. T. Kovanen, G. David, and T. Hamalainen. Survey: Intrusion detection systems in encrypted traffic. In *Internet of Things, Smart Spaces, and Next Generation Networks and Systems*, pages 281–293. Springer, 2016.
15. A. Pastor, A. Mozo, D.R. Lopez, J. Folgueira, and A. Kapodistria. The Mouseworld, a security traffic analysis lab based on NFV/SDN. In *Proceedings of the 13th International Conference on Availability, Reliability and Security*, page 57. ACM, 2018.
16. W. Wang et al. HAST-IDS: Learning hierarchical spatial-temporal features using deep neural networks to improve intrusion detection. *IEEE Access*, 6:1792–1806, 2018.
17. L. Fernandez Maimo, A. Huertas Celdran, A.L. Perales Gomez, F.J. Clemente Garcia, J. Weimer, and I. Lee. Intelligent and dynamic ransomware spread detection and mitigation in integrated clinical environments. *Sensors*, 19(5):1114, 2019.
18. L. Fernandez Maimo, A.L. Perales Gomez, F.J. Garcia Clemente, M. Gil Perez, and G. Martinez Perez. A self-adaptive deep learning-based system for anomaly detection in 5G networks. *IEEE Access*, 6:7700–7712, 2018.
19. A.J. Aviv and A. Haeberlen. Challenges in experimenting with botnet detection systems. In *4th Conference on Cyber Security Experimentation and Test*, pages 6–6, 2011.
20. J. M. Jorquera Valero, P. M. Sanchez Sanchez, L. Fernandez Maimo, A. Huertas Celdran, M. Arjona Fernandez, S. de los Santos Vilchez, and G. Martinez Perez. Improving the security and QoE in mobile devices through an intelligent and adaptive continuous authentication system. *Sensors*, 18(11), 2018.
21. G. Vormayr, T. Zseby, and J. Fabini. Botnet communication patterns. *IEEE Communications Surveys & Tutorials*, 19(4):2768–2796, Fourthquarter 2017.
22. P. Garcia-Teodoro, J. Diaz-Verdejo, G. Macia-Fernandez, and E. Vazquez. Anomaly-based network intrusion detection: Techniques, systems and challenges. *Computers & Security*, 28(1-2):18–28, February 2009.
23. E. Hodo, X. Bellekens, A. Hamilton, C. Tachtatzis, and R. Atkinson. Shallow and deep networks intrusion detection system: A taxonomy and survey. *arXiv preprint*, arXiv:1701.02145, 2017.
24. S. M. Erfani, S. Rajasegarar, S. Karunasekera, and C. Leckie. High-dimensional and large-scale anomaly detection using a linear one-class SVM with deep learning. *Pattern Recognition*, 58:121–134, October 2016.

25. Q. Niyaz, W. Sun, and A.Y. Javaid. A deep learning based DDoS detection system in software-defined networking (SDN). *EAI Endorsed Transactions on Security and Safety*, 4(12), 12 2017.
26. S. Garcia, M. Grill, J. Stiborek, and A. Zunino. An empirical comparison of botnet detection methods. *Computers & Security*, 45:100–123, September 2014.
27. A. Huertas, M. Gil, F.J. Garcia, F. Ippoliti, and G. Martinez. Dynamic network slicing management of multimedia scenarios for future remote healthcare. *Multimedia Tools and Applications*, 78(17):24707–24737, September 2019.
28. A. Huertas, M. Gil, F.J. Garcia, and G. Martinez. Enabling highly dynamic mobile scenarios with software de_ned networking. *IEEE Communications Magazine*, 55(4):108–113, 2017.
29. A. Huertas, M. Gil, F.J. Garcia, and G. Martinez. Sustainable securing of medical cyber-physical systems for the healthcare of the future. *Sustainable Computing: Informatics and Systems*, 19:138–146, 2018.
30. A. Huertas, M. Gil, F.J. Garcia, and G. Martinez. Policy-based management for green mobile networks through software-defined networking. *Mobile Networks and Applications*, 24(2):657–666, April 2019.
31. A. Huertas, M. Gil, F.J. Garcia, and G. Martinez. Towards the autonomous provision of self-protection capabilities in 5G networks. *Journal of Ambient Intelligence and Humanized Computing*, 10(12):4707–4720, December 2019.
32. M. Gil Perez, A. Huertas Celdran, F. Ippoliti, P.G. Giardina, G. Bernini, R.M. Alaez, E. Chirivella-Perez, F. Garcia Clemente, G. Martinez Perez, and E. Kraja. Dynamic reconfiguration in 5G mobile networks to proactively detect and mitigate botnets. *IEEE Internet Computing*, 21(5):28–36, 2017.
33. C.C. Machado, L.Z. Granville, and A. Schaeffer-Filho. ANSwer: Combining NFV and SDN features for network resilience strategies. In *IEEE Symposium on Computers and Communication*, pages 391–396, 2016.
34. K. Cabaj and W. Mazurczyk. Using software-defined networking for ransomware mitigation: The case of cryptowall. *IEEE Network*, 30(6):14–20, 2016.
35. G. Cusack, O. Michel, and E. Keller. Machine learning-based detection of ransomware using SDN. In *Proceedings of the 2018 ACM International Workshop on Security in Software Defined Networks & Network Function Virtualization*, SDN-NFV Sec'18, pages 1–6, New York, NY, USA, 2018. ACM.
36. ETSI NFV ISG. Network functions virtualisation (NFV) network operator perspectives on NFV priorities for 5G. Technical report.
37. M. S. Siddiqui et al. Hierarchical, virtualised and distributed intelligence 5G architecture for low-latency and secure applications. *Transactions on Emerging Telecommunications Technologies*, 27(9):1233–1241, September 2016.
38. P. Neves et al. Future mode of operations for 5G-The SELFNET approach enabled by SDN/NFV. *Computer Standards & Interfaces*, 54(4):229–246, November 2017.
39. L. Fernandez Maimo, A. Huertas Celdran, M. Perez Gil, F.J. Garcia Clemente, and G. Martinez Perez. Dynamic management of a deep learning-based anomaly detection system for 5G networks. *Journal of Ambient Intelligence and Humanized Computing*, 10(8):3083–3097, 2019.
40. A. Huertas, F.J. Garcia, J. Weimer, and I. Lee. ICE++: Improving security, QoS, and high availability of medical cyber-physical systems through mobile edge computing. In *Proceedings of the IEEE 20th International Conference on e-Health Networking, Applications and Services (Healthcom)*, pages 1–8, September 2018.

CHAPTER 8

Deep Learning for Network Intrusion Detection: An Empirical Assessment

Arnaldo Gouveia and Miguel Correia

CONTENTS

8.1 INTRODUCTION

With the continuous growth in a number and impact of network attacks, *network intrusion detection systems* (NIDSs) are increasingly becoming critical security components. From a research standpoint, although investigated for many years [1], NIDSs continue to get a lot of attention due to their practical interest and to challenges such as the need to handle large data volumes [2, 3].

Anomaly-based network intrusion detection is a particularly promising approach as it allows detecting previously unknown attacks. This approach resorts to machine learning to create models of normal behavior, then detecting deviations from this behavior. Most of the research in this area uses standard predictors known for decades, such as naive Bayes, thus supervised machine learning.

Deep learning is currently a hot topic in machine learning [4, 5]. This kind of technique has been achieving excellent results in the recognition of speech, faces, and images, to name a few [4]. Schmidhuber has a recent survey on major work in the area [6].

The term deep learning first appeared in the late 2000s associated with recent work in *neural networks* with many layers, or *deep belief networks*. However, with time its semantics expanded to a variety of other methods as neural networks are equivalent to other statistical and machine learning approaches in structural terms. Recursive generalized linear models [7] have the same structure as deep belief networks [6]. Gradient boosting machines [8] leverage performance via boosting with an ensemble of weak prediction models, typically decision trees.

This chapter explores the use of these three deep learning techniques in the context of anomaly-based network intrusion detection: *generalized linear models* (GLM), *gradient boosting machines* (GBM), and *deep learning* (DL), using H2O's denominations [9]. We executed both standard predictors and these advanced techniques with a recent, carefully crafted, data set of network traffic containing known attacks. The goal is to understand if deep learning algorithms have superior performance over standard predictors and at what cost.

The data set is composed of sequences of records that represent traffic flows. Each recorded is labeled as normal or attack (i.e., as being an attack or part of an attack). In terms of machine learning, the intrusion detection problem can be modeled as a regression problem. This means that we use a machine learning predictor (e.g., GBM) configured with the train data set to predict if the every record of the test data set is normal or attack. We assess the quality of the prediction using standard metrics such as the mean square error.

The main contributions of this chapter are: (1) a comparison of intrusion detection with three deep learning algorithms versus three classical machine learning algorithms; (2) showing that deep learning can be used for intrusion detection and that it achieves better results than classical algorithms (lower mean square error, higher R squared, lower analysis time).

The experimental results show that the deep learning algorithms provide a better fit than older algorithms, taking advantage of their superior modelling capacities.

8.2 METHOD

The chapter presents an experimental comparison of deep mining algorithms with more classical mechanisms in the context of network intrusion detection. The deep mining algorithms selected were those more popular today and with implementations available. For the classical algorithms there was a vast choice, so we selected a few of different types among those known to provide better results.

The experimental comparison method used was:

1. select an adequate data set of network traffic containing attacks;
2. tune the detection algorithms;
3. run the algorithms with the data set;
4. analyze the results.

We start with the first and leave the rest for Section 8.3.

8.2.1 Experimental Data Set

Data sets have been widely adopted as a means to benchmark NIDSs (or intrusion detection systems (IDSs) in general). Two of the first data sets were DARPA1998 and DARPA1999, developed under the auspices of DARPA. These data sets have been widely used for several years [10, 12]. However, they were also much criticized because they were synthetic, so could lead to overoptimistic results; the attack characterization was not realistic; and they did not allow false alarm evaluation [11, 12]. The KDD99 data set was developed in an attempt to improve these two data sets. The definition of this data set involved identifying parameters that would allow effective detection while minimizing processing costs. The objective was to allow real-time detection. Extensive work has been done over this data set, but many criticisms were also made [13]. However, an extensive study of the KDD99 data set argues that this data set is still valid within the following scope of objectives [14]: dealing with high dimensional data; learning from a very large imbalanced data set; feature selection research; detecting new intrusions, and simulating intrusion detection of encrypted data.

The UNB ISCX NSL-KDD data set. With such a scarcity of adequate data sets, researchers from the University of New Brunswick (UNB) created a new data set known as the UNB ISCX NSL-KDD data set or NSL-KDD data set for short [15]. The NSL-KDD data set is fairly recent and aims at solving, among others, the most hindering problems identified by Travalle et al. in KDD99 [13]. The authors explain that their data set may still suffer from some of the problems discussed by McHugh [12]. McHugh claims that a data set should reflect as much as possible live data, at risk of bias toward unrealistic results with respect to true detection, false alarms, or both. However, the UNB researchers claim that processed data sets have benefits in terms of redundancy reduction in both the train and test sets. Also unbalance issues of the KDD99 data set have been minimized by redistributing the sample proportions of classes. The cardinality of the records corresponding to each difficulty level group is inversely proportional to the percentage of records in the original KDD99 data set. As a consequence, the prediction rates of different machine learning methods vary in a wider range arguably allowing a more precise assessment of distinct learning techniques. In this regard, the NSL-KDD aims at a greater degree of balancing.

The UNB ISCX NSL-KDD data set is composed of sequences of entries or records. Each entry represents a *flow*, i.e., a sequence of IP packets starting at a time instant and ending at another, between which data flows to and from a source IP address to a target IP address using a transport-layer protocol (TCP, UDP) and an application-layer protocol (HTTP, SMTP, SSH, IMAP, POP3, or FTP). The data set has around 2.5 million flows and 76.6 GB, so it is reasonably large.

Each flow is labeled as either *normal* or *attack*. In the version of the data set used, no specifics on the attack types are included, although there were several types of attacks. The attacks fall into four main classes:

- Probing—scan-based attacks, e.g., network scans; port sweep;
- Denial of service (DoS)—network and systems denial of service, e.g., syn flood;
- Remote to local (R2L)—unauthorized access from a remote machine, e.g., password guess attacks;
- User to remote (U2R)—unauthorized access to local superuser (root) privileges, e.g., buffer overflow attack on a program running as root.

The data set is composed of two subdata sets: *train data set*, used for training a NIDS, and *test data set*, used for testing the NIDS. Both have the same structure and contain attacks of the four classes as illustrated in Table 8.1. It is important to note that the test data does not have the same probability distribution as the training data, and it includes specific attack types not in the training data. A total of 37 types of attack are included in contrast with only 23 in the train data set. Table 8.1 summarizes this information. Each record of the data set is characterized by features that fall into three categories: basic, features, and traffic features. These features are represented in Table 8.2. Also in Figure 8.1 it is visible the correlation structure among the data set features. The degree of correlation among the distribution of the various features is clearly visible with the most significant groups of correlation underlined with square boxes. The train data set has around 19.8 GB of data, whereas the test data set has 3.5 GB.

Cost-based models have often been used regarding feature definition in fraud-based detection. Stolfo [16] demonstrated how cost-based assertions developed for fraud detection can be generalized and applied to the area of network intrusion detection as a criteria for feature finding. With this approach in mind this author defined a set of features intrinsically related to Probing, Remote to Local, Denial of Service, and User to Remote

TABLE 8.1 Attacks in the UNB ISCX NSL-KDD Train and Test Data Sets (All Attacks from the Train Exist Also in the Test Data Set)

Class	Train Data Set Attacks	Attacks in Test Data Set Only
Probe	portsweep, ipsweep, satan, guesspasswd, satan, spy, nmap	snmpguess, saint, mscan, xsnoop
DoS	back, smurf, neptune, land, pod, teardrop, buffer overflow, warezclient, warezmaster	apache2, worm, udpstorm
R2L	imap, phf, multihop	snmpget, httptunnel, xlock
U2R	perl, loadmodule, ftp write, rootkit	sqlattack, mailbomb, processtable

TABLE 8.2 Features Used to Characterize Each Flow in the Data Set: Basic (Top), Content (Middle), Traffic (Bottom)

Feature	Detail
Duration	Length of the flow in seconds
Protocol-type	Type of the protocol, e.g., TCP, UDP, ICMP
Service	Network service, e.g., HTTP, telnet
src-bytes	Num. of data bytes from source to destination
dst-bytes	Num. of data bytes from destination to source
Flag	Status of the flow, normal or error
Lang	1 if flow is from/to the same host/port; 0 otherwise
Wrong-fragment	Num. of erroneous fragments
Urgent	Num. of urgent packets
Hot	Num. of hot indicators
num-failed-logins	Num. of failed login attempts
Logged-in	1 if successfully logged-in; 0 otherwise
num-compromised	Num. of compromised conditions
Root-shell	1 if root shell is obtained; 0 otherwise
su-attempted	1 if su root command attempted; 0 otherwise
num-root	Num. of root accesses
num-file-creations	Num. of file creation operations
num-shells	Num. of shell prompt
num-access-files	Num. of operations on access control files
num-outbound-cmds	Num. of outbound commands in an ftp session (obliterated)
is-host-login	1 if the login belongs to the hot list; 0 otherwise
is-guest-login	1 if the login is a guest login; 0 otherwise
Count	Num. of connections to the same host as current connection
serror-rate	Perc. of connections that have SYN errors
rerror-rate	Perc. of connections that have REJ errors
same-srv-rate	Perc. of connections to the same service
diff-srv-rate	Perc. of connections to different services
srv-count	Num. of connections to the same service as current connection
srv-serror-rate	Perc. of connections that have SYN errors
srv-rerror-rate	Perc. of connections that have REJ errors
srv-diff-host-rate	Perc. of connections to different hosts
dst-host-count	Num. of connections to the same destination host
dst-host-srv-count	Num. of connections to the same service as current connection
dst-host-same-srv-rate	Perc. of connections to the same service
dst-host-diff-srv-rate	Perc. of connections to different services
dst-host-same-src-port-rate	Perc. of connections from same source and port (from dst to src)
dst-host-srv-diff-host-rate	Perc. of connections to different services (from dst to src)
dst-host-serror-rate	Perc. of connections that have SYN errors (from dst to src)
dst-host-srv-serror-rate	Perc. of connections that have SYN errors per service (from dst to src)
dst-host-rerror-rate	Perc. of connections that have REJ errors(from dst to src)
dst-host-srv-rerror-rate	Perc. of connections that have REJ errors per service (from dst to src)

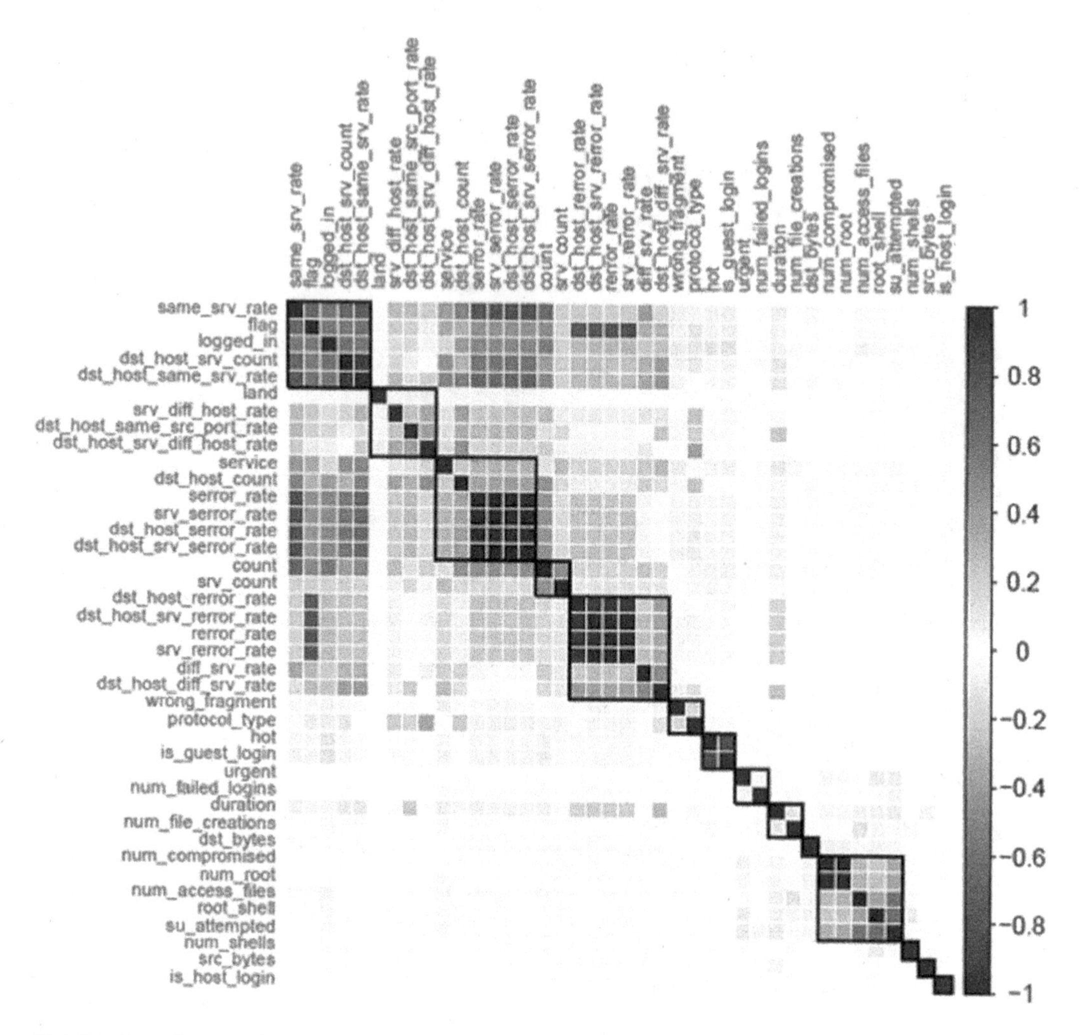

FIGURE 8.1 Plot of the statistical correlation between features distributions obtained with the "Pearson" method.

as classes of intrusions considered to maximize specific cost models. As such a number of features have been identified and applied to the predecessor of the UNB ISCX NSL-KDD data set, the KDD99 data set. We used this set of features in the work we present here.

8.2.2 Analysis Tools

All experiments were performed with R, which is a programming language and software environment for statistical computing and graphics supported by the R Foundation for Statistical Computing. The R language is widely used among statisticians and data miners for developing statistical software and data analysis [17]. It is a GNU project that provides a wide variety of statistical (linear and nonlinear modeling, classical statistical tests, time-series analysis, KNN, clustering, etc.) and graphical techniques, and is highly extensible. R provides tools for data preprocessing, classification, regression, clustering, association rules, and visualization.

The R packages H2O [9], kknn [18], rpart [19], and e1071 [20] were used in the analysis process. The main package was H2O that implements deep learning algorithms for R. It calls the H2O software package (Apache 2.0 License) produced by H2O.ai (previously 0xdata). The H2O R package provides a number of implementations of machine learning algorithms: GLM, GBM, deep learning (multilayer neural network models), K-means, naive Bayes, principal components analysis, principal components regression, and random forest.

8.2.3 Performance Metrics

To assess our results we use a set of metrics adequate for regression problems: mean square error, root mean square error, and R square (also named coefficient of determination). The commonly used error indicator is not used since it is not adequate for regression models.

Accuracy. Accuracy is a metric commonly used to evaluate IDS. Accuracy is also used as a general statistical measure of how well a classification test correctly identifies or not a given variable. The accuracy is the sum of true results (both true positives and true negatives) divided by the total number of cases (sum of all true positives, true negatives, false positives, and false negatives); in terms of numerical value its most favorable figure would be 1.

Mean square error. The mean squared error of an estimator measures the average of the squares of the errors or deviations between the estimated values ($\hat{y}_i$) and the real values (y_i). In other words, it gives the average of the squared differences between the estimator and what is estimated: $MSE = \frac{1}{n}\sum_{i=1}^{n}(y_i - \hat{y}_i)^2$.

Root mean square error. The square root of MSE. The use of RMSE is very common as it makes an excellent general purpose error metric for numerical predictions. Compared to the similar mean absolute error, RMSE amplifies and severely punishes large errors. $RMSE = \sqrt{\frac{1}{n}\sum_{i=1}^{n}(y_i - \hat{y}_i)^2}$.

Root square error. The square root of the square of the errors. It provides a measure of error without error signal bias. $RSE = \sqrt{\sum_{i=1}^{n}(y_i - \hat{y}_i)^2}$.

R square. R square or the coefficient of determination R^2 is the proportion of variability in a data set that is accounted for by a statistical model. In other terms, it is an indicator of how well data fits a statistical model. It varies between 0 and 1. In this definition, the term variability is defined as the sum of squares: $R^2 = 1 - \frac{SS_{res}}{SS_{tot}} = 1 - \frac{\sum_i (y_i - \bar{y})^2}{\sum_i (f_i - \bar{y})^2}$. The values of f_i are the predicted values in the model. Notice than on the contrary of the errors that should be as small as possible, the best value for this metric is 1. There are, however, many examples where the correlation coefficient is close enough to one but the model is still not appropriate, so in these cases the mean residual deviance should be used.

Mean residual deviance. In statistics, deviance is a quality-of-fit statistic for a model. It is a generalization of the idea of using the sum of squares of residuals in ordinary least squares to cases where model fitting is achieved by maximum likelihood. Formally, one can views deviance as a sort of distance between two probabilistic models. The residual deviance is the sum, over all the observations, of terms which vary according to type (regression/classification) of tree. The residual mean deviance is then obtained after dividing by the degrees of freedom (number of observations minus the number of terminal nodes).

TABLE 8.3 Results for KNN Tuning Phase with the Train Data Set

Mean error	MSE	Best kernel	Best k
0.003	0.003	Triangular	6

8.3 RESULTS

This section presents the study itself. The section starts with the older algorithms (Section 8.3.1), then presents the advanced, deep learning, and schemes (Section 8.3.2). For each algorithm we explain briefly how it works, how it was tuned in order to obtain results as good as possible, and its results for intrusion detection.

8.3.1 Classical Algorithms

This section explores three algorithms: k-nearest neighbors, decision trees, and support vector machines.

k-nearest neighbors. The k-nearest neighbors algorithm (KNN) is a nonparametric method used for classification and regression. In both cases, the classes are grouped according with the k closest training examples in the feature space. Here we use KNN for regression. Kernel methods introduce nonlinearity through implicit nonlinear or linear transforms, often local in nature. The main objective is to facilitate separability, i.e., the closer two points are, the larger the kernel value. In general terms, a kernel is transformation that takes place before the model is processed.

Tuning. For tuning the predictor we used the training data set, which has the same shape and features of the testing data set, as previously explained. The algorithm has been tested against a set of possible kernel types, namely triangular, rectangular, and Epanechnikov, and optimal with a maximum value for k of 15 and a distance of 3. The best value obtained for k was 6 and the best kernel was triangular. The metrics obtained are shown in Table 8.3.

Results. We run the tuned predictor, i.e., KNN with $k = 6$ and triangular kernel, with the test data set. The results are presented in Table 8.4. Notice that the errors are considerable and R square very low (it varies from 0 to 1 and 1 is the best).

Decision trees. Decision trees (DTs) are a nonparametric supervised learning method used for classification and regression. The goal is to create a model that predicts the value of a target variable by learning simple decision rules inferred from the data features. The deeper the tree, the more complex the decision rules and the fitter the model. The decision trees implementation used was the one in R's *rpart* package.

Tuning. While running the *rpart* algorithm, the number of cross-validations was set to 10 and the complexity parameter (*cp*) was *null*. This essentially meant that the optimization has been left to the cross-validation process without any tentative to spare computational time. With such a configuration, any split that does not decrease the overall lack

TABLE 8.4 Results for KNN with the Test Data Set

MSE	RMSE	RSE	R square
0.228	0.447	0.929	0.071

TABLE 8.5 Results for Decision Trees (*rpart*) with the Test Data Set

MSE	RMSE	RSE	R square
0.2027	0.477	0.929	0.070

of fit by a factor of *cp* is not attempted. The main role of this parameter, if different from *null*, is to save computing time by pruning-off splits that are obviously not worthwhile. Essentially, any split which does not improve the fit by the figure of *cp* will likely be pruned off by cross-validation, and that hence the program need not pursue it. In our case, it was left for the cross-validation process the responsibility of optimization.

Results. The results of applying the predictor with the test data set are presented in Table 8.5. The errors are slightly lower than KNN's, and R square is slightly higher, so this predictor provided better results.

Support vector machines. SVMs are a group of algorithms usually considered to be among the best out of the box algorithms for classification and regression. There are several types of SVMs. The simplest is the maximum margin classifier (MMC), but it cannot be applied to our most data sets, including ours, since it requires the classes to be separable by a linear boundary. The support vector classifier (SVC) is an extension of the MMC that can be applied in a broader range of cases. Here we use the SVM algorithm implemented in the *e1071* package [20], which is a further extension to accommodate nonlinear class boundaries. This algorithm builds a model defining a set of support vectors, i.e., of vectors that model the data.

Tuning. Our preliminary experiments with SVM have shown that the quality of the prediction (i.e., the error) depends strongly on the number of support vectors, therefore tuning was particularly important in this case. The tuning phased involved using the *tune.svm* primitive of the *e1071* R package. That phase aimed at choosing good values for the cost (C) and gamma (γ) parameters and using a Gaussian radial kernel. The ranges of values considered were from 0.0001 to 1000 for cost, and from 10^{-6} to 10^{-3} for gamma. The best values were around $C = 1000$ and $\gamma = 0.0001$. C and gamma are the parameters for a nonlinear support vector machine (SVM) with a Gaussian radial basis function kernel. The C parameter tells the SVM optimization how much you want to avoid misclassifying each training example. For large values of C, the optimization will choose a smaller-margin hyperplane if that hyperplane does a better job of getting all the training points classified correctly. On the other hand, a very small value of C will cause the optimizer to look for a larger-margin separating hyperplane, even if that hyperplane misclassifies more points. For small values of C, you should get more misclassified examples, often even if your training data is linearly separable. On its turn, gamma is the free parameter of the Gaussian radial basis function. A small gamma shapes a Gaussian kernel with a large variance. If gamma is large, then variance is small implying the support vector does not have wide-spread influence. Large gamma leads to high bias and low variance models, and vice versa. Figure 8.2 shows a heat graph of the performance for different combinations of the parameters, with better performance represented in darker blue.

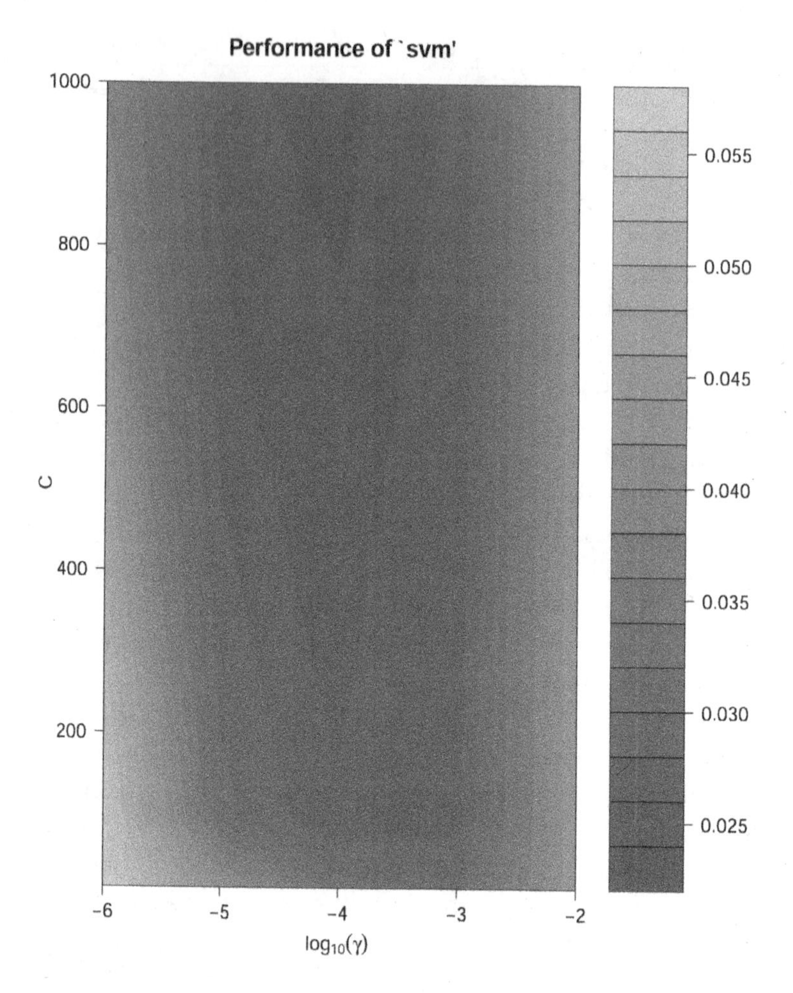

FIGURE 8.2 SVM optimization heat graph for gamma and cost.

Results. The results for SVM with the test data set for $C = 1000$ and $\gamma = 0.0001$ are presented in Table 8.6. These values are very similar to those obtained with decision trees. However, the SVM has shown to be extremely inefficient with analysis times of hours that is much longer that the other algorithms.

8.3.2 Advanced Algorithms

There has been much excitement about deep learning techniques due to their ability large improvements in terms of predictive power. Unlike the neural networks of the past, deep learning has improved training stability, generalization, and scalability, so it is quickly becoming the approach of choice for prediction. Here we consider three of the most popular algorithms of the area: GLM, GBM, and DL. All implementations are from R's *H2O*

TABLE 8.6 Results for SVM with the Test Data Set

MSE	RMSE	RSE	R square
0.200	0.447	0.817	0.0948

package. Modern deep learning algorithms have proven great training stability and generalization and scaling on big data. It is also the algorithm of choice for highest predictive accuracy. One of the root analysis to explain why deep learning works is deep rooted in the theory of deep Boltzmann machines and will not be pursued in this chapter.

Generalized linear models. Generalized linear models (GLMs) are a generalization of linear regression. Linear regression models the dependency of a response y on a vector of predictors $x(y: x^T \beta + \beta - 0)$. These models are built with the assumptions that y has a Gaussian distribution with a variance of σ^2 and the mean is a linear function of x with an offset of some constant $\beta - 0$, i.e. $y = \mathrm{N}(x^T \beta + \beta - 0; \sigma^2)$. These assumptions can be overly restrictive for real-world data that does not necessarily has a Gaussian distribution. GLM generalizes linear regression in the following ways:

- adds a nonlinear link function that transforms the expectation of response variable, so that $link(y) = x^T \beta + \beta - 0$
- allows variance to depend on the predicted value by specifying the conditional distribution of the response variable or the family argument.

Tuning. For the tuning phase of the GLM algorithm, a number of n folds for cross-validation have been chosen and four probability distributions have been tested (Tweedie, Gaussian, Poisson, gamma) against two types of regularization types: ridge and lasso.

The best model was elected to be the one with Poisson as probability distribution with ridge regularization as it exhibited the most balanced set of performance parameters.

Results. The assessment used the *H2O.glm* function of the *H2O* package. Performance indicators have been extracted with the training data and with the test data. Table 8.7 shows these results.

Gradient boosting machines. Gradient boosting machines (GBM) is an algorithm designed to produce a prediction model in the form of an ensemble of weak prediction models. It encloses the concepts of gradient descent and boosting, hence the name. It builds the model in a stage-wise fashion and is generalized by allowing an arbitrarily differentiable loss function. GBM fits consecutive trees where each solves for the net error of the prior. The idea of gradient boosting originated in Breiman's observation that boosting can be interpreted as an optimization algorithm using a suitable cost function [21].

Tuning. In general, three types of parameters are subject to choice in order to optimize GBM:

- Tree-specific parameters: these affect each individual tree in the model;
- Boosting parameters: these affect the boosting operation in the model;
- Miscellaneous parameters: other parameters for overall functioning.

TABLE 8.7 Results for GLM

Performance Metrics	Results with Training Data	Results with Test Data
MSE	0.046	0.177
R square	0.815	0.276
Mean residual deviance	0.031	0.117

TABLE 8.8 Results for GBM

Performance Metrics	Results with Training Data	Results with Test Data
MSE	0.051	0.153
R square	0.794	0.373
Mean residual deviance	0.051	0.153

The maximum depth of a tree is a tree-specific parameter used to control over-fitting as higher depth will allow model to learn relations very specific to a particular sample. In the tuning phase, a set of maximum depth of 2, 3, and 4 have been used.

A nonnegative integer that defines the number of trees has been tested with two values, 5 and 10.

The learning rate, as a boosting parameter, has been set to 0.1. This determines the impact of each tree on the final outcome. Lower values are generally preferred as they make the model robust to the specific characteristics of tree and thus allowing it to generalize well, although with an expense in processing time.

Several good and approximate models have been obtained. We presented the results for one with a number of trees = 10, mean depth = 4, min leaves = 15, and max leaves = 16.

Results. The results were obtained with the *H2O.gbm* function of the *H2O* package. Performance indicators have been extracted for both data sets (Table 8.8). The results are worse than those obtained with GLM.

Deep learning. Deep learning is a form of machine learning that uses a model of computing that's very much inspired by the structure of the brain. Hence we call this model a neural network. The basic foundational unit of a neural network is a node usually compared to a neuron. This is the most common description of deep learning although a number of other algorithms are also described as deep learning approaches. Each node or neuron has a set of inputs, each of which is given a specific weight. The node computes some function on these weighted inputs: in general terms a linear node takes a linear combination of the weighted inputs. A neural network morphs hooking up neurons to each other, usually arranged in layers, to the input data, and to the "outlets," which correspond to the network's answer to the learning problem.

H2O deep learning algorithm in unsupervised mode currently supports only (deep) autoencoding, where all layers are trained at once, just like a regular feed-forward neural network with the output layer using the MSE loss function in training.

Tuning. Tuning in deep learning was centered in the number of layers and respective dimensions. As such nets with three layers with (10,5,10) nodes, five layers with (10,5,10,5,10) nodes, and seven layers with (10,5,10,5,10,5,10) nodes. Best results have been achieved with three layers with (10,5,10) nodes. Fifty training epochs have been defined for all cases and the value for l1 (or lasso) regularization penalty of 0.1.

Results. The results were obtained with the *H2O.deeplearning* function of the *H2O* package, again for both data sets (Table 8.9). The results are very similar to those obtained with GBM.

TABLE 8.9 Results for Deep Learning

Performance Metrics	Results with Training Data	Results with Test Data
MSE	0.104	0.166
R square	0.581	0.320
Mean residual deviance	0.104	0.166

TABLE 8.10 Summary of Test Set Prediction Results for the Classical Algorithms

Regression Metrics	KNN	Dec. Trees	SVMs
Mean square error	0.230	0.203	0.222
Root mean square error	0.480	0.447	0.471
Root square error	0.940	0.929	0.906
R square	0.186	0.070	0.094

8.4 ANALYSIS AND DISCUSSION

The results from the previous section are presented in Table 8.10 for the classical algorithms and Table 8.11 for the advanced algorithms.

In relation to the classical algorithms, all the three show rather close figures for all the performance indicators extracted. According to what could be expected all the three extract a model that allows rather close predictions for the test data. In itself this is a good behavior that justifies using the same algorithms comparatively.

The results of the advanced algorithms are clearly better. The results for the *test data set* are the ones that are more relevant, as the others were used for training the algorithms (in italics at the table). For the test data set, *the best results were those of GBM*, with the lowest MSE and the highest R square. In comparison with the classical algorithms, the error in terms of MSE reduced 25% and R square increased by one order of magnitude.

When comparing the predictive results for GBM and DL it is noticeable a degree of overfitting when the performance metrics are extracted over the training data. This is to expect since the model is built on the same data. This shows that the existence of a test data set is important, as there is no overfitting with that data for the obvious reason that the model is not trained with it.

8.5 RELATED WORK

A number of anomaly-detection techniques related with malicious activity have been studied in the fields of computing, communications, and networking [22]. Some examples are: anomaly detection in data in computers and systems, anomaly detection in network traffic, detection of network configuration and vulnerabilities, assessing network and data integrity, recognition of attack patterns, tracking policy violations, and analysis of abnormal activities in logs. A particularly exhaustive review has been done by Shan et al. [23].

Anomaly detection systems were first described by Denning in her seminal work on the host-based intrusion-detection expert system (IDES) that described a model of a real-time expert system capable of detecting break-ins, penetrations, and other forms of breach [24].

TABLE 8.11 Summary of Prediction Results for the Advanced Algorithms (*tr* Stands for Train Data Set, *test* for Test Data Set)

Regression Metrics	DL(tr)	DL (test)	GLM(tr)	GLM (test)	GBM(tr)	GBM (test)
Mean square error	0.104	0.169	0.046	0.177	0.051	0.153
R square	0.581	0.320	0.815	0.276	0.794	0.373
Mean residual deviance	0.104	0.166	0.031	0.117	0.051	0.153

Denning's model describes profiles for representing the behavior of subjects with respect to objects in terms of metrics and statistical models, and rules for acquiring knowledge about this behavior from audit records and for detecting anomalous behavior. This model lay the ground on modeling data sets identifying anomalies to be used for machine learning IDS, NIDS included. Subsequent machine learning approaches have been tested against attacks and intrusions of several types, e.g., attacks against web servers [25]. Several angles were also investigated, e.g., the problem of labelling samples from real applications [26].

Deepa and Kavitha [27] surveyed a comprehensive set of machine learning approaches for intrusion detection, namely association rules, data clustering, Bayesian networks, hidden Markov models, decision trees, support vector machines [28], genetic algorithms, and honey pots. Garcia-Teodoro et al. [29] have produced another survey of anomaly-based intrusion detection techniques, outlining the main challenges facing wide scale deployment of anomaly-based intrusion detectors, with special emphasis on assessment issues. Sommer and Paxson [30] have noted a gap between the amount of research papers and the number of implementations of IDSs based on machine learning and identified several difficulties.

There is one previous work on applying deep learning to network intrusion detection [31]. This work used the same data set as us. However, it considered a single detection algorithm, the one we call DL, instead of providing a comparison of several as we do. Interestingly, we obtained better results with another of the algorithms, GBM, nevertheless close to the results obtained by deep learning.

8.6 CONCLUSIONS

From the comparative analysis of the various algorithms several conclusions can be extracted. The set of classical algorithms proved to convey equivalent models and results under the same test conditions. This supports the perspective conveyed by Engen et al. [14] regarding the quality of the UNB ISCX NSL-KDD data set. We recall that one of the main objections regarding KDD99 was related to the disparity of results obtained. In this regard, the results obtained for the UNB ISCX NSL-KDD data set seem to indicate a similitude of results given the same experimental conditions.

Overfitting can be observed in GLM, GBM, and deep learning when testing with training data (not with the test data set). This is a usually observed property regarding learning algorithms pointing that in this particular case and for algorithm evaluation purposes the best approach is to test the models obtained with a distinct test set. Given the similar values for R square and MSE obtained for GBM and deep learning with the test set we may assume a good degree of similitude for prediction between these two algorithms. Taken at face value the results seem to indicate good these are good candidates for ensemble regression.

When compared with the classical cases we may notice a better performance judging from the values of MSE and R square obtained: MSE is consistently lower for MSE and deep learning and the bigger values of R square obtained seem to indicate that the models do a better job at assimilating the variance of the train data. GBM was clearly the best algorithm, closely followed by DL. The performance of the deep learning algorithms was also better than what we observed for the classical algorithms.

REFERENCES

1. B. Mukherjee, L. T. Heberlein, and K. N. Levitt. Network intrusion detection. *IEEE Network*, 8(3):26–41, 1994.
2. H. A. Nguyen and D. Choi. Application of data mining to network intrusion detection: Classifier selection model. In *Challenges for Next Generation Network Operations and Service Management, 11th Asia-Pacific Network Operations and Management Symposium*, pages 399–408, 2008.
3. T.-F. Yen, A. Oprea, K. Onarlioglu, T. Leetham, W. Robertson, A. Juels, and E. Kirda. Beehive: large-scale log analysis for detecting suspicious activity in enterprise networks. In *Proceedings of the 29th ACM Annual Computer Security Applications Conference*, 2013.
4. G. Anthes. Deep learning comes of age. *Communications of the ACM*, 56(6):13–15, 2013.
5. I. Arel, D. C. Rose, and T. P. Karnowski. Deep machine learning: A new frontier in artificial intelligence research. *IEEE Computational Intelligence Magazine*, 5(4):13–18, 2010.
6. J. Schmidhuber. Deep learning in neural networks: An overview. *Neural Networks*, 61:85–117, 2015.
7. P. McCullagh and J. A. Nelder. *Generalized linear models (Second edition)*. London: Chapman & Hall, 1989.
8. J. H. Friedman. Greedy function approximation: A gradient boosting machine. *Annals of Statistics*, 29:1189–1232, 2000.
9. S. Aiello, T. Kraljevic, and P. Maj. *h2o: R Interface for H2O*, 2015. R package version 3.6.0.8.
10. C. Brown, A. Cowperthwaite, A. Hijazi, and A. Somayaji. Analysis of the 1999 DARPA/Lincoln laboratory IDS evaluation data with NetADHICT. In *Proceedings of the 2nd IEEE International Conference on Computational Intelligence for Security and Defense Applications*, pages 67–73, 2009.
11. M. V. Mahoney and P. K. Chan. An analysis of the 1999 DARPA/Lincoln laboratory evaluation data for network anomaly detection. In *Proceedings of the 6th International Symposium on Recent Advances in Intrusion Detection*, pages 220–237. Springer, 2003.
12. J. McHugh. Testing intrusion detection systems: A critique of the 1998 and 1999 DARPA intrusion detection system evaluations as performed by Lincoln laboratory. *ACM Transactions on Information and System Security*, 3(4):262–294, Nov. 2000.
13. M. Tavallaee, E. Bagheri, W. Lu, and A. A. Ghorbani. A detailed analysis of the KDD CUP 99 data set. In *Proceedings of the 2nd IEEE International Conference on Computational Intelligence for Security and Defense Applications*, pages 53–58, 2009.
14. V. Engen, J. Vincent, and K. Phalp. Exploring discrepancies in findings obtained with the KDD Cup'99 data set. *Intelligent Data Analysis*, 15(2):251–276, Apr. 2011.
15. A. Shiravi, H. Shiravi, M. Tavallaee, and A. A. Ghorbani. Toward developing a systematic approach to generate benchmark datasets for intrusion detection. *Computers & Security*, 31(3):357–374, May 2012.
16. S. J. Stolfo, W. Fan, W. Lee, A. Prodromidis, and P. K. Chan. Cost-based modeling for fraud and intrusion detection: Results from the jam project. In *In Proceedings of the 2000 DARPA Information Survivability Conference and Exposition*, pages 130–144. IEEE Computer Press, 2000.
17. R Foundation. R Project, 2015. https://www.r-project.org/about.html.
18. K. Schliep and K. Hechenbichler. *kknn: Weighted k-Nearest Neighbors*, 2015. R package version 1.3.0.
19. T. Therneau, B. Atkinson, and B. Ripley. rpart: Recursive partitioning and regression trees, 2015. R package version 4.1-10.
20. D. Meyer, E. Dimitriadou, K. Hornik, A. Weingessel, and F. Leisch. e1071: Misc functions of the department of statistics, probability theory group, 2015. R package version 1.6-7.
21. L. Breiman. Arcing the Edge. Technical report, Statistics Department, University of California, Berkeley, June 1997.

22. A.-S. K. Pathan, editor. *The State of the Art in Intrusion Prevention and Detection*. Auerbach Publications, 2014.
23. A. A. Shah, M. S. H. Khiyal, and M. D. Awan. Analysis of machine learning techniques for intrusion detection system: A review. *International Journal of Computer Applications*, 119(3):19–29, June 2015.
24. D. E. Denning. An intrusion-detection model. *IEEE Transactions on Software Engineering*, 13(2):222–232, Feb. 1987.
25. C. Kruegel and G. Vigna. Anomaly detection of web-based attacks. In *Proceedings of the 10th ACM Conference on Computer and Communications Security*, 2003.
26. Y. Park, I. M. Molloy, S. N. Chari, Z. Xu, C. Gates, and N. Li. Learning from others: User anomaly detection using anomalous samples from other users. In *Computer Security—ESORICS 2015*, pages 396–414. Springer, 2015.
27. A. Deepa and V. Kavitha. A comprehensive survey on approaches to intrusion detection system. *Procedia Engineering*, 38:2063–2069, 2012.
28. L. Khan, M. Awad, and B. Thuraisingham. A new intrusion detection system using support vector machines and hierarchical clustering. *The VLDB Journal*, 16(4):507–521, Oct. 2007.
29. P. García-Teodoro, J. Díaz-Verdejo, G. Maciá-Fernández, and E. Vázquez. Anomaly-based network intrusion detection: Techniques, systems and challenges. *Computers & Security*, 28(1–2):18–28, Feb. 2009.
30. R. Sommer and V. Paxson. Outside the closed world: On using machine learning for network intrusion detection. In *Proceedings of the 2010 IEEE Symposium on Security and Privacy*, pages 305–316, 2010.
31. M. Z. Alom, V. Bontupalli, and T. M. Taha. Intrusion detection using deep belief networks. In *Proceedings of the 2015 National Aerospace and Electronics Conference*, pages 339–344, June 2015.

CHAPTER 9

SPATIO: end-uSer Protection Against ioT IntrusiOns

Gil Mouta, Miguel L. Pardal, João Bota, and Miguel Correia

CONTENTS

9.1 INTRODUCTION

The Internet of Things (IoT) is one of the fastest growing paradigms today. It consists of many objects of daily use with embedded smart sensors and computational resources, connected to the Internet. In 2018, there were 7×10^9 connected IoT devices (IoT Analytics 2018), and it is estimated that the number of IoT devices will surpass the number of non-IoT active device connections by 2020. With IoT becoming more pervasive, it becomes a target for malicious agents. According to an intelligence report (NetScout 2018), IoT devices are attacked within 5 minutes of being plugged in, and targeted by specific exploits in 24 hours, and a survey (Ponemon Institute 2018) reveals that 21% of respondent organizations had a cyber-attack or data breach caused by an unsecured IoT device in 2017.

IoT devices often present certain characteristics, such as reduced dimensions, low energy consumption, limited network (Internet) connection, limited processing and storage resources, and low cost. These characteristics help disseminate the IoT paradigm to consumers, but they also make traditional security tools less effective. For example, traditional cryptography is harder to apply in IoT devices, due to the heavy use of resources that such algorithms put on the device (Trappe et al. 2015). Another aggravating factor in IoT security is the quantity of devices, many of which are of different types, from sensors to actuators, and from different vendors, highly increasing the chance that at least one of these devices will have a vulnerability or be misconfigured. Even if the IoT device being targeted is not the target desired by the attacker, a compromised device can act as a foothold on a network and allow it to launch more attacks.

Intrusion Detection Systems (IDSs) are a common component of network security systems. They help discover, determine, and identify unauthorized use, duplication, alteration, and destruction of information and information systems (Mukkamala et al. 2005). IDS can monitor activity on the network, or run on individual hosts or devices, using data collected locally about events taking place, such as activity logs and system calls. They can also employ different detection methods, such as recognizing known attacks, or measuring the normal behavior of a device, and labeling any behavior differing from it as anomalous.

While IDSs are one of the primary tools used for protection of traditional networks and information systems, some characteristics of the IoT make traditional IDS solutions inadequate (Zarpelao et al. 2017). The high number of small devices with constricted resources makes host-based detection harder, while the high volume of traffic generated by these devices, on a high bandwidth network running on technology such as the recent 5G cellular network, makes near real-time network level analysis more difficult. It is also important that modern systems be able to deal with the growing concern about data privacy.

The fog computing paradigm evolved to deal with some of these issues, where focus is shifted away from a centralized cloud computing environment toward the edge of the network. The main idea behind this paradigm is that gateway devices such as routers or base stations in an IoT network utilize their processing and storage resources to process data closer to its source, reducing latency and processing cost.

9.1.1 Objectives

We propose an architecture to deal with the challenges of designing an IDS for the IoT, namely, the limited characteristics of the IoT devices, the high amount of data generated by the large amount of devices, most of which are sensors, and the need for data privacy. We compare different types of machine learning approaches to intrusion detection and analyze the advantages of offloading work closer to the edge of the network.

To this end, we developed a prototype of our architecture, SPATIO (end-uSer Protection Against ioT IntrusiOns), which uses outlier detection algorithms to detect anomalies in IoT networks. The core machine learning model works in a streaming paradigm, analyzing data as it comes and producing alerts in a timely manner.

We also compared a fog computing approach versus a more centralized one, by distributing some processing of data to the edge of the network, where gateway devices monitor the IoT device network traffic and transform it into flow metrics, which are consumed by the machine learning algorithm.

In developing SPATIO, we analyzed available machine learning frameworks and discussed which is best for our case. We also compare several public datasets on IoT intrusions. We evaluate our prototype by measuring its accuracy in detecting intrusions, as well as counting the number of false alarms produced. We also evaluate the advantages of offloading work to devices in the edge of the network, by comparing its performance to a centralized approach. We do this evaluation on a public dataset of IoT intrusions.

9.1.2 Context

Telecommunication networks play an important role: they connect the smart objects to the Internet, and are interested in deploying IoT networks in industrial contexts, or for large cultural events such as concerts. As such, it is in their interest to protect clients from potential network intrusions that can cause damage to their clients or their brand. We developed this work in collaboration with Vodafone Portugal, a big Portuguese telecom company and internet service provider, to better understand the current trends in IoT networks, such as what types of devices or architectures are being used.

9.2 BACKGROUND AND RELATED WORK

We start by presenting the IoT, discuss the fog computing paradigm, and review research by other authors in IDSs for the IoT.

9.2.1 Internet of Things

The IoT is the ever-growing network of physical objects that are connected to the Internet. It consists of many objects of daily use with embedded smart sensors and computational resources. These objects make possible new and innovative applications (Al-Fuqaha et al. 2015).

In 2018, there were 7 billion connected IoT devices, and by 2020, it is expected that the number of IoT devices will surpass the number of non-IoT active device connections worldwide (IoT Analytics 2018).

With the ubiquity of the IoT, it makes sense to separate an industrial focused use case from a consumer use case.

The Industrial IoT is the concept of IoT applied to the industry. For example, IoT has been applied to the mining industry, for the safety of miners. Environmental sensors are deployed to detect flooding, fires, gas or dust explosions, cave-ins, toxic gas, and other risk factors. They also allow precise location and identification of miners. By using wireless communication technology, such as Wi-Fi and RFID (Radio-frequency Identification) to send the collected information to the surface, mining companies can keep track of the miners and analyze critical safety data to enhance safety measures (Qiuping et al. 2011).

On the consumer side of the IoT, smart home technology uses sensors and actuators in common household objects to provide a better quality of living. Home parts are equipped with different technologies for more intelligent monitoring and remote control, as well as allowing interaction between each smart component, such that everyday house activities are automated without user intervention, or with more convenient remote control (Kadam et al. 2015). Some of its applications include lighting, air conditioning, energy management, security, and control of other home appliances. For example, a kitchen can have refrigerators, microwaves, coffee makers, and dishwashers with IoT capabilities. One example is the Internet Refrigerator, as developed by LG Electronics, an Internet enabled refrigerator, which lets the users download recipes and display them on a liquid-crystal display (LCD) screen. The refrigerator is also able to keep track of its inventory automatically, and alert the user to what is there and what is missing.

9.2.2 Fog Computing

Fog computing is a distributed computing paradigm that acts as an intermediate layer between services provided by the cloud and the edge of the network, where the IoT devices are (Mahmud et al. 2017). The concept was introduced by Cisco in order to address the challenges of IoT applications in conventional cloud computing, such as the high amount of nodes, with a wide geo-distribution, making a traditional geographically centralized cloud solution not ideal, especially when dealing with latency-sensitive operations.

A fog computing environment typically consists of gateway devices, such as routers and base stations, placed in proximity to the IoT devices and sensors. As opposed to the more resource constrained IoT devices, these intermediary devices have higher processing, storage, and networking resources to be able to do some preprocessing of the data coming from the IoT devices before being sent to the cloud.

A fog computing approach can bring the following advantages (Dastjerdi et al. 2016): reduction of network traffic by doing preprocessing of data before being sent to the cloud, better serve local events by using locally stored information to act on requests, lower latency by doing processing closer to source, instead of having to do a round-trip to the cloud and back, better scalability in the cloud by reducing the amount of work, and better control of data privacy, as data are processed closer to its owner.

These advantages make fog computing a better fit for IoT applications when compared to a traditional centralized approach.

An example of a fog computing application is FogLearn (Barik et al. 2018), a fog architecture framework which, among other scenarios, was used in detecting diabetes patients. It leverages data collected from smart wearables in patients, using fog devices to reduce latency and increase throughput. Data collected from patients in the edge layer are first preprocessed in the fog layer, before being sent to the cloud layer for further analysis and long term storage.

9.2.3 Machine Learning for Intrusion Detection

Security is a growing concern in the IoT. IoT inherits the traditional network security problems and adds its own set of problems. For example, IoT devices tend to have limited computational resources and a limited battery life. This makes conventional cryptography harder to apply to such devices, due to the heavy use of resources such algorithms put on the device (Trappe et al. 2015).

IDSs are a common component of network security systems. They help discover, determine, and identify unauthorized use, duplication, alteration, and destruction of information and information systems (Mukkamala et al. 2005).

When dealing with anomaly detection, for the purpose of network intrusion detection, a fast solution with as little human effort required as possible is preferred, to ensure timely detection of any possible attack. This is where machine learning solutions come in.

Machine learning consists of automated detection of meaningful patterns in data (Shalev-Shwartz and Ben-David 2014). It is an emerging field of data analysis, which became common standard in solving most problems that require analyzing a large amount of data, finding complex patterns that a human programmer cannot explicitly define.

In this section, we look at some of the machine learning solutions for anomaly detection proposed by other authors.

Gonçalves et al. (2015) proposed a machine learning and data mining approach for analyzing log data and semi-automatically discovering misbehaving hosts, without having to instruct the system about how hosts misbehave. A probability-based clustering algorithm, the Expectation-Maximization algorithm, is used, which has the benefit of not requiring prior knowledge of the distribution of each feature and other parameters, making it adequate for clustering large datasets. Classification algorithms, a method of supervised machine learning, are then used to assign clusters to classes, such as "normal" and "abnormal." This is the phase that requires the most human effort, but after the creation of the classification model, human intervention is only needed if the classification results are wrong, to update the model. The authors concluded that while the presented process did extract relevant security information from the logs that otherwise is not directly observable, it is not accurate enough to take automatic actions to deal with the misbehaving host.

Another example of machine learning applied to IoT intrusion detection, Meidan et al. (2018) propose a network-based approach to anomaly detection using autoencoders to detect botnets. The use of autoencoders has the advantage of not needing manual classification of what is malicious or benign, meaning unknown attacks can be detected.

Also, by profiling each device with a separate autoencoder, this method addresses the growing heterogeneity of IoT devices. For experimental evaluation, the authors created and used a dataset out of devices purposefully compromised to the Mirai and BASHLITE botnets. They were successful in detecting every attack launched by each compromised IoT device, that is, true positive rate of 100%. Also, compared to other algorithms commonly used in anomaly detection (Local Outlier Factor [LOF], One-Class Support Vector Machine, and Isolation Forest), this method got the fewest false alarms, while also being the fastest method at detecting the attacks.

Chawathe (2018) attempted to achieve similar accuracy, and potentially better performance, than the system proposed by Meidan et al. (2018). The author proposes a network-based method for monitoring botnets on IoT devices, requiring no access to the device other than the ability to monitor network traffic from some other hosts on the network. The author suggests the use of a simpler method based on supervised machine learning through the use of classifiers alone. The author concludes that simple classification methods can achieve high accuracy in classifying network activity into 11 classes, as well as achieving those high accuracy values by using only a small fraction of the original attributes (over 99% accuracy using only 20 of the 115 attributes on one of the devices in the dataset).

Outlier detection algorithms work by finding values which deviates from other observations on data. This type of algorithm lends itself to intrusion detection, since an attack is by definition an outlier, as it deviates from the normal behavior of a system. Auskalnis et al. (2018) proposed the usage of the LOF algorithm to detect anomalies in a computer network. The LOF algorithm detects outliers by evaluating each event's uniqueness based on the distance from the k-nearest neighbors. The expectation is that the density of events around an outlier is significantly different than the density of its neighbors. The authors evaluate their solution on an existing dataset, the NSL-KDD dataset[1]. An accuracy rate of 0.84 was obtained. The results are also compared to the accuracy rates of classification methods, and obtain the same values for accuracy; however, the authors conclude using LOF or other unsupervised outlier detection methods have the advantage of not needing labeled datasets for learning.

9.3 SPATIO

In this chapter, we present the SPATIO system and its prototype implementation. We start with an overview of the architecture and data pipeline in Section 9.3.1, and then a more detailed explanation of method and tool selection for each step, starting with the machine learning framework in Section 9.3.2, feature extraction in Section 9.3.3, and feature selection in Section 9.3.4. We conclude with a summary of the SPATIO system.

9.3.1 Architecture

The goal of SPATIO is to detect anomalies in an IoT network. For this, we analyze network traffic, as the heterogeneity of IoT devices, with varied characteristics, capacities, and

[1] https://www.unb.ca/cic/datasets/nsl.html

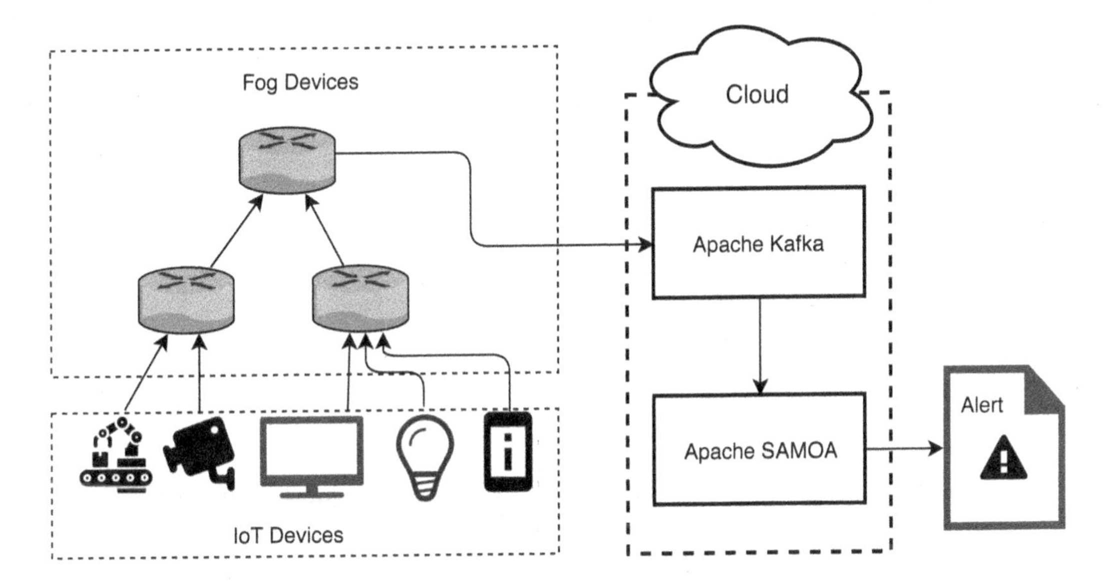

FIGURE 9.1 Overview of the SPATIO architecture.

goals, makes host-based detection difficult. We also expected to obtain network captures from our collaboration with a telecom company, which deploys IoT networks. Figure 9.1 gives an overview of the SPATIO architecture.

Devices in the fog of the network, such as routers, collect incoming and outgoing network traffic of the IoT devices (left side of Figure 9.1). We then take a fog computing approach, as described in Section 9.2.2, by having these devices also process the raw data traffic into flow features. This process is described in Section 9.3.3. By doing this transformation closer to the data source, we try to reduce network traffic, as only the reduced flow features need to be sent to the cloud, instead of the entire raw traffic. We also reduce the work needed to be done in the cloud, increasing scalability.

Another advantage of the fog computing approach is increased data privacy. By summarizing the whole traffic into flow features, the actual content of communication is not sent. Additionally, information identifying the devices, such as the IP address, can be transformed into an alias to identify the device.

The flow features are sent to the machine learning algorithm through Apache Kafka[2], a distributed streaming platform which works like a publish-and-subscribe system, acting as a message queue. We choose Kafka for its ability to be distributed, preventing a single point of failure, as well as its interoperability with our chosen machine learning framework, Apache SAMOA, described in Section 9.3.2.

The end result of SPATIO is alerts that indicate an anomaly in the network, affecting a certain device. This anomaly might or might not be the result of an ongoing attack. Human validation is required to differentiate a benign anomaly from an intrusion. This validation can later be done to teach a classification algorithm to be able to differentiate benign from malicious anomalies, although that was not explored in this research.

[2] https://kafka.apache.org/

9.3.2 Streaming and Machine Learning Framework

When choosing a machine learning framework, we looked at available data analysis frameworks which featured machine learning or integrated with a framework that did.

Streaming data analysis has the advantage of not requiring data to be stored in memory and providing quicker results. In an IoT environment, the large number of devices can generate a stream of data that is not feasible to be stored, and dealing with intrusion detection requires timely detection, which makes a streaming approach a good fit. As such we took a streaming approach for SPATIO, using Apache SAMOA.

Apache SAMOA and Apache MOA[3] are open source frameworks for data stream mining written in the Java programming language. They both function very similarly in terms of programming interface. However, SAMOA was built to be used in a distributed environment. To this end, it can integrate with Apache Storm, Apache S4, or Apache Samza, other stream processing frameworks that do not support machine learning natively, but are made to be distributed over several nodes. Although in our prototype for evaluation, we do not use this feature, opting for Apache SAMOA's local mode, in a real-world scenario having the IDS be distributed is advantageous as it prevents a single point of failure and spreads computation allowing for faster response times.

SAMOA is an Apache Incubator project, meaning it is still a developing project not ready to enter the Apache Software Foundation. This reflects on the fact that it lacks diversity in available machine learning algorithms and a somewhat lacking documentation. To solve the former problem of lack of machine learning algorithms, we used MOA's algorithms in SAMOA. MOA is SAMOA's centralized counterpart. It is a more mature project, leading it to have more machine learning algorithms available, in several categories from basic classification and clustering tasks to outlier detection and active learning. To export MOA's algorithms to SAMOA, we used an existing library[4], although this library had not been updated in 5 years and required some modifications before it could be used.

SAMOA defines a topology using the following basic elements: Processor, Content Event, Stream, Task, Topology Builder, and Learner. Figure 9.2 gives an overview of the topology we used for SPATIO. A Processor is the basic logical processing unit. It is where the processing logic for the data is written. A Processor can be an Entrance Processor if it generates its own stream of data. For SPATIO, we create an Entrance Processor, which reads incoming data from the gateway devices collecting flow information. A Stream is a physical unit of the SAMOA topology, which represents the flow of information and connects different Processors. Information goes through a Stream in the form of Content Events. When the Entrance Processor reads the flow data, it converts each flow, represented in a CSV (comma-separated value) format, into a Content Event object, through a deserializer. The Entrance Processor connects to a Learner. A Learner is the unit in the topology, which runs a machine learning model. The resulting output of the Learner is connected to a final Processor, where alerts are generated. Finally, a Task in SAMOA is

[3] https://moa.cms.waikato.ac.nz/

[4] https://github.com/samoa-moa/samoa-moa

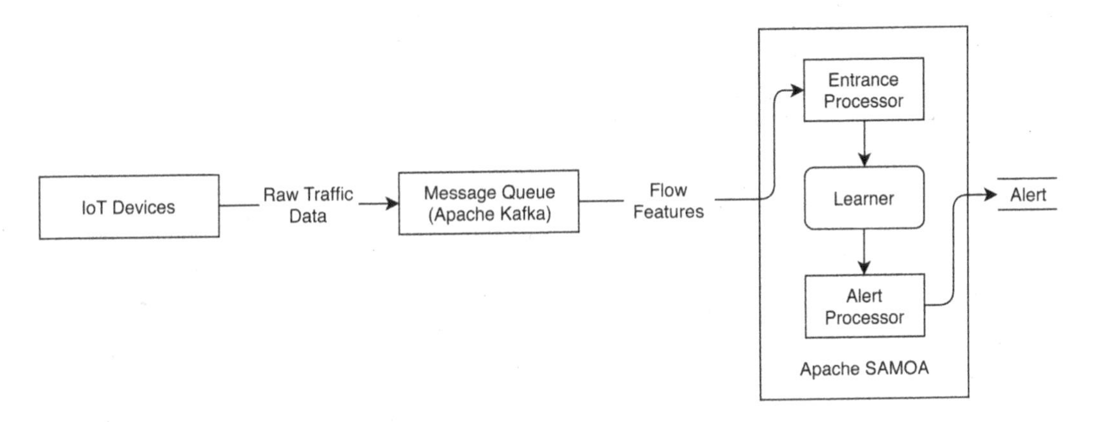

FIGURE 9.2 Overview of the SAMOA topology used for SPATIO.

an execution entity, as in, the class that is executed by SAMOA. It is where the topology is defined, through a Topology Builder.

We use outlier detection algorithms, which are a good fit for intrusion detection since by definition, an attack is an anomaly. They are an unsupervised algorithm and do not require a training period. SAMOA and MOA include the following outlier detection algorithms, which we test in Section 9.4.2.2: ExactSTORM, ApproxSTORM, AbstractC, and MCOD.

9.3.3 Feature Extraction

SPATIO monitors the network traffic to detect anomalies. To this end gateways, such as routers, to which the IoT devices are connected are constantly capturing network traffic using tools such as tcpdump[5]. While most of these tools allow for filtering packets by port, address or protocol, or more detailed filters, for SPATIO, we do not use this filtering feature and capture every packet arriving at the network card.

For SPATIO, feature extraction is done in IoT devices on the fog layer. Since it is important for an attack to be detected as quickly as possible, and these types of device can be resource constrained, we focus on analyzing the header of the packages, instead of analyzing the contents of each packet. Packet header anomaly detection also has the added advantage of remaining valid when the payload is encrypted, such as in an SSL session.

For our approach, we use single connection derived features, features derived from a single connection, or flow. A flow consists of all traffic with similar characteristics, usually source and destination address and ports.

To transform a series of network packets, which is what we are collecting with tcpdump, into flow records, we chose the NetMate-flowcalc[6] tool. NetMate (Network Measurement and Accounting Meter) is a network measurement tool used to generate statistics of network traffic. We chose NetMate for its ability to operate in real time, which is necessary for our streaming approach, for its ease of configurability, where each configuration, such as features extracted and the timeout period for flow creation, is easily changed in a rule file,

[5] https://www.tcpdump.org/

[6] https://github.com/DanielArndt/netmate-flowcalc

and because it, by default, outputs the features in a CSV format which makes no further parsing required before sending the flow data to the machine learning algorithm.

NetMate captures 38 features by default plus the (source address, source port, destination address, and destination port) tuple used to identify the flow, that is, all traffic which match these same characteristics are grouped into a single flow, as long as the time between packets does not cross a certain threshold, in which case a new flow is created. We use those features as the base features for SPATIO, before feature selection described in the next section.

9.3.4 Feature Selection

Feature selection consists of selecting a subset of features in order to reduce the dimension of the problem. This technique brings several advantages, such as a better generalization of the machine learning model to the problem, reducing overfitting; reduction of network bandwidth used on sending the flow features to the nodes running the machine learning model; avoiding the "curse" of dimensionality, in which more features make the algorithm less efficient (Verleysen and François 2005). For anomaly detection, the more features we consider in a flow, the more likely flows are to be different from each other, making outlier detection more difficult. As such, we can consider feature selection to be the process of identifying and removing as much irrelevant and redundant information as possible.

For feature selection we use WEKA[7], a suite of machine learning tools and algorithms written in Java. It contains several feature selection algorithms. For our purposes, we need a feature selection algorithm, which runs on unlabeled data. We chose to use Principal Component Analysis (PCA) as a feature evaluator. PCA works by transforming the current set of features into a new reduced set of features, known as the Principal Components. This new set is chosen such that each component is orthogonal to each other, that is, uncorrelated. Two features are correlated when a change in one leads to a change in the other. An example of two correlated features are total fpackets (total packets in the forward direction) and total fvolume (total bytes in the forward direction), since one can expect the number of bytes to increase with more packets sent. This does not always need to be observed for the features to be correlated, but the more present it is, the more strongly correlated the features are.

The new set of features produced by PCA is essentially linear combinations of the original features, where each feature is given a weight. Although new set of features can be used for machine learning directly, we thought it would be interesting to analyze, which features of the original set extracted would be more relevant to intrusion detection in our dataset. To this end, we summed the weights given to each feature in each Principal Component, considering that the higher this sum, the more relevant the feature is.

9.3.5 Summary

This chapter presented SPATIO, a system to detect anomalies in an IoT network. Figure 9.3 summarizes the data flow between the components of the SPATIO architecture. Network traffic is collected from IoT. This traffic gets processed into flows through feature extraction.

---

[7] https://www.cs.waikato.ac.nz/ml/weka/

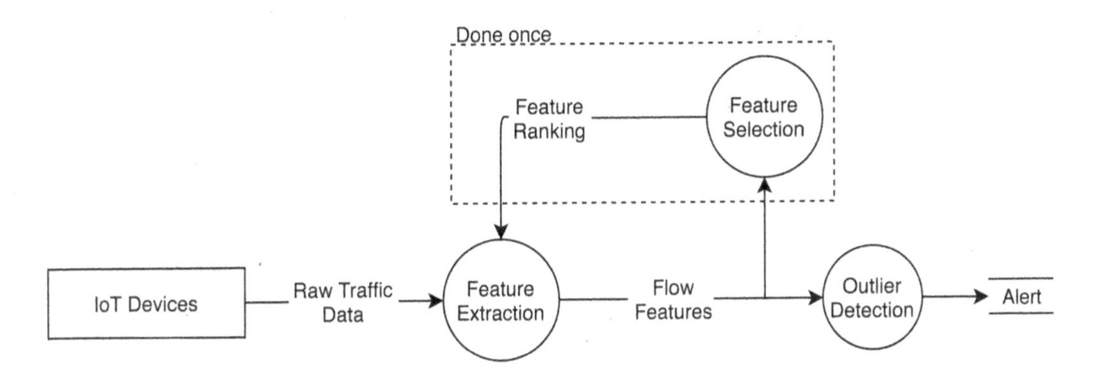

FIGURE 9.3 Data-flow diagram for the SPATIO architecture.

The system can then use these flows, or further refine them by going through feature selection. Feature selection analyses the extracted flows, and through correlation analysis returns the n-sized subset of features that more accurately represent the dataset. This process only has to be done once. After feature extraction, the flows are sent to the machine learning framework, running an outlier detection method. This outputs alarms, which indicate an anomaly in a certain device.

9.4 EVALUATION

In this chapter, we present the experiments that were done to evaluate the SPATIO prototype using a public dataset. The experiments assessed the accuracy of the prototype and the choice of architecture. We were interested in answering the following questions about our system:

- What percentage of attacks can our system detect?
- How many false alarms will be generated?
- What features are relevant for IDS, and how does the quantity of considered features affect the results?
- What are the advantages of offloading work to fog devices?

9.4.1 Dataset

To evaluate SPATIO, we tested our prototype on an existing dataset. We were originally planning on capturing our own set of data from a real IoT network in collaboration with a telecom company, but we were not able to find a proper IoT environment currently running and get the required authorizations for data capture in time. Since we are interested in testing the feature extraction from the raw data traffic, we look exclusively at datasets that provided the entire network capture, in a pcap format. The datasets we considered are summarized in Table 9.1.

To choose a dataset, we focused on certain qualities: the dataset had to be labeled, so we could easily test the accuracy of our solution, it had to be IoT related, and have as many

TABLE 9.1 Overview of the Studied Datasets

Dataset	Source	Size	Attacks	Duration	Labeling	IoT
CTU Honeypot Capture—Stratosphere Lab 2019	Collected traffic from an IoT connected camera	7 MB	Unspecified	1 day	None	Yes
CTU IoT Malware Capture—Stratosphere Lab 2019	Collected traffic from an IoT host infected with a botnet	1 MB	IoT botnet	1 day	None	Yes
Sivanathan et al. 2018	Testbed with 28 different IoT devices	6 GB	None	2 weeks	None	Yes
Hamza et al. 2019	Testbed with 27 different IoT devices	60 GB	Denial-of-service	1 month	Attack timeframe	Yes
UNSW-NB15—Moustafa and Slay 2015	Simulated by traffic generator	100 GB	Fuzzers, analysis, backdoors, DoS, exploits, generic, reconnaissance, shellcode, worms	16 hours	Attack timeframe	No
UNSW BoT-IoT—Koroniotis 2018	Virtual machines simulation IoT services and attacks	69 GB	DDoS, DoS, OS and service scan, keylogging, and data exfiltration attacks	Unspecified	Files divided by attack type	Yes
Bezerra et al. 2018	Single device simulating different IoT scenarios	1.5 GB	IoT botnets	Per scenario, 1 hour not infected, 1 hour infected	File separated by benign/infected	Yes
CICIDS 2017—Sharafaldin 2018	Empirical network with 12 victims and 4 attackers	50 GB	Brute force, heartbleed, botnet, DoS, DDoS, web attack, infiltration attack	5 days, one attack type per day	Attack frame and attack type per day	No

devices as possible. We decided on using a dataset collected for ACM SOSR 2019 (Hamza et al. 2019). Although this dataset covers less devices and less attack types, focusing only on Denial of Service (DoS) attacks, it was the IoT dataset with the best labeling for our purposes.

The chosen dataset consists of network traffic collected in a lab. The lab comprised a TPLink gateway with OpenWrt firmware serving 27 IoT devices, of which 10 are used as victims for attacks. The network traffic is captured at the gateway. Benign traffic is collected by simulating possible user interaction with each IoT device, while malicious traffic is captured by running a script exploiting known vulnerabilities to carry out both reflected and direct DoS attacks. Traffic is captured over a month, with two periods of around a week where attacks happen, which we name Period 1 and Period 2. The attacks present in the dataset consist of direct DoS attacks, in which the targeted devices are the victims, and reflected DoS attacks, in which the targeted device is used to attack another outside device. All of the attacks are sustained for 10 minutes, at various packet rates, and in total there are 200 attacks.

9.4.2 Accuracy Evaluation

The first set of tests measures the accuracy of the system. We take into consideration the following metrics:

- Total alarms, the total number of alarms generated.
- Attacks detected, the number of attacks detected. To calculate this metric, we consider any attack during which an alarm is generated, linked to the device being attacked, as being detected. Multiple alarms during the same attack do not further increase this number.
- Percentage of Attacks detected, the percentage of attacks detected, calculated by dividing Attacks detected by the total number of attacks as stated by the ground truth of the dataset.
- Alarms during attack, the number of alarms generated during attacks. This includes multiple alarms generated in a single attack.
- Percentage of Alarms during attack, the percentage of alarms, which fall inside an attack window for that device. This is calculated by dividing Alarms during attack by Total alarms. The remaining percentage is the false alarm percentage.
- False Alarms (ALL), the number of false alarms. A false alarm is an alarm, which happens while there is no attack occurring on the related device.
- False Alarms (1 minute). This metric works like the one above, but after an alarm occurs, no more alarms can occur for 1 minute.
- False Alarms per day, the amount of false alarms generated per day, based on the results of the metric False Alarms (1 minute).

9.4.2.1 Evaluation Environment

Since these tests do not consider the performance of the solution, and only the accuracy of the machine learning model, they were run on a dedicated machine with the pipeline simplified to SAMOA reading the previously generated flow metrics directly from a file. This machine has 16 GB of RAM, 1 TB of disk storage, and is running Debian GNU/Linux Buster.

After collecting the alerts in a file, a Python script is run which compares them with the ground truth provided by the dataset, providing the metrics described in the previous section.

9.4.2.2 Outlier Detection Algorithms

We start by measuring how the considered metrics vary by using different outlier detection methods. The results are shown in Tables 9.2 and 9.3. We defer the discussion of these results to Section 9.4.2.4.

9.4.2.3 Feature Selection

As explained in Section 9.3.4, we use PCA in WEKA to rank our features, by considering the sum of the weights of each Principal Component. We take 1 day of Period 1 where attacks happen as the training dataset to run PCA. Selecting a day with attacks ensures we capture the features more relevant to detect attacks, while using only 1 day ensures we do not overfit our selection.

With the feature ranking in mind, we tested how the number of considered features affects the metrics described in Section 9.4.2. We run five tests for each period, considering the 6, 12, 18, 24, and 30 most relevant features. We also compare with the base 36 feature results obtained in the previous section.

These tests use the same environment as the one done in the previous section and run the MCOD outlier detection algorithm. The results are shown in Tables 9.4 and 9.5.

9.4.2.4 Discussion

We can see all the tested outlier detection algorithms give similar results. In Period 1, we were able to detect 79.62% of the attacks in all cases. However, a difference can be seen when comparing false alarms. The MCOD algorithm had the lowest amount of generated

TABLE 9.2 Evaluation of the Accuracy Metrics According to Tested Outlier Detection Algorithm for Period 1

Algorithm	MCOD	ExactSTORM	ApproxSTORM	AbstractC
Total alarms	2,18,203	2,22,147	2,18,912	2,19,045
Attacks detected	43	43	43	43
% Attacks detected	79.63%	79.63%	79.63%	79.63%
Alarms during attack	5,411	5,459	5,441	5,438
% Alarms during attack	2.48%	2.46%	2.49%	2.48%
False alarms (all)	2,12,792	2,17,238	2,13,334	2,13,539
False alarms (1 min)	6,759	6,759	6,759	6,759
False alarms per day	563	563	563	563

TABLE 9.3 Evaluation of the Accuracy Metrics According to Tested Outlier Detection Algorithm for Period 2

Algorithm	MCOD	ExactSTORM	ApproxSTORM	AbstractC
Total alarms	1,30,312	97,216	95,841	95,443
Attacks detected	61	58	58	58
% Attacks detected	75.31%	71.60%	71.60%	71.60%
Alarms during attack	22,681	20,341	20,228	20,203
% Alarms during attack	17.41%	20.92%	21.11%	21.17%
False alarms (all)	1,07,631	97,216	95,842	95,443
False alarms (1 min)	7,999	5,957	5,953	5,950
False alarms per day	667	496	496	496

TABLE 9.4 Evaluation of the Accuracy Metrics According to Changing Number of Features for Period 1

Number of Features	6	12	18	24	30	36
Total alarms	50,999	2,04,649	2,04,819	2,11,459	2,18,221	2,18,203
Attacks detected	38	39	39	43	43	43
% Attacks detected	70.37%	72.22%	72.22%	79.63%	79.63%	79.63%
Alarms during attack	2,069	4,640	4,686	4,717	5,413	5,411
% Alarms during attack	4.06%	2.27%	2.29%	2.23%	2.48%	2.48%
False alarms (all)	40,042	2,04,814	2,04,477	2,12,563	2,12,792	2,12,792
False alarms (1 min)	10,115	5,549	5,186	6,759	6,759	6,759
False alarms per day	920	504	471	614	614	614

TABLE 9.5 Evaluation of the Accuracy Metrics According to Changing Number of Features for Period 2

Number of Features	6	12	18	24	30	36
Total alarms	22,967	28,423	82,541	90,330	92,554	1,30,312
Attacks detected	53	55	56	58	58	61
% Attacks detected	65.43%	67.90%	69.14%	71.60%	71.60%	75.31%
Alarms during attack	17,001	18,537	19,020	19,302	20,663	22,681
% Alarms during attack	74.02%	65.22%	23.04%	21.37%	22.33%	17.41%
False alarms (all)	21,142	26,164	75,983	78,334	80,444	1,07,631
False alarms (1 min)	6,079	6,001	5,931	6,070	5,957	7,999
False alarms per day	553	546	539	552	542	727

alarms, and one of the highest percentage of alarms that were generated during an attack, leading it to have the least amount of false alarms. In a real life situation, a false alarm requires human intervention, wasting time. It also potentially leads to alert fatigue, a situation in which so many false alerts are being generated that the human operator becomes desensitized to them, leading to a longer response time to real attacks, or even missing them completely (Sasse 2015). As such, a lower number of false alarms, and of alarms in general, is desirable. To this extent, we measured the number of false alarms generated if we blocked alarm generation for 1 minute after an alarm. The principle behind this idea is that when an anomaly occurs, whether it is benign or not, it can have an impact on multiple connections, generating several alerts in a short amount of time. This does seem to be the case, as this change reduces the number of false alarms considerably. A similar idea can be applied to real attack alarms, to reduce the number of alarms in general. In Period 2, we observe a smaller percentage of detected attacks and a smaller percentage of total alarms. This can occur because of the different devices being tested in the different periods, where attacks on this new set of devices are harder to distinguish from normal benign traffic.

In Section 9.4.2.3, we tested how the considered metrics change as the number of features change. A noticeable pattern is that as less features are considered, less alarms are generated, but the number of attacks detected also lowers. This can be explained by the loss of information when removing features. Outlier flows could be outliers by varying a lot on a certain feature, which could then be removed in feature selection, thus this outlier is no longer classified as such. We can see that when we use only six features, the metrics of false positives start to get worse due to too much loss of information. It is also interesting to note that the number of false alarms tends to reduce as less features are used, but only because the total alarms is also reducing, as the percentage of false alarms actually increases with less features. Still, using a lower number of features is something to be considered because although the detection rate of the system is lower, the bandwidth used to transmit the flows will also be lower, and less alarms in total will be generated, which can be useful in a situation where dealing with alarms has some resource cost associated. In Period 2, we see a bigger drop off in accuracy with less metrics. This can be caused by the fact that feature selection was done over the dataset of Period 1, making those features more accurate at representing traffic from Period 1.

9.4.3 Architecture Comparison

The second set of tests compares the chosen architecture of SPATIO, where work is offloaded to fog devices, with a centralized approach, that is, an architecture that skips this step, choosing instead to transmit the captured packets directly to a central server running the machine learning algorithm, which then also becomes responsible for feature extraction.

9.4.3.1 Evaluation Environment

To test our architecture in comparison with a centralized approach, we set up two scenarios using virtual machines. In the first scenario, testing the fog computing approach, two VMs act as gateway devices, collecting the IoT traffic and running the feature extraction, and then sending the flow features to a VM representing the cloud, running Apache Kafka and Apache SAMOA. In the second scenario, testing the centralized approach, the two gateway VMs still collect traffic from the IoT devices, but this traffic is sent directly to the cloud VM, with no feature extraction. The cloud VM is then responsible for doing feature extraction.

The gateway VMs were modeled after specifications of gateways being used in real IoT networks, given to us by a telecommunications company. As such, it contains 512 MB of RAM and 4 GB of disk storage. The cloud VM has 8 GB of RAM and 32 GB of disk storage. Both VMs run Debian GNU/Linux Jessie 64 bit.

To emulate the IoT device traffic, tcpreplay[8] was used to replay the dataset in a dummy network interface.

9.4.3.2 Network Bandwidth Usage

In this section, we consider the differences of network bandwidth usage between sending the raw traffic in the network, or only sending the extracted flow records. We measure the size of the raw traffic for each period and the size of the flow records. We then calculate the average bandwidth in bps by dividing the total size by the duration of the experiment. The results are shown in Table 9.6.

We can see using flow metrics results in much lower bandwidth requirements, reaching a 90% reduction in the case of Period 1.

[8] https://tcpreplay.appneta.com/

TABLE 9.6 Network Bandwidth Difference between Centralized and Fog Approach

Period	Total Size Raw Traffic	Total Size Flows	Bandwidth Avg. Sending Raw Traffic	Bandwidth avg. Sending flow Features
1	28.2 GB	1.72 GB	29,209 bps	1,787 bps
2	25.6 GB	4.50 GB	28,998 bps	5,092 bps

9.4.3.3 Latency

In this section, we compare the different architectures by measuring the latency in anomaly detection. To do this, we use ten benchmark alerts and measure how fast each solution creates that alarm. We repeat each test five times and take the average as our value. Since the dataset contains traffic captured over a month, to reduce the time it takes to perform the test and to emulate more network traffic, 1 day was chosen from the dataset and data were replayed at a rate of 100 MB/s. The results are shown in Figure 9.4, where each data point represents a benchmark anomaly.

9.4.3.4 Discussion

IoT networks tend to have many connected devices, and use lower bandwidth technologies such as Lowpower wide-area networks for long-range communications at low power consumption and low data rate. As such, minimizing network traffic is advantageous. Transforming raw traffic into flow records can reduce network usage by a lot, as would be expected since an entire connection of packets is transformed into a single line of features. An interesting fact to note is that despite having lower traffic size, Period 2 flow records are bigger. This can be explained by the fact that a flow can cover an entire connection, regardless of number/size of packets. For example, if a device was to download a video of size 1 GB from the Internet, then this entire 1 GB worth of traffic would be transformed in a single flow. In fact, by further analyzing which devices are being used in Period 2, we can

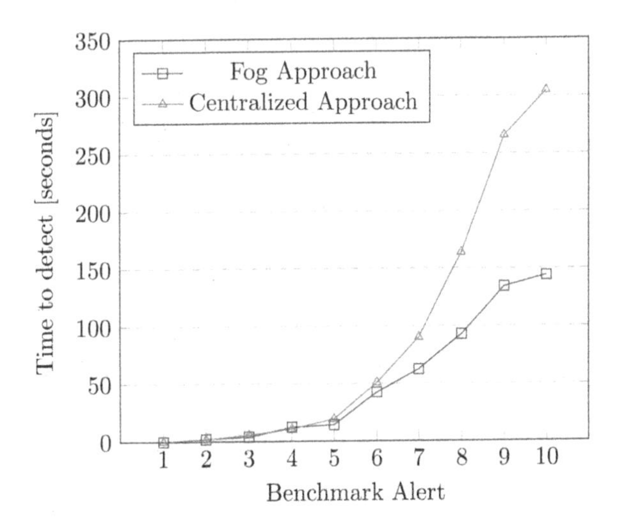

FIGURE 9.4 Detection latency in fog versus centralized approach.

see devices such as Chromecast, used to stream video, and Amazon Echo, a virtual assistance with features such as playing music from the Internet.

In the latency test, we can observe both approaches get worse latency as time goes by. This indicates data are being backlogged somewhere. In the centralized approach, this latency increase is more accentuated, which could indicate feature extraction on the server accounts for a big portion of the backlog, showing the advantage of distributing this task to the gateway devices. It is important to note that this test is done under a constant rate of 100 MB/s of traffic. In a situation with lower data rates, the difference between the two approaches is likely to reduce.

9.4.4 Summary

The experimental results of the SPATIO prototype regarding accuracy and false positive rate allow us to reach a few conclusions. First, the choice of algorithm is not as important as the features selected to represent the data. We can observe that although not much changes in accuracy between algorithms, selecting different amounts of features lead to a big change. When developing an IDS, one must consider the trade-off between having many features, leading to a higher network and computing cost, and having fewer features, leading to a lower accuracy. We can also observe how proper feature selection impacts the results. In our prototype, feature selection was done based on data from Period 1 alone. This leads to a situation where, when using a lower amount of features, the system performs better on Period 1 than on Period 2.

We also observed the advantages of the fog computing approach taken in SPATIO. We were able to reduce network bandwidth usage by 90% by preprocessing the data closer to the data source. We also reduced load on the cloud server, leading to a lower alarm latency. This means that alarms are delivered faster, which can bring benefits when responding to attacks.

9.5 CONCLUSION

In this chapter, we analyzed the IoT as a growing market and identified the challenges it brings. We contribute with a solution for better security in IoT, as several reports show it to be a growing concern. Specifically, we address how to design an IDS for the IoT, having to deal with the IoT device's constrained processing and storage resources, the high amount of traffic in an IoT network, and the timely response time needed when dealing with potential security alerts.

Our presented solution is called SPATIO, a system to detect intrusions in the IoT. SPATIO finds intrusions in a network by monitoring traffic of the IoT devices, and feeding it to a machine learning model running outlier detection algorithms, which are a good fit to this problem as attacks are outliers by nature. To deal with the high amount of traffic in the network, while providing timely alerts, we adopted a streaming solution, which processes the incoming data flows in a continuous manner. This means we do not need to wait for a window of processing, as would be the case in batch processing, providing a faster generation of alerts. This streaming approach also has the advantage of not requiring the storage of data, which would not be practical in the IoT devices due to storage constraints, nor in the cloud, due to the high amount of devices generating high amounts of traffic.

We also take a fog computing approach in SPATIO. Feature extraction, that is, generating useful metrics based on flows of data, is done on devices closer to the edge of the network. Using the IoT devices themselves, that is, edge computing would not be practical due to the constrained processing resources they possess. Instead, by using gateways such as routers and base stations, we reduce workload on the cloud (central servers), increasing scalability, while also reducing overall network traffic, since only the transformed flow features are sent, instead of the entire raw traffic. Additionally, fog computing allows for better data privacy control, as data are processed closer to the owner of the data. This allows the removal of sensitive data and any personally identifiable information before it is sent to the cloud.

9.5.1 Achievements

We developed a prototype of SPATIO and used it to run tests on a public IoT network intrusion dataset. In the best case, we manage to detect 79.63% of the attacks, while producing on average 614 false alarms per day, over an 11-day period. While this amount of false alarms is somewhat high, we believe it can be much further reduced and discuss a possible way in Section 9.5.2.

In our evaluation, we observed a trade-off between having a higher amount of features, resulting in a better accuracy, and a lower amount of features, reducing network traffic and requiring less computational power.

We also observed the advantages of taking a fog computing approach. SPATIO reduces traffic in network by 90% by sending flow features to the cloud instead of the entire raw traffic. It also shows a clear advantage in timely processing of the information when offloading work to the fog devices, especially in a period of high constant throughput, where a backlog of data starts forming.

9.5.2 Future Work

While evaluating our prototype, it became clear it produced too many false alerts. This can lead to alert fatigue, a situation in which so many false alerts are being generated that the human operator becomes desensitized to them, leading to a longer response time to real attacks, or even missing them completely. To deal with this issue, a classification algorithm can be added to the SPATIO machine learning pipeline, after the outlier detection. The idea would be to further classify alerts into malicious and nonmalicious. While this would require initial human effort, the system would learn from each false alarm and gradually produce fewer and fewer false alarms.

Something that is difficult to our work when evaluating SPATIO was the lack of a good IoT dataset containing traffic from a high number of devices in a real scenario. Although the lack of such public datasets might be understandable, by the fact that real traffic captures might contain private data and therefore not many entities currently running IoT networks would be willing to part with their captures. Nonetheless, SPATIO would benefit from a dataset containing more throughput and devices, mirroring a real life scenario.

Finally, as computational resources become cheaper, more work can be offloaded to fog devices. An interesting path for future work is to further research what can be offloaded to these devices. A possible approach would be using federated learning, as recently proposed by Google (Bonawitz et al. 2019), where the fog devices would download the machine learning model from the cloud, update it with local data, and send back the update to the cloud, allowing distributed learning from the collected data, without the data ever leaving the devices.

REFERENCES

Al-Fuqaha, A., Guizani, M., Mohammadi, M., Aledhari, M., & Ayyash, M. (2015). Internet of Things: a survey on enabling technologies, protocols, and applications. *IEEE Communications Surveys & Tutorials*, 17(4), 2347–2376.

Auskalnis, J., Paulauskas, N., & Baskys, A. (2018). Application of local outlier factor algorithm to detect anomalies in computer network. *Elektronika ir Elektrotechnika*, 24, 96–99.

Barik, R. K., Priyadarshini, R., Dubey, H., Kumar, V., & Mankodiya, K. (2018). FogLearn: leveraging fog-based machine learning for smart system big data analytics. *International Journal of Fog Computing*, 1, 15–34.

Bezerra, V. H., Costa, V. G. T., Barbon, S., et al. (2018). Providing IoT hostbased datasets for intrusion detection research. In *Proceedings of Simpósio Brasileiro em Segurança da Informação e de Sistemas Computacionais (SBSeg).*

Bonawitz, K., Eichner, H., Grieskamp, W., et al. (2019). Towards federated learning at scale: system design. arXiv.

Chawathe, S. 2018. Monitoring IoT networks for botnet activity. In *Proceedings of the 17th IEEE International Symposium on Network Computing and Applications.*

Dastjerdi, A. V., Gupta, H., Calheiros, R. N., Ghosh, S. K., & Buyya, R. (2016). Fog computing: principles, architectures, and applications. In *Internet of Things: Principles and Paradigms*, Chapter 4. Morgan Kaufmann.

Gonçalves, D., Bota, J., & Correia, M. (2015). Big data analytics for detecting host misbehavior in large logs. In *Proceedings of the 14th IEEE International Conference on Trust, Security and Privacy in Computing and Communications (TrustCom).*

Hamza, A., Gharakheili, H. H., Benson, T., & Sivaraman, V. (2019). Detecting volumetric attacks on IoT devices via SDN-based monitoring of MUD activity. In *Proceedings of the 2019 ACM Symposium on SDN Research (ACM SOSR).*

IoT Analytics. (2018). State of the IoT & short term outlook. https://iot-analytics.com/product/state-of-the-iot-2018/ (accessed May 10, 2020).

Kadam, R., Mahamuni, P., & Parikh, Y. (2015). Smart home system. *International Journal of Innovative Research in Advanced Engineering (IJRAE)*, 2(1), 126–130.

Koroniotis, N., Moustafa, N., Sitnikova, E., et al. (2018). Towards the development of realistic botnet dataset in the Internet of Things for network forensic analytics: Bot-IoT dataset. *Future Generation Computer Systems*, 100, 779–796.

Mahmud, R., Kotagiri, R, & Buyya, R. (2017). Fog computing: a taxonomy, survey and future directions. In *Internet of Everything: Algorithms, Methodologies, Technologies and Perspectives*, eds. DiMartino, B., Li, K., Yang, L. T., & Esposito, A. Springer, Singapore.

Meidan, Y., Bohadana, M., Mathov, Y., et al. (2018). N-BaIoT: network-based detection of IoT botnet attacks using deep autoencoders. *IEEE Pervasive Computing*, 17(3), 12–22.

Moustafa, N., & Slay, J. (2015). UNSW-NB15: a comprehensive data set for network intrusion detection systems. In *Proceedings of Military Communications and Information Systems Conference (MilCIS).*

Mukkamala, S., Sung, A., & Abraham, A. (2005). Cyber security challenges: designing efficient intrusion detection systems and antivirus tools.

NetScout. (2018). Dawn of the Terrorbit era. https://www.netscout.com/blog/dawn-terrorbit-era (accessed May 10, 2020).

Ponemon Institute. (2018). Second annual study on the Internet of Things (IoT): a new era of third-party risk. https://sharedassessments.org/wp-content/uploads/2018/04/2018-IoTThirdPartyRiskReport-Final-04APR18.pdf (accessed May 10, 2020).

Qiuping, W., Shunbing, Z., & Chunquan, D. (2011). Study on key technologies of Internet of Things perceiving mine. *Procedia Engineering*, 26, 2326–2333.

Sasse, A. 2015. Scaring and bullying people into security won't work. *IEEE Security & Privacy*, 13 (3): 80–83.

Shalev-Shwartz, S., & Ben-David, S. (2014). *Understanding Machine Learning: From Theory to Algorithms*. Cambridge University Press New York, NY.

Sharafaldin, I., Lashkari, A. H., & Ghorbani, A. A. (2018). Toward generating a new intrusion detection dataset and intrusion traffic characterization. In *Proceedings of the 4th International Conference on Information Systems Security and Privacy (ICISSP)*.

Sivanathan, A., Gharakheili, H., H., Loi, F., et al. (2018). Classifying IoT devices in smart environments using network traffic characteristics. In *Proceedings of IEEE Transactions on Mobile Computing*.

Stratosphere Lab, CTU University of Prague, Czech Republic. (2019). CTU HoneyPot and Malware Capture datasets. https://www.stratosphereips.org/datasets-iot (accessed May 10, 2020).

Trappe, W., Howard, R., & Moore, R. S. (2015). Low-energy security: limits and opportunities in the Internet of Things. *IEEE Security & Privacy*, 13(1): 14–21.

Verleysen, M. & François, D. (2005). The curse of dimensionality in data mining and time series prediction. In *Proceedings of the 8th International Conference on Artificial Neural Networks: Computational Intelligence and Bioinspired Systems (IWANN)*.

Zarpelão, B. B., Miani, R. S., Kawakani, C. T., & de Alvarenga, S. C. (2017). A survey of intrusion detection in Internet of Things. *Journal of Network and Computer Applications*, 84, 25–37.

CHAPTER 10

Low Power Physical Layer Security Solutions for IoT Devices

Chithraja Rajan, Dheeraj Sharma,
Dip Prakash Samajdar and Jyoti Patel

CONTENTS

10.1 INTRODUCTION

The global connectivity of the mobile devices over a network called Internet of Things (IoT) was an attractive breakthrough for the general public as well as the industrialists since its evaluation in the 1990s [1–3]. This picture becomes more fascinating by the enormous growth of smart devices (e.g., phones, wearables, kitchen wares, medical equipment, automobiles, etc.), which provoked on the go demand and provided the feasibility for the realization of consumer web services through the distributed computation of data and information over a "cloudy" cyber-space [2]. Nevertheless, while enjoying this charm of communication freedom every other person is compromising with the need of data privacy and protection. So, here comes the issues and challenges [3] associated with the security concerns in IoT generally, confined in three paradigms, namely, confidentiality, authentication, and integrity of device, data, and domain. When we talk about security in IoT, it must be noted here that unlike a PC or general purpose gadget the IoT devices are resource constrained and dedicated in nature [2]. Hence, the straightforward acceptance of cyber defense solutions in IoT platform is not as easy as it is predicted and defiantly, the nasty cyber invaders might not step out to show their brutality to these naïve devices through several invasive and non-invasive attacks [1]. Also, the software security solutions are based on the blind assumption that the physical layer is restricted of any kind of security breaches. However, in actual scenario the hardware module can either be a real culprit or a life savior. Fortunately, a number of hardware protection schemes are in trend, which promise to provide high level of cyber space security insurance to IoT by providing safe locker, fingerprinting and camouflaging techniques [4]. In this context, it is important to highlight here that the traditional cryptographic algorithms implemented over hardware are impractical in IoT due to its inadequate computational and secure key storage capacity. Therefore, it is high time to introduce alternative ultralow power, fast, and small self-defensive IoT modules. In this contest, Physical Unclonable Functions (PUFs) [5–8] perfectly match the concept of reproducing secret keys as and when required without storing in any physical means. Alternatively, PUFs are coined as IP (Intellectual Property) fingerprint [8] as each IP is born with time and space independent uniqueness, which are random in nature and can neither be predicted nor be duplicated The coming subsections discuss the various security primitives, constraints that the IoT facing and the importance of physical layer security in IoT perspective.

10.2 SECURITY PRIMITIVES

The issues and challenges associated with the security concerns in IoT are generally, confined in three paradigms namely, confidentiality, authentication, and integrity of device, data, and domain [9].

10.2.1 Confidentiality

International Standard Organization (ISO) has defined the term confidentiality as, "the sophisticated information must be confidential and accessible to only those who are authorized to access". For example, the login details like username and password provided by the bank need to be confidential and not to be shared with anybody.

10.2.2 Authentication

Authentication is a bidirectional communication procedure where both the customer and service provider need to be brought under the confidentiality for smooth exchange of critical information. For example, while doing a digital transaction on an online shopping portal, it is necessary that you confirm that you are dealing with an authentic pay channel of your bank, as well as the bank needs your proof of identity for which a One Time Password (OTP) has been generated and sent to your registered mobile number or e-mail id. The high level authentication can even check the biometric details of the person.

10.2.3 Integrity

Data integrity is another pillar of hardware security, which means that data should not be manipulated while processing and should remain original.

10.3 IoT CONSTRAINTS

When considering security paradigms for billions of heterogeneous devices connected in IoT network, there are many factors [10–12] that need to be recalled and become the reason for the security vulnerability sources. The security concerns related to IoT devices are not separate from rest of the electronic industry because from design to final IC packaging, devices are passed through a number of parties, which are dedicated and involved to bring the best commodity to market. Broadly, they are design owner, system integrator, tester, IP provider, and finally manufacturer [10]. However, not everybody play fairly and become "unfair competitors" willing to earn extra pie through unwanted means. Apart, from these communities there are "brokers" who buy and sell the electronic parts to integrators and users directly which promote counterfeit. The honest conduction of electronic industry is insured through "Trustworthy party" who has no direct involvement in production but certifies delivery of legitimate products in market. Lastly, the "waste collector" whose role is electronic waste recycling but play major part in counterfeiting. Apart from these traditional reasons of security breaches, several other shortcomings prevail in IoT, which need to be addressed to maintain ethical environment. The shortage of energy supply to IoT devices is one of the disadvantages because the prime application includes operation at remote locations where they lack availability of constant power source [11]. Second limitation is the smaller area constraint of IoT devices, which privileges portability and concinnity. Then comes, heterogeneity as devices with different nature and constraints can communicate with each other in order to provide better and wider throughput [12]. Consequently, these devices become less intelligent in terms of security as their computational capability is victimized by the IoT resource constraints.

10.4 PHYSICAL SECURITY

Hardware security [13] is otherwise known as physical layer security as the security is provided by and to the hardware rather than the software, which is installed on the hardware in the application layer. This initial definition of hardware security has itself drawn a broad line between hardware and software security. The traditional hardware security modules (HSMs) [14] are dedicated crypt processors that generate keys, which later are encrypted or decrypted for authentication purposes whereas software share the system resources and hence, they slow down the overall performance while the crypt processors boots up as soon as system is turned on and operate independently. Hardware security can withstand cold boot attacks [15], but if there is only software security then hackers can easily recover the data deleted even from the memory. In spite of these advantages, HSMs are preferred only in applications where cost is least significant factor, which indicates that their direct adoption is not feasible in the miserly atmosphere of IoT where the cryptographic keys have to be kept in custody of nonvolatile memory (NVM) [16]. Also, the bit stream of digital keys can be easily recovered from the NVM even after their erasure as illustrated by FBI with the crack of iPhone using a "secure golden key". But even this does not favor cost-effective software-based security solutions for IoT nodes because they can be cheaply reverse engineered and scripts can be easily decoded by any professional programmer. So, the mystery remains weather there exists possible security technique for such adverse and incredible IoT. Of course, the answer is "yes". As we proceed in this chapter we disclose how the physical disorders, which are usually unfortunate in chips arrived during manufacturing become source of distinct security proposal for delicate IoT devices. Before exploring aforementioned security solution, a deep understanding of various security threats and available solutions is required, which are explained in next sections.

10.5 SECURITY THREATS

Generally, a HSM is like a black box that generates, stores, and provides the key whenever required [17]. So from single transistor to entire chip a hacker can make an entry and can steal, manipulate, or even destroy the information. Broadly, attacks are classified as invasive, semi-invasive, and noninvasive [18]. During invasive attacks [19] the IP structure can be easily mapped by taking high resolution photographs of chip surface using optical microscope with Charge Couple Device (CCD) camera after removing layer by layer. On the other hand, during noninvasive attacks [20], there is no sign of physical harm to chip but the power and clock signals are varied which provoke system to operate different instructions and hence, generate entirely misleading outputs. Semi-invasive attacks [21] are close to invasive attacks but they do not require depassivation as no probing is targeted but information can be gathered using electromagnetic field ionization methods like UV and X-rays. Therefore, based on the kind of attacks, five basic threats are identified and explained in details in the following text.

10.5.1 Reverse Engineering

Reverse engineering [22] is most common threat to investigate device working, which can be performed by either invasive or noninvasive way. As mentioned earlier, during invasive attack the reverse engineering is performed by taking layer by layer photographs of the chip surface, which are then pattern recognized in order obtain the final picture of the RTL. This way of reverse engineering may even damage the chip. On the other hand, the noninvasive way of attack simply discovers the chip behavior through finite state machine and truth tables by subjecting it to most possible cases of tests.

10.5.2 Counterfeiting

Counterfeiting [23] is the most serious issue of the electronics industry. Counterfeiting can be done in any of the three ways namely: relabeling, refurbishing, and repackaging. As the name indicates, the relabeling is simply erasing company identifiers of the original one and replacing it with another one which they publish thereafter as product of same make or higher version. This act of hypocrisy leads to customer dissatisfaction and flow of corrupted product in the market. Refurbishing is the act of selling esthetically improved but discarded electronic parts, which hold no value. Repackaging means original IC has been repackaged in order to modify the order of pins. This act of removing original package alters the circuit resulting in improper working condition. Definitely, all these misconducts consequences to heavy loss to business.

10.5.3 Theft and Cloning

It happens when foundry persons do overbuilding of the products or clone the design data without designers' permission, which later can be used to manufacture the products or can be sold illegally [24]. Similarly, the manufacturing resources like masks can be stolen and reused to create duplicate copies of device. Of course, any such unrightful behaviors can neither be justified nor be tolerated in any domain of life.

10.5.4 Hardware Trojans

Hardware Torjan (HT) [25] is hot topic of discussion for electronic security engineers as HTs are miniature hardware units, which are very hard to detect but can easily leak the cypto keys and send it to outside sources without affecting system performance.

10.6 HARDWARE SECURITY SOLUTIONS

Threats and reasons discussed in previous sections give a clear overview that how difficult and complex is the situation in IoT security. Nevertheless, there are many existing hardware-based protection schemes, which have to be familiarized that can be or cannot be a life savior at certain times. Therefore, in this section various protection schemes are discussed mentioning their advantages as well drawbacks.

10.6.1 Hardware Obfuscation

Obfuscation is the term used by the software programmers where a program is transformed to another program difficult to reverse engineer as it won't produce the same response to the test as produced by the original program [26]. Same technique can be deployed in hardware parts through locking/unlocking schemes using encrypted keys. Locking has been done immediately after IC built and can be unlocked by the trustworthy party. Example of secure digital circuits include additional blocking states are inserted in FSM which remain in that state unless a sequence of specified key is applied at input. Hence, this prevents IC or IP stealing and cloning.

10.6.2 Hardware Camouflaging

Camouflaging [27] prevents reverse engineering based on optical means of image processing. The technique is to hide the system details at layout level designing using nonstandard library contents, which make it impossible to distinguish the individual transistor functionality by optical means. Split manufacturing scheme seems to be the future hardware camouflaging technique where designer needs a trustworthy local manufacturer for final packaging of the IC.

10.6.3 Reconfigurable Hardware

Reconfigurable hardwares [28] such as Field Programmable Gate Arrays (FPGA) can be inserted at critical area of chip like a bus or instruction decoder, which can change its nature timely and can stop the non-invasive reverse engineering attacks. Only the authentic person with correct keys can force the reconfigurable part to be functional. However, this is an expensive technique, which increases the area along with that keys need to be safe through bitstream encryption.

10.6.4 Hardware Identification

An identification ID can be either stored in NVM or can be generated from PUF for each IC [29]. Watermarking [30] and fingerprinting [31] are two types of hardware identification technique. Watermark can be applied to each IC pre-synthetically, in-synthetically or post-synthetically to check its authenticity. Pre-synthetic watermarking is suitable for DSPs only as they are algorithm dependent and hence, of high cost. Comparatively, in-synthetic watermarking insertion of digital signature through automatic tool so is fast implementation but is not favorable for area, power, and timing. Finally, post-synthetic watermarking are device dependent handmade technique and hence, time consuming. Therefore, none of the watermarking methods is compatible for IoT devices.

Fingerprinting is to find refurbished devices detected through difference in delay between two chips on yearly basis. This does not require additional circuitry and is easy to find old ICs but is difficult to implement in IPs because they are manufactured on different platforms with different time and area constrains. Hash functions are the fingerprints used for IP as they can differentiate the original IPs from fake one as they return correlated results from similar IPs irrespective of space and time variations. Apart from these, PUF is an evolving fingerprinting technique, which generate challenge response pairs (CRP) based on the manufacturing mismatches prevailing due to nonideal fabrication process. Some of the basic known PUFs are delay and memory-based PUFs, which are explained in details in next sections.

10.6.5 IP Licensing

Inspired from the software licensing technique the user has to pay only for the copies of use and cryptography, which facilitates the right user and owner to exchange the signatures between each other with the help of a trustworthy middleman. Otherwise, IP licensing [32] promote another way of business by which the companies can offer the demo version of the IP to the customer for prepurchase testing. The performance of such IP is low, but it attracts customer satisfaction before final license purchase.

10.7 PHYSICAL UNCLONABLE FUNCTIONS (PUFs)

In the previous sections, various attacks, threats, and protection schemes have been demonstrated well. But when we come to IoT devices the picture becomes complicated because these miniature resource constraint equipment may not be able to tolerate the extra burden of these power and area thirsty security mediums, which not only slows down system performance but also are incapable of providing 100% security. Alternatively, PUF [33] is the only solution, which is capable to "protect", "limit", and "detect" the threats altogether. PUF which is not only capable to detect the threat but also stop it or reduce the attack in most effective way so that the attacker has to think twice before making next attempt. Therefore, the concept of PUF, which has been discussed in many literatures is elaborated here in terms of IoT security.

It was in year 2002, the term PUF was first coined, which like a mathematical function generates the measurable output when queried with particular input [34]. Here, the unique keys related to the PUF are collected as the challenge-response pairs where challenge means the input which when applied to the PUF will generate unique output called as *response*. PUF extracts the unique fingerprint of the IoT device that has been acquired by them during manufacturing time due to process variations. These process variations [35], random and irreproducible in nature, are intrinsic to the device properties, which can neither be detected nor be hypothetically assumed by the mathematical models. Even the right manufacturer with the original layout and mask cannot replicate the same disorder in the system as it is an uncontrollable phenomenon of the fabrication process. The designers and the manufacturer are even though targeted to reduce these imperfections so that devices manufactured through similar procedures are same in nature. Still their characteristics slightly fluctuate from each other and become source of uniqueness, which favored PUF. The next subsections deal with the elaborated description regarding sources of process variations, types of PUF, and their figure-of-merits.

10.7.1 Sources of Variations

The heart line of the PUF depends only on the percentage of variations IC possesses. These variations deviate the device and circuit designed values, which become source of PUF characteristics. Therefore, this section reveals variability sources and their impact on device properties. Broadly, variations can be classified as environmental and manufacturing or process variations [36].

10.7.1.1 Environmental Variations

As the name indicates that environmental variations show up whenever there is fluctuation in operating conditions such as temperature, power supply, noise, aging, radiations, etc. [37]. They are unpredictable and uncontrolled in nature. Since, they depend on operating medium; these variations affect PUF robustness and are harmful for system integrity. Therefore, for reliable security applications the side effects of these variations should be negligible.

10.7.1.2 Process Variations

These variations are result of manufacturing or fabrication imperfections. The imperfect masks, processes, and wear outs develop permanent back end of line (BOEL) and front end of line (FOEL) variations [38, 39]. The transistor level variations such as dimensional variations, doping related physical fluctuations are major source of FEOL. On the other hand, BEOL sources include variations in metal and dielectric interconnects results in thickness and space variations, which change the path resistivity and hence, delay and power fluctuations. These variations can be either inter die or intra die variations. The identical dies on same or different wafers or on separate lots can be different due to fabrication process dependencies and they are often less severe as any abnormalities existing as compared to the nominal parameters can be compensated by adapting proper biasing and dynamic voltage sources. However, the intra die variations exist within the die are design and fabrication dependent and no counteract techniques are available. Intra die variations can be either systematic or random in nature. Systematic variations are in general deterministic and can be modelled as they have been sourced by chemical mechanical polishing (CMP) and lithography, which depend on layout density and dimensions. Random variations are unpredictable caused by random doping fluctuations (RDF) [40]. These variations directly affect the timing performance and hence, leverage PUF performance in security terms. The variations can be modeled either as Gaussian or non-Gaussian distribution. Most process variations are assumed to be Gaussian but some are particularly non-Gaussian like via resistance variations. In simulation set up Monte Carlo analysis is in the close proximity to experimental level variations to compute their after effects on system performance.

10.7.2 PUF Classification and Examples

PUF can be classified based on the construction and the working principles. In real-world scenario, there are many nonelectronic PUFs that are existing from many years and had been utilized for security purposes. In this subsection, we give an idea about various PUFs for clear understandability of dealing with them in various applications.

10.7.2.1 Nonelectronic PUFs

The absence of any electronic component for PUF construction but later processing and measurements are done electronically constitute the class of nonelectronic PUFs [41]. It includes optical, paper, CD, and RF-DNA PUFs. Optical PUFs are evolved well before electronic PUF and are used in World War-II to identify weapons. They produce unique pattern on reflection by light. The orientation of light incidence on optical PUF is the challenge and the resultant pattern is the response. Paper PUFs are the concept incorporated in currency note printing to avoid counterfeiting by linking the digital signatures to that the unique UV fibers are introduced explicitly during manufacturing. CD PUFs are often used to protect compact disk piracy through scanning the lands and pits length in CD, which varies during manufacturing. RF-DNA PUFs are just like optical PUFs; instead of light, a copper wire is embedded in a silicon shield for EM field scattering and RF waves are observed for security purposes. Magnetic PUFs are often used in credit cards, where the unique particle patterns have been scanned and used for identification and authentication purposes.

10.7.2.2 Electronic PUFs

Inspired from the traditional way of security, the PUF had found its place in electronic world as well. It includes analog PUFs, delay-based and memory-based PUFs. These are explained in detail one by one in the following sections.

10.7.2.3 Analog Electronic PUF

The analog behavior of the PUFs that can be measured electronically is analog electronic PUFs. It comprises V_T, power distribution, coating, and LC PUFs. In the V_T PUFs, a large number of similar transistors are connected to an addressable array and whenever any one of them is addressed using a challenge, it will drive a load and voltage across the load, which is digitized using a comparator. Due to fabrication imperfections the voltage on the load varies every time. Power distribution PUFs [42] are proposed here due to manufacture variability in which the equivalent resistance across the power grid varies that can be measured and treated as CRP. Instead of depending on the natural manufacture variability, in coating PUF [43] dielectric coating is spread over the comb-shaped sensors whose capacitances are measured. A glass plate embedded between two metal plates form a capacitor that are chained using a metal wire act as a inductor forms the basic structure of a LC PUF. LC PUF is passive in nature which when placed in an RF field results in occurrence of resonance frequency sweep based on the inductance and capacitance value, which varies with the manufacture variability.

10.7.2.4 Delay-Based PUFs

They are intrinsic PUFs which stay inside the chip integrally and can measure their own value. Also, they can be constructed with the same steps of chip manufacturing and no overhead is needed. Here, the delay of the signal path is the measurable quantity, which varies because of manufacture variability. Again, there are two popular delay-based PUFs, arbiter [44] and ring oscillator (RO) PUFs [45]. The arbiter PUF works on the principle of digital race between two symmetrically designed lines, which might have different delays due to manufacture variability. The switch blocks or flip-flops have been used to construct delay lines. The challenge over these blocks will either pass the signal straight or switch the path. The CRP so generated will be exponential to the number of switch blocks. In case of RO PUFs, a number of RO stages have been constructed, which due to manufacture variations oscillate with slightly different frequencies and when selected and compared, the bits have been generated. The resultant CRPs are stored in CRT.

10.7.2.5 Memory-Based PUFs

Another kind of intrinsic PUFs are based on memory elements, which has tendency to be in one of its stable states. But, due to manufacture mismatch it is unpredictable to find out which one will be the initial stable state of the memory when switched on and this randomness becomes blessing for PUF purposes. SRAM, butterfly, flip-flops, and latches are the memory-based PUFs.

10.7.2.6 SRAM PUFs

Static random access memory (SRAM) consists of two inverters connected back to back has four transistors plus two transistors for reading and writing purpose. Due to manufacturing variability, it is unknown that which SRAM cell takes 1 or which cell takes 0 at power up time. As the numbers of SRAM cells increase, the unpredictability also enhances which motivates PUF behavior [46].

10.7.2.7 Butterfly PUF

These are two cross-coupled data latches on FPGA with two stable states just like SRAM but need not power up every time as the preset/clear functions do the same for the PUF. Again, the preferred stable state of the butterfly cell is determined by the manufacturing mismatch [47].

10.7.2.8 Latch PUF

Similar to SRAM and butterfly PUF, two NOR gates are cross-coupled to form a NOR latch, which when reset will occupy one of the stable states based on the component mismatch in the internal circuit [48].

10.7.2.9 Flip-Flop PUF

Flip-flops are cross-coupled in similar fashion like that of SRAM for implementation on FPGA or as ASIC [49].

10.7.3 PUF Performance Metrics

After elaborating overview of various PUFs and their concepts in previous sections, this section discusses the performance metrics of the PUF, which decide the quality and their range of applications.

The quality of the PUF can be evaluated by following metrics:

Uniqueness: It determines the similarity factor among various PUFs and is calculated using inter-hamming distance (HD). In order to differentiate between various PUFs, it is important to maintain a HD of at least 50%. The uniqueness can be obtained from eq. (10.1)

$$Uniqueness = \frac{2}{j(j-1)} \sum_{i=1}^{j-1} \sum_{k=i+1}^{j} \frac{HD(R_i, R_k)}{n} \times 100\% \quad (10.1)$$

where Ri and Rk are n-bit response of chips i and k ($i \neq k$) and j is the number of chips.

$$HD(R_i, R_k) = \sum_{i=1}^{n} (r_{i,t} \oplus r_{k,t})$$

Reliability: It determines that the PUF response should remain consistent with the changing environmental conditions such as power supply, temperature, or aging. It is obtained using intra HD. The HD should be closed to 0%. It can be calculated from the formula in eq. (10.2)

$$Reliability = \frac{1}{s} \sum_{m=1}^{s} \frac{HD(R_i, R'_{i,m})}{n} \times 100\% \quad (10.2)$$

10.7.4 PUF Security Applications

System identification and authentication are the two applications of PUF.

10.7.4.1 Chip Identification and Authentication

Similar to biometric identification, PUF can be used to identify the chip, which avoids counterfeiting. During enrollment phase, a large set of CRPs are collected and stored at verifier database such that whenever the system identification is required it can just apply a challenge to the PUF. If the generated response is closely matched to the CRT table, then the legal system can be identified. In order to avoid repay attack, the used CRP has been discarded.

10.8 NANOTECHNOLOGY FOR IoT SECURITY

So far it is clear that among various classical ways of protection, PUF is the practical security solution for the IoT application because of the number of the limitations that restrict the direct adaptability of hardware security solution in the IoT platform. However, the PUF performance is also under attackers' observation and has been challenged through reliability issues, modeling attacks (MA), side channel attacks (SCAs), and cryptanalysis attack (CA) [50]. However, researchers have identified that a complicated behavioral pattern is tolerated against SCA and machine learning algorithms as much training and computational capacity are needed by the adversary. Here, comes the role of eminence possibilities of emerging semiconductor devices that had unfolded new opportunities for

PUF designs as novel devices have their own advantages, which make them most immune to the existing threats. Besides these facts that the ever scaling mature CMOS technology is facing leakage problems like short-channel effects (SCEs), subthreshold slope degradation, velocity saturation, and drain-induced barrier lowering (DIBL) [51]. Also, due to insufficient channel control the subthreshold swing (SS) of CMOS is restricted to 60 mV/decade, which increases the OFF current exponentially. Alternatively, devices like FinFET, Tunnel FET (TFET), memirstor, and negative capacitance FET (NCFET) are well-studied low-power transistors, which leave their separate footmark in security applications as well [52, 53]. In this chapter, we focus is on TFET-based PUF design as it is the next-generation low-power device, which has been better explained in the next subsection and its role in security domain is also discussed.

10.8.1 Tunnel FET: Opportunities and Challenges

According to experiments performed in ref. [52], there exists a clear trade-off in between performance improvement and variability control due to a dominant source of variation existing at the tunneling region, which attracts PUF designers. Additionally, TFET that works on the principle of band-to-band tunneling of carriers from source valence band (VB) to channel conduction band (CB) has unlimited SS, immune to SCEs, and least leakage current. However, TFET suffers from certain drawbacks like any emerging technologies such as low ON current and ambipolar conduction [53]. Therefore, the unidirectional current flow, ambipolarity, and effect of Miller capacitance have made TFET the best possible alternative for PUF realization. Further, the nanodimension devices are difficult to form as to obtain an absolute doping profile, fixing exact metal work function values and creating ultra-sharp geometric profile is a tough job, which enhances system mismatch and encourages PUF features. Therefore, in IoT perspective, TFET opens the ocean of opportunities for low power, compact hardware security solution as nano TFET PUF.

10.8.2 Case Study: Implementation of Security Module Using TFET for Low-Power IoT Applications

As we discussed in previous subsection that the silicon TFET has better SS and OFF current but suffers from low ON current and ambipolarity. Therefore, a III–V-based heterojunction TFET can be the best possible solution for better ON current and reduced ambipolarity. In this regard, we have made a comparative analysis among various III–V-based heterojunction nanowire TFET at source/channel and drain regions in a nanowire. Figure 10.1 shows

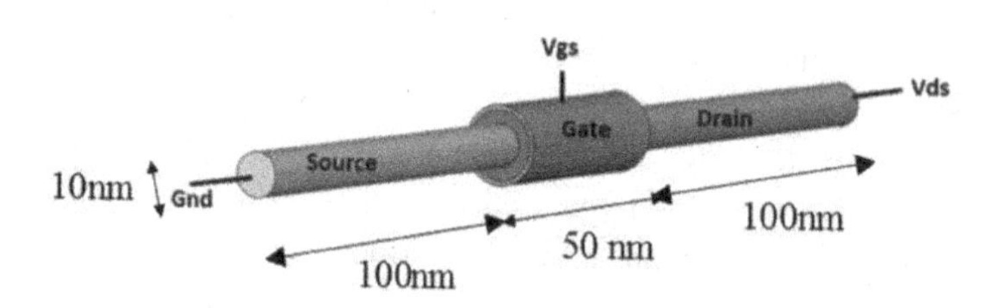

FIGURE 10.1 3-D structural view of a nanowire TFET.

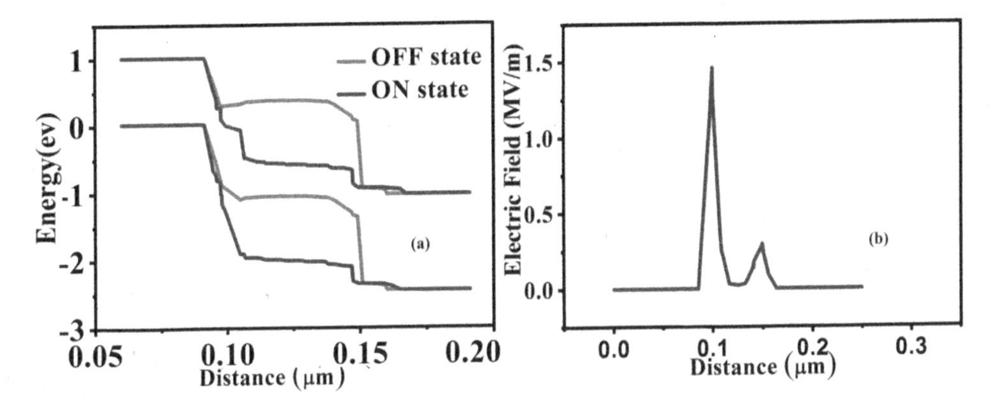

FIGURE 10.2 (a) Energy band diagram and (b) electric field of nanowire TFET [54].

the 3-D structural view of the proposed device. Here, a low band gap material at source/channel and a wide band gap material at the drain region shows better band bending at source/channel junction than at drain/channel junction, resulting in better ON current and low ambipolarity.

Various material combination tested here are Ge/GaAs, Si/GaAs, InAs/GaAs, InAs/Si, $In_{0.3}Ga_{0.7}As/Si$, silicon, $Si_{0.5}Ge_{0.5}/Si$, and InAs/GaSb. It has been found that Ge/GaAs provides the best device characteristics. Figure 10.2 shows the energy band diagram and the electric field (EF) characteristics of the Ge/GaAs TFET. There is enough band bending during ON condition, which results in higher EF at source/channel junction than at drain/channel junction. Figure 10.2 shows that the Ge/GaAs provides better drain current and transconductance (g_m) in comparison to other III–V materials. Figure 10.3(a) shows that these devices have parasitic capacitance in the range of "atto" Farad, which results in high gain bandwidth product (GBP) in "Tera" Hz range and maximum is provided by Ge/GaAs.

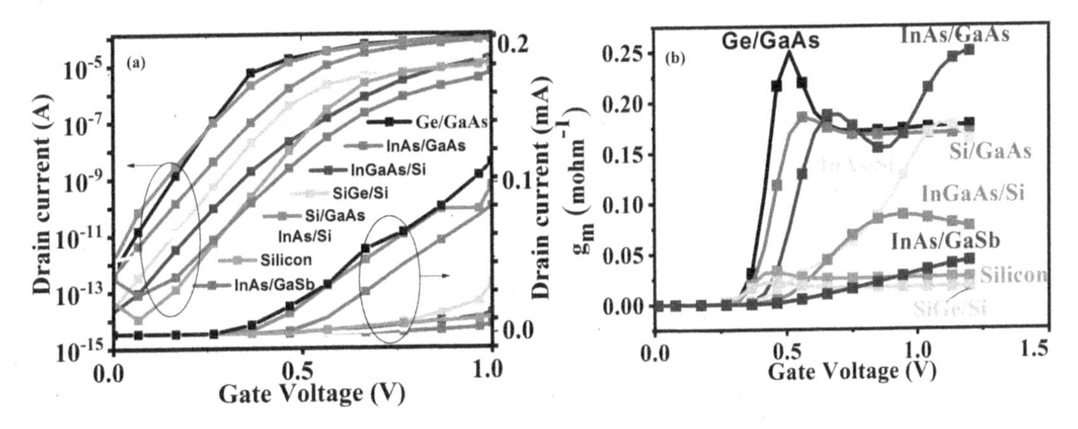

FIGURE 10.3 Heteromaterial nanowire TFET comparison of (a) drain current and (b) transconductance [54].

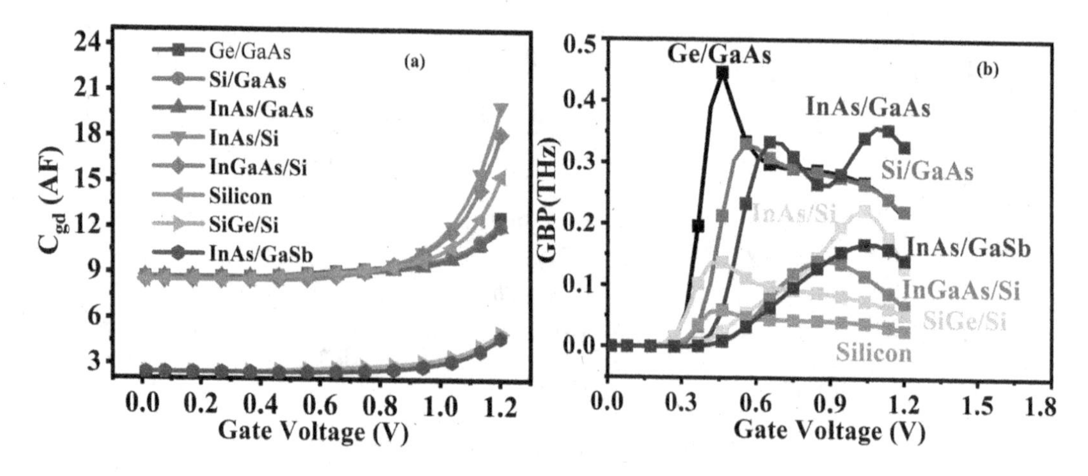

FIGURE 10.4 Heteromaterial nanowire TFET comparison of (a) parasitic capacitance and (b) gain bandwidth product [54].

The device has been implemented in Sentaurus tool, the device characteristics are then exported in look up tables, which are used in Cadence Virtuoso using Verilog-A code. In cadence, we have developed a 32-bit RO PUF, which has been implemented using 64 RO stages, as shown in Figure 10.4. The rest of the calculations are done in MATLAB®, which provided 48.9% inter HD (Figure 10.5) and 2.03% intra HD (Figure 10.6a) for temperature variations and 3.8% intra HD (Figure 6) for supply voltage variation. Finally, Figure 10.7 shows leakage power comparison between MOS and TFET. Both n and p TFETs yield power in "pico" Watt range, while MOS exhibits leakage power in the range of "nano" Watts, which is a significant difference when IoT

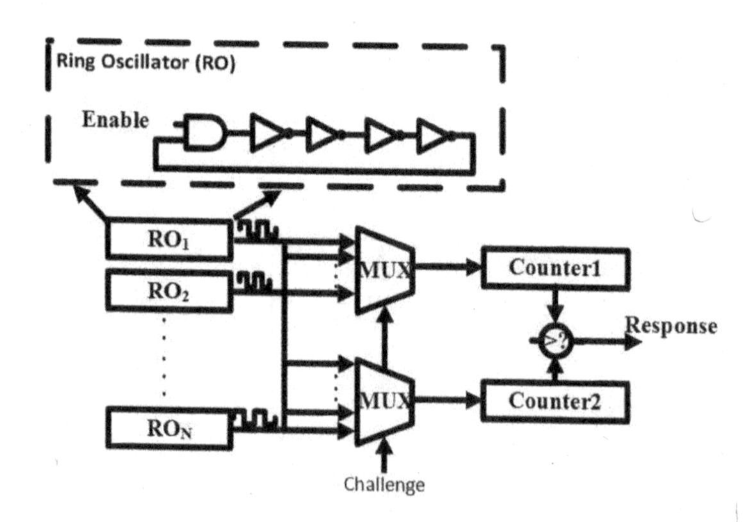

FIGURE 10.5 RO PUF architecture.

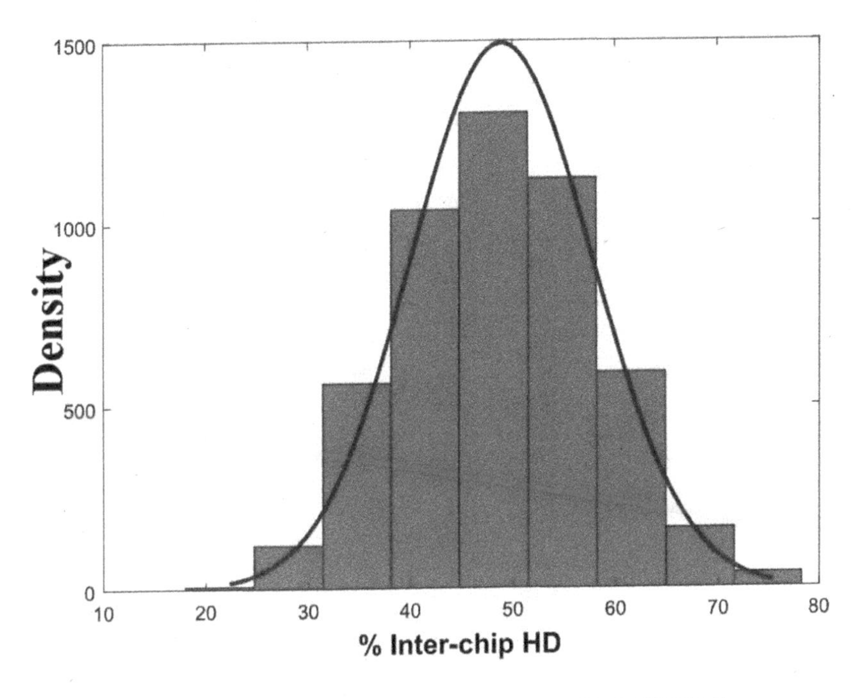

FIGURE 10.6 Inter hamming distance of nanowire TFET-based RO PUF [54].

is considered. Therefore, these statistics show that the proposed device can be a strong alternative in PUF for IoT applications.

10.9 CONCLUSION

In this chapter, the importance of hardware security in IoT devices is portrayed as they are the most power and area constraint modules, and the situation becomes more complex when invaders try to sneak into their heterogeneous atmosphere. Therefore, direct adaptation of conventional hardware security solutions such as cryptographic methods in IoT is impossible. Alternatively, PUFs had emerged as a strong IoT security solution as they generate secure CRPs through random manufacture variations and need not to be kept in separate memory in chip. This chapter summarizes various PUFs and their importance in electronics world. PUF applications are also emphasized and its performance metrics are also discussed in detail. But, the traditional PUFs are CMOS-based, which reached at scaling limit and great source of power loss due to SCEs and of limited SS of 60 mV/decade. Therefore, low power heterojunction nanowire structure of novel TFETs can replace the CMOS technology in existing PUF architectures. In this chapter, various III–V semiconductor-based heterojunction nanowire TFETs have been compared because Si TFET lacks sufficient ON current to drive the consecutive stages and suffers from higher ambipolarity. Therefore, III–V semiconductor-based heterojunctions at source/channel and drain regions provide

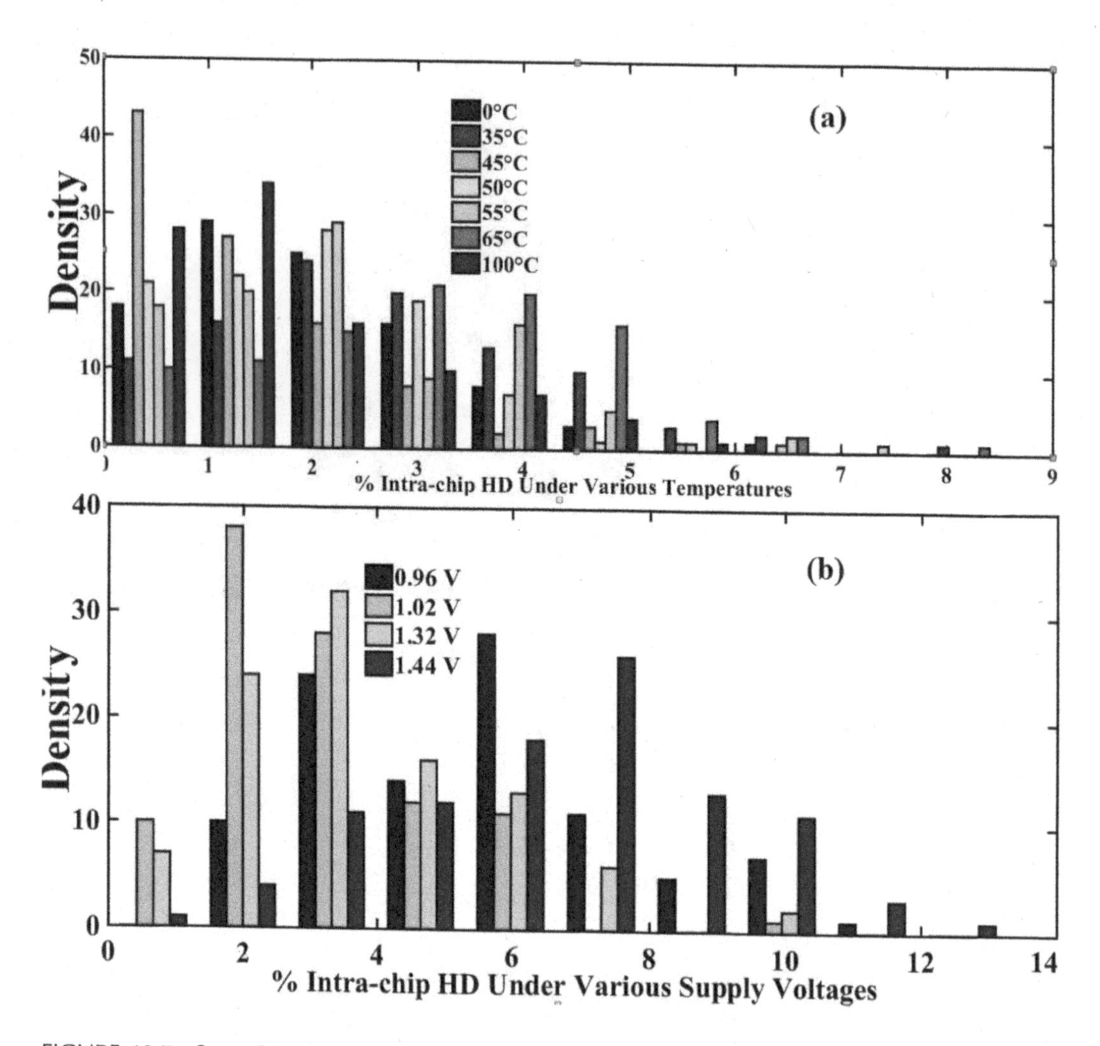

FIGURE 10.7 Intra Hamming Distance of Nanowire TFET-Based RO PUF for (a) Temperature Variations and (b) Voltage Variations [54].

better DC/AC characteristics using band gap engineering. We found that Ge/GaA-based nanowire TFET produced 1 mA current has 0.29 V threshold voltage and of 0.25 $m\Omega^{-1}$ transconductanc. Also, it sheds only pico Ampere leakage current during OFF condition and exhibits very small parasitic capacitance, which favors Ge/GaA-based nanowire TFET for ultra-low power security utilizations in IoT (Figure 10.8). Additionally, the novel device is novel because fabrication methods generate novel process variations and hence, strengthen PUF qualities. Therefore, in this chapter, the 10% Monte Carlo simulation of RO PUF investigated using Ge/GaA-based nanowire TFET exhibits 48.9% inter HD, while 3.792% and 2.03% intra HD for temperature and voltage variations, respectively.

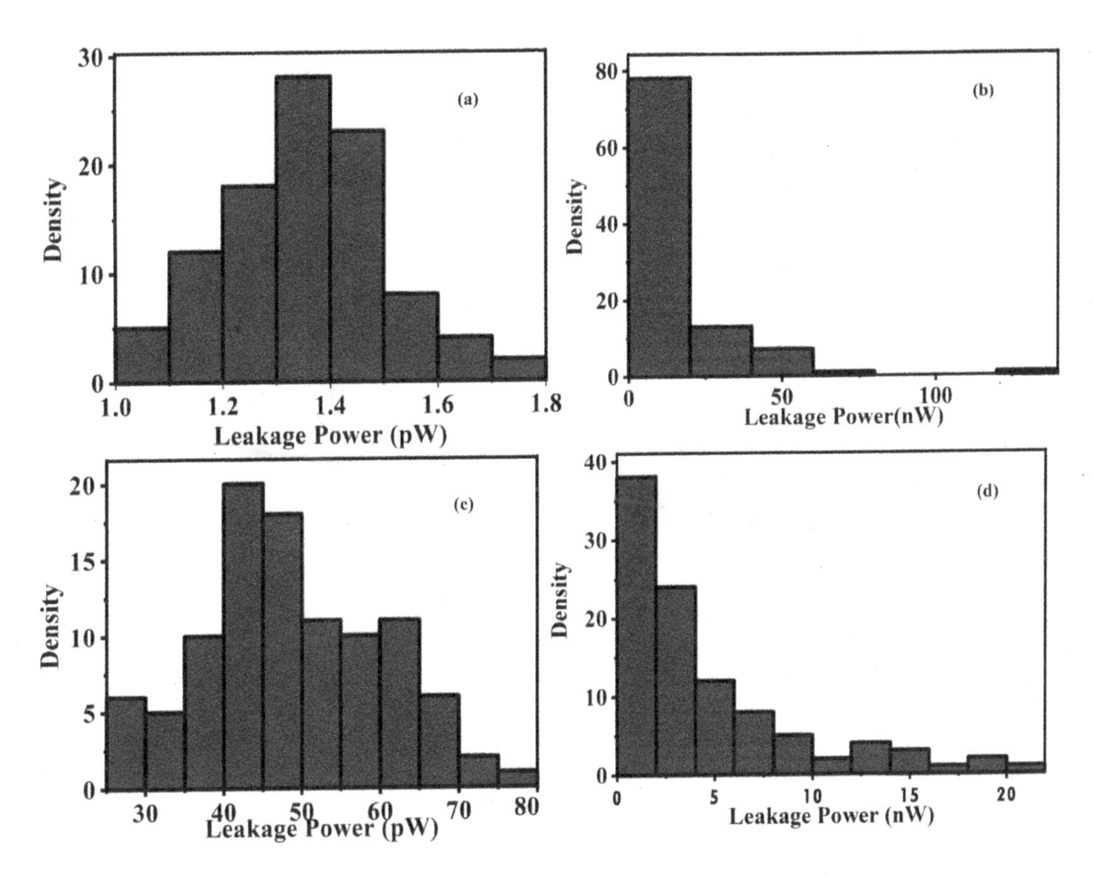

FIGURE 10.8 Caparison of leakage power for (a) n-TFET, (b) n-MOS, (c) p-TFET, and (d) p-MOS.

REFERENCES

1. Xu, Teng, James B. Wendt, and Miodrag Potkonjak. "Security of IoT systems: Design challenges and opportunities". In 2014 IEEE/ACM International Conference on Computer-Aided Design (ICCAD), IEEE, San Jose, CA, pp. 417–423, 2014.
2. Ammar, Mahmoud, Giovanni Russello, and Bruno Crispo. "Internet of Things: A survey on the security of IoT frameworks". *Journal of Information Security and Applications* 38 (2018): 8–27.
3. Aman, Muhammad N., Kee Chaing Chua, and Biplab Sikdar. "Position paper: Physical unclonable functions for IoT security". In Proceedings of the 2nd ACM international workshop on IoT privacy, trust, and security, pp. 10–13, 2016.
4. Papaspiliotopoulos, Vasileios A., George N. Korres, Vasilis A. Kleftakis, and Nikos D. Hatziargyriou. "Hardware-in-the-loop design and optimal setting of adaptive protection schemes for distribution systems with distributed generation". *IEEE Transactions on Power Delivery* 32, no. 1 (2015): 393–400.
5. Idriss, Tarek, Haytham Idriss, and Magdy Bayoumi. "A PUF-based paradigm for IoT security". In 2016 IEEE 3rd World Forum on Internet of Things (WF-IoT), IEEE, Reston, VA, pp. 700–705, 2016.
6. Halak, Basel, Mark Zwolinski, and M. Syafiq Mispan. "Overview of PUF-based hardware security solutions for the Internet of Things". In 2016 IEEE 59th International Midwest Symposium on Circuits and Systems (MWSCAS), IEEE, Abu Dhabi, UAE, pp. 1–4, 2016.

7. Wallrabenstein, John Ross. "Practical and secure IoT device authentication using physical unclonable functions". In 2016 IEEE 4th International Conference on Future Internet of Things and Cloud (FiCloud), Vienna, Austria, IEEE, pp. 99–106, 2016.
8. Aman, Muhammad N., Kee Chaing Chua, and Biplab Sikdar. "Position paper: Physical unclonable functions for IoT security". In Proceedings of the 2nd ACM international workshop on IoT privacy, trust, and security, pp. 10–13, 2016.
9. Cao, Yuan. "Design of security primitives for trustworthy integrated circuits". PhD diss., 2015.
10. Haroon, Asma, Munam Ali Shah, Yousra Asim, Wajeeha Naeem, Muhammad Kamran, and Qaisar Javaid. "Constraints in the IoT: The world in 2020 and beyond". *Constraints* 7, no. 11 (2016): 252–271.
11. Zhang, Zhi-Kai, Michael Cheng Yi Cho, Chia-Wei Wang, Chia-Wei Hsu, Chong-Kuan Chen, and Shiuhpyng Shieh. "IoT security: Ongoing challenges and research opportunities". In 2014 IEEE 7th International Conference on Service-Oriented Computing and Applications, IEEE, Matsue, Japan, pp. 230–234, 2014.
12. Mukherjee, Amitav. "Physical-layer security in the Internet of Things: Sensing and communication confidentiality under resource constraints". *Proceedings of the IEEE* 103, no. 10 (2015): 1747–1761.
13. Tehranipoor, Mohammad, and Cliff Wang, eds. *Introduction to Hardware Security and Trust*, Springer Science & Business Media Verlag New York, 2011.
14. Rostami, Masoud, Farinaz Koushanfar, Jeyavijayan Rajendran, and Ramesh Karri. "Hardware security: Threat models and metrics". In 2013 IEEE/ACM International Conference on Computer-Aided Design (ICCAD), San Jose, CA, IEEE, pp. 819–823, 2013.
15. Sklavos, N., K. Touliou, and C. Efstathiou. "Exploiting cryptographic architectures over hardware vs. software implementations: advantages and trade-offs". *Memory* 13 (2006): 18.
16. Skorobogatov, Sergei P. "Semi-invasive attacks - A new approach to hardware security analysis. University of cambridge." computer laboratory: Technical report (2005).
17. Chin, Wen-Long, Wan Li, and Hsiao-Hwa Chen. "Energy big data security threats in IoT-based smart grid communications". *IEEE Communications Magazine* 55, no. 10 (2017): 70–75.
18. Zhang, Zhi-Kai, Michael Cheng Yi Cho, and Shiuhpyng Shieh. "Emerging security threats and countermeasures in IoT". In Proceedings of the 10th ACM Symposium on Information, Computer and Communications Security, pp. 1–6. 2015.
19. Wan, Meilin, Zhangqing He, Shuang Han, Kui Dai, and Xuecheng Zou. "An invasive-attack-resistant PUF based on switched-capacitor circuit". *IEEE Transactions on Circuits and Systems I: Regular Papers* 62, no. 8 (2015): 2024–2034.
20. Šimka, Martin, and Park Komenského. "Active non-invasive attack on true random number generator". In 6th PhD Student Conference and Scientific and Technical Competition of Students of FEI TU Košice, Košice, Slovakia, pp. 129–130, 2006.
21. Merli, Dominik, Dieter Schuster, Frederic Stumpf, and Georg Sigl. "Semi-invasive EM attack on FPGA RO PUFs and countermeasures". In Proceedings of the Workshop on Embedded Systems Security, pp. 1–9. 2011.
22. Torrance, Randy, and Dick James. "The state-of-the-art in IC reverse engineering". In International Workshop on Cryptographic Hardware and Embedded Systems, Springer, Berlin, Heidelberg, pp. 363–381, 2009.
23. Devadas, Srinivas, Edward Suh, Sid Paral, Richard Sowell, Tom Ziola, and Vivek Khandelwal. "Design and implementation of PUF-based 'unclonable' RFID ICs for anti-counterfeiting and security applications". In 2008 IEEE International Conference on RFID, IEEE, Las Vegas, NV, pp. 58–64, 2008.
24. Skorobogatov, Sergei. "Physical attacks on tamper resistance: progress and lessons". In Proceedings of the 2nd ARO Special Workshop on Hardware Assurance, Washington, DC, 2011.

25. Chakraborty, Rajat Subhra, Seetharam Narasimhan, and Swarup Bhunia. "Hardware Trojan: Threats and emerging solutions". In 2009 IEEE International High Level Design Validation and Test Workshop, IEEE, San Francisco, CA, pp. 166–171, 2009.
26. Forte, Domenic, Swarup Bhunia, and Mark M. Tehranipoor, eds. *"Hardware Protection Through Obfuscation*, Springer International Publishing Switzerland AG., 2017.
27. Rajendran, Jeyavijayan, Michael Sam, Ozgur Sinanoglu, and Ramesh Karri. "Security analysis of integrated circuit camouflaging". In Proceedings of the 2013 ACM SIGSAC Conference on Computer & Communications Security, pp. 709–720, 2013.
28. Kastner, Ryan, and Ted Huffmire. "Threats and challenges in reconfigurable hardware security". California Univ. San Diego La Jolla Dept. of Computer Science and Engineering, 2008.
29. Tuyls, Pim. "Hardware intrinsic security". In *International Workshop on Radio Frequency Identification: Security and Privacy Issues*, Springer, Berlin, Heidelberg, pp. 123–123, 2010.
30. Noman, Ali Nur Mohammad, Kevin Curran, and Tom Lunney. "A watermarking based tamper detection solution for RFID tags". In 2010 Sixth International Conference on Intelligent Information Hiding and Multimedia Signal Processing, IEEE, Darmstadt, pp. 98–101, 2010.
31. Kheir, Mohamed, Heinz Kreft, and Reinhard Knöchel. "UWB on-chip fingerprinting and identification using carbon nanotubes". In 2014 IEEE International Conference on Ultra-WideBand (ICUWB), IEEE, Paris, pp. 462–466, 2014.
32. Bossuet, Lilian, and Brice Colombier. "Comments on 'A PUF-FSM binding scheme for FPGA IP protection and pay-per-device licensing'". *IEEE Transactions on Information Forensics and Security* 11, no. 11 (2016): 2624–2625.
33. Herder, Charles, Meng-Day Yu, Farinaz Koushanfar, and Srinivas Devadas. "Physical unclonable functions and applications: A tutorial". *Proceedings of the IEEE* 102, no. 8 (2014): 1126–1141.
34. Suh, G. Edward, and Srinivas Devadas. "Physical unclonable functions for device authentication and secret key generation". In 2007 44th ACM/IEEE Design Automation Conference, IEEE, San Diego, CA, pp. 9–14, 2007.
35. Rührmair, Ulrich, Jan Sölter, and Frank Sehnke. "On the foundations of physical unclonable functions". *IACR Cryptology ePrint Archive* 2009 (2009): 277.
36. Xu, Sarah Q., Wing-kei Yu, G. Edward Suh, and Edwin C. Kan. "Understanding sources of variations in flash memory for physical unclonable functions". In 2014 IEEE 6th International Memory Workshop (IMW), IEEE, Taipei, pp. 1–4, 2014.
37. Günlü, Onur, Onurcan İşcan, and Gerhard Kramer. "Reliable secret key generation from physical unclonable functions under varying environmental conditions". In 2015 IEEE International Workshop on Information Forensics and Security (WIFS), IEEE, Rome, pp. 1–6, 2015.
38. Wang, Xiaoxiao, and Mohammad Tehranipoor. "Novel physical unclonable function with process and environmental variations". In 2010 Design, Automation & Test in Europe Conference & Exhibition (DATE 2010), IEEE, Dresden, pp. 1065–1070, 2010.
39. Zhang, Le, Zhi Hui Kong, Chip-Hong Chang, Alessandro Cabrini, and Guido Torelli. "Exploiting process variations and programming sensitivity of phase change memory for reconfigurable physical unclonable functions". *IEEE Transactions on Information Forensics and Security* 9, no. 6 (2014): 921–932.
40. Stanzione, Stefano, Daniele Puntin, and Giuseppe Iannaccone. "CMOS silicon physical unclonable functions based on intrinsic process variability". *IEEE Journal of Solid-State Circuits* 46, no. 6 (2011): 1456–1463.
41. Maes, Roel, and Ingrid Verbauwhede. "Physically unclonable functions: A study on the state of the art and future research directions". In *Towards Hardware-Intrinsic Security*, Springer, Berlin, Heidelberg, pp. 3–37, 2010.
42. Helinski, Ryan, Dhruva Acharyya, and Jim Plusquellic. "A physical unclonable function defined using power distribution system equivalent resistance variations". In 2009 46th ACM/IEEE Design Automation Conference, IEEE, San Francisco, CA, pp. 676–681, 2009.

43. Skoric, Boris, Geert-Jan Schrijen, Wil Ophey, Rob Wolters, Nynke Verhaegh, and Jan van Geloven. "Experimental hardware for coating PUFs and optical PUFs". In *Security with Noisy Data*, Springer, London, pp. 255–268, 2007.
44. Hori, Yohei, Takahiro Yoshida, Toshihiro Katashita, and Akashi Satoh. "Quantitative and statistical performance evaluation of arbiter physical unclonable functions on FPGAs". In 2010 International Conference on Reconfigurable Computing and FPGAs, IEEE, Quintana Roo, pp. 298–303, 2010.
45. Maiti, Abhranil, Jeff Casarona, Luke McHale, and Patrick Schaumont. "A large scale characterization of RO-PUF". In 2010 IEEE International Symposium on Hardware-Oriented Security and Trust (HOST), IEEE, pp. 94–99, 2010.
46. Garg, Achiranshu, and Tony T. Kim. "Design of SRAM PUF with improved uniformity and reliability utilizing device aging effect". In 2014 IEEE International Symposium on Circuits and Systems (ISCAS), IEEE, Melbourne, pp. 1941–1944, 2014.
47. Kumar, Sandeep S., Jorge Guajardo, Roel Maes, Geert-Jan Schrijen, and Pim Tuyls. "The butterfly PUF protecting IP on every FPGA". In 2008 IEEE International Workshop on Hardware-Oriented Security and Trust, IEEE, Anaheim, CA, pp. 67–70, 2008.
48. Habib, Bilal, Jens-Peter Kaps, and Kris Gaj. "Implementation of efficient SR-Latch PUF on FPGA and SoC devices". *Microprocessors and Microsystems* 53 (2017): 92–105.
49. Maes, Roel, Pim Tuyls, and Ingrid Verbauwhede. "Intrinsic PUFs from flip-flops on reconfigurable devices". In 3rd Benelux Workshop on Information and System Security (WISSec 2008), vol. 17, p. 2008, 2008.
50. K. Shamsi, W. Wen and Y. Jin, "Hardware security challenges beyond CMOS: Attacks and remedies", Ultimate Integration of Silicon, IEEE Computer Society Annual Symposium on VLSI, Pittsburgh, PA, 2016.
51. H. Liu and S. Datta and V. Narayanan, "Steep switching tunnel FET: A promise to extend the energy efficient roadmap for post-CMOS digital and analog/RF applications", International Symposium on Low Power Electronics and Design (ISLPED), IEEE, Beijing, pp. 145–150, Sept. 2013.
52. A. M. Ionescu and H. Riel, "Tunnel field-effect transistors as energy efficient electronic switches", *Nature*, vol. 479 no. 7373 pp. 329–337, Nov. 2011.
53. A. R. Trivedi and S. Carlo and S. Mukhopadhyay, "Exploring Tunnel-FET for ultra low power analog applications: A case study on operational transconductance amplifier", 50th ACM/EDAC/IEEE Design Automation Conference (DAC), IEEE, Austin, TX, pp. 1–6, May 2013.
54. C. Rajan, D. Sharma, and D. P. Samajdar. "Implementation of physical unclonable functions using hetero junction based GAA TFET". *Superlattices and Microstructures* 126 (2019): 72–82.

CHAPTER 11

Some Research Issues of Harmful and Violent Content Filtering for Social Networks in the Context of Large-Scale and Streaming Data with Apache Spark

Phuc Do, Phu Pham and Trung Phan

CONTENTS

11.1 INTRODUCTION

Nowadays, it is become easier for people to share all kinds of information to others through social networks, thanks to the development of Internet (Fire et al., 2014; Zhang et al., 2015a; Ismailov, 2017). However, besides its advantages, harmful and violent contents in social networks lead to dangerous effects on those who lacking proper knowledge of the social networks. During the last few years, the popularity of content sharing on Internet, like websites, forums, and social networks, has given considerable rise in digital contents such as news, comments, posts, video clips, etc. that can be uploaded as well as accessed by a large number of the people around the world, including sensitive social groups such as children, teenagers, etc. Comments, posts, images, or videos shared on social networks lead to severe problems such as the possibility of individuals being tremendously obsessed due to violent and dangerous things on the Internet. Therefore, in this chapter we propose a novel approach of applying Apache Spark for solving problems of harmful and violent contents such as comments, posts, images, etc. in popular social networks like as Facebook, Twitters, YouTube, etc. and for protecting sensitive social groups, such as children. Content filtering (Golbeck, 2016; Vanetti et al., 2010; Steinberg et al., n.d.) for comments, posts, and images on the social networks is considered as a very challenging task not only because the harmful contents are formed in varied and highly complex structures but also the large amount of data from popular social networks also cause challenges for a traditional standalone-based system to handle. The literature related to harmful and violence content detection is limited and, in most of the cases, it examines only visual writing styles and features of data to decide whether the content is harmful or not. Moreover, most of data from social networks are continuously generated as rapid data streams (Ediger et al., 2010; Kim et al., 2013; Riedy & Bader, 2013; Nair & Shetty, 2015) which are considered as highly computationally challenging tasks for standalone-based systems. Recently, demands for new efficient methods for processing large-scale data on social networks in real time are growing (Ferrara et al., 2016; Bányai et al., 2017; Peel & Clauset, 2015). In normal operation, the continuous high-speed data streams which are generated from large social networks are extremely high, for example, over 510K comments/posts are posted to Facebook every minute, more than 500 million tweets (including text-based contents and images) are sent to Twitter every day, or averagely 400 hours of video clips are uploaded to YouTube per minute. Task of analyzing and mining data streams is unique from the other static mining approaches by being adaptable to the data-transferring rate as well as the available computing resources of memory and time constraints. The challenges of mining data streams are unbounded computer's memory requirements as well as storage resources. In data-streaming analysis and mining, the performance of response time and throughput are the most important and critical criteria. The streaming-based mining systems must ensure that all arriving data are processed on time and correctly. In data-stream mining, current mining tasks need to be completed before the other data batch comes. From the past, approaches of mining streaming data have been recognized due to its potential application in several areas. However, the size of data's volume is not taken into consideration. However, numerous applications have been developed for solving the problems of detecting harmful and violent contents in social networks. They lack the process of thorough evaluations on how to deal with large-volume and high-velocity dataset such as Facebook or YouTube social networks. Mining data streams from social networks is one of the current interesting research areas due to its valuable applications in marketing, social network analysis, social trends, or events recognition, etc. From social networks, analysis systems conduct process of extracting interesting patterns/keywords/phrases, or trends from a sequence of comments, post, video, images, etc. that arrive continuously in a rapid speed. Besides retrieving knowledge from social networks, people also want to automatically detect and prevent harmful, dangerous, and violent content from social networks. Typically, to deal with these large-scale continuous datasets, the Apache Spark Streaming framework is designed for batch processing of high-velocity traffic data. Spark streaming is a recent and most well-known approach for tackling the high-speed and large-scale data streams using the distributed processing baselines of Apache Spark. In order to adopt streaming-based approach for our custom application, we need to develop a platform based on Spark Streaming. A Spark Streaming-based program is constructed as a packaged application and runs on the Apache ecosystem. For every incoming data from the outside, the application will fetch the data into multiple micro-batches and transfer to all Sparks' nodes following the resilient distributed datasets (RDDs) architecture. This is called as discretized stream, or DStream. Then, parallel map and reduce operations can be applied in these discretized stream, also called DStream, like other Spark-based applications. Therefore, Spark Streaming-based application is based on the level of data's volume as well as velocity, which are needed to process. Based on the architecture of Spark Streaming, we develop a novel approach that can detect harmful and violent content for large-scale and high-speed data, which are collected from social networks. To collect large content datasets from social networks, we apply Apache Nutch, which is a distributed web crawler on the Apache Hadoop ecosystem. We also build custom natural language processing (NLP) tools and knowledge bases to support the process of detecting harmful and violent contents, which are all developed on the Spark Streaming frameworks. This chapter presents three main contributions, which are as follows:

- First, we develop an approach of using Apache Nutch as a distributed web crawler to collect data from popular social networks like as Facebook, Twitter, etc. The crawled contents are then stored to HDFS and the next processing batches are arranged.

- Second, we present the proposed novel NLP tools and construct the knowledge base to support the harmful and violent content detection model with high performance in accuracy.
- Finally, we introduce a novel mechanism for detecting harmful and violent content from content dataset (comments, posts, tweets, etc.) of large social networks on the Spark Streaming framework. Our novel detection models include:
    - For text-based contents such as comments, posts, news, etc., we apply our proposed text-processing model to analyze harmful or violent key phrases from given text corpus, then we apply text embedding model to learn the representation of these key phrases, which then are used to recognize pattern of harmful contents in the future.
    - For images/videos that contain harmful or violent scenes, we apply the OpenCV[1] tool combining with convolutional neural network (CNN) approach to extract distinctive features from image-based data. By using combining these extracted features with our labeled violent knowledge-based image dataset we able to decide that these images or videos might contain harmful or violent content or not.
    - All harmful or violence detection models for both text-based and image-based contents are developed on the Spark Streaming framework in order to ensure the capability of handling large-scale and high-velocity data streams from multiple data source, such as forums, social networks, etc.

11.2 RELATED WORK

The harmful and violent contents are mostly considered as the emerging topics and most of research studies are still in consideration. In fact, analyzing, detecting, and filtering the cyberbullying, harmful, and violent contents from large social networks such as Facebook, Twitter, Instagram, YouTube, etc. are difficult and challenging due to the multityped format of the content. With Facebook, beside text-based contents such as comments, posts, news, etc., we also have images, video clips, etc. Building a combined system for detecting, analyzing, and validating both image-based and text-based contents is difficult and takes a lot of efforts. There is a multitude of research contributions in the area of detection and validation of the abuse, cyberbullying, harmful and violent content that are based on different aspects. In recent years, large social network such as Facebook, Twitter, YouTube, etc. attract more and more people to share and post their own contents like feelings via comments, posts, status, etc. or images and video clips, etc. These activities have gradually become the ultimate targets for harmful, abused as well as violent contents to be popular. Therefore, we develop a protection mechanism to

[1] OpenCV: https://opencv.org/

protect users against these kinds of harmful, abusive, and violent contents on the social network, as it is extremely necessary. Most recent research studies have come up with the solution of building automated system for detecting harmful and violent contents or activities on the social networks like violent or aggressive posts, comments or abusive text messages or sexual or terrorist images, video clips, etc. There are various research studies to investigate the negative effects and challenges of spams in social networks and propose the methods for preventing these kinds of contents. From a Machine Leaning perspective, the detection of harmful and violent content in social networks can be considered as the classification task which helps in categorizing the text-based and image-based content that is harmful or malicious to the user. There are several researches (Ferrara et al., 2016; Bányai et al., 2017; Peel & Clauset, 2015) straightforwardly focused on this approach and primarily collecting and selecting unique features that harmful and violent contents should have. Then, these selected features are used to train machine learning classification models such as Logistic Regression, Support Vector Machine, Decision Tree, etc. for analyzing and classifying contents on the social networks. This approach is also popular in other social network-controlling aspects, such as spam filtering mentioned in the works of (Heymann et al. 2007). It proposed an approach of defeating spam in social networks like Twitter by applying trained classification model to identify spammers in the social network. There are several advanced studies that proposed the approaches (Chen et al. 2009) of evaluating the relationship between social network groups to identify the root source of violent, harmful, or spam contents such as common groups of people usually posting harmful contents in the social networks. Similar to the approach of (Wang et al. 2011 and Wang 2010), they proposed novel techniques to detect spam contents on the social networks based on the training of the classification learning models. Another approaches proposed a scalable and online harmful, violent content analysis, and detection based on the data analysis and mining techniques (Jin et al., 2011; Chowdury et al., 2013; Yue et al., 2007). These approaches use the client's feedback, reporting activities on social networks to build and extract features for enriching the classification model. These proposed models are designed to detect harmful and violent contents depending on the user's behaviors and reactions on uncomfortable contents or messages from other users but largely ignore the relationships between users and the social groups on the social networks. The other approaches of applying similarity measure and ranking techniques to detect and filter the harmful and violent contents on social networks. (Noll et al. 2009) proposed a graph-based model to calculate the similarity and rank the contents that users would like to read. (Noll et al. 2009) and (Yamaguchi et al. 2010) proposed the approach, called TuRank, which is an authoritative ranking depending on user's flow information on Twitter, or (Das Sarma et al. 2010) proposed a model for comparing the score-based ranking mechanism to gain the ranking score of user's feedback. Similar to spam content analyzing and filtering in social networks, cyberbullying content filtering is also a hot topic in recent times. There are researches that indicated the negative influences of cyberbullying on social networks nowadays. The online social cyberbullying (Dadvar & De Jong 2012); (Reynolds et al. 2011) is considered as the shared text-based or text-based contents on the social networks or forums, which might cause depression or other tremendous life-threatening problems for people who read or see these contents. Similar to the analyzing and detecting techniques of harmful/violent or spamming contents (Chen et al., 2012); Li & Tagami, 2014), some researches demonstrated the uses of machine learning-based techniques such as classification, data analyzing and mining, or ranking to validate the cyberbullying contents on the social networks. Moreover, post or comment of a user is normally about 5–100 words and it might be not enough to extract distinctive feature from these kinds of contents. Therefore, we need to build more complex models, which must be suitable for classifying short-length text-based contents like as posts or comments. There are combined methods between machine learning and user's feedback reaction on bad contents in social network to build more effective systems for preventing cyberbullying. These hybrid-systems help to analyze and investigate the characteristics and behaviors of clients through their reactions or reporting activities in social networks. Then the information is applied to build the more accurate analysis and detection mechanism for preventing harmful or cyberbullying content on the social networks. These advanced techniques have performed better in the detection of cyberbullying contents even in the short-length text-based content like comments, posts, etc. with only few words. Short-length contents on the social networks like posts or comments are also considered as major challenges for analysis and detection of harmful and violent content in social networks. (Vanhove et al., 2013) proposed a flexible system architecture for combined methods that can be quickly applied to multiple harmful and violent content filtering purposes (Xiang et al., 2012; Benevenuto et al., 2009; Diakopoulos & Naaman, 2011). The combined system can be changeable for all text-based and image-based content analysis and categorization. However, most of recent studies rarely mentioned about challenges related to the volume size and velocity of the large-scale social network such as Facebook, Twitter, Instagram, etc. In the context of Big Data, the needed handling datasets from online resources (social networks, forums) are not only big in size but also coming very fast. Therefore, the designed system needs to be capable of handling large scale with high-velocity datasets. Recent studies on large-scale social network processing have indicated challenges of handling large-scale and high-velocity incoming of data from social networks (Sosic & Leskovec, 201; Bischke et al., 2019; Wang et al., 2018; Eirinaki et al., 2018; Lumsdaine et al., 2007). The advanced researches in the large-scale and high-velocity data processing, which are covered in these researches, have placed strong baselines for efficient knowledge upon key requirements and challenges for current issues of social network analysis tasks. Combining with advances in deep learning approach, we apply the CNN for solving the harmful and violent content analysis and filtration (Mu et al., 2016; Song et al., 2019; Ali & Senan, 2016; Dong et al., 2016). The CNN architecture is used to scan and extract distinctive features from text-based and image-based contents, which are collected from multiple online resources such as social networks (Facebook, Twitter, and YouTube), forums, online news, etc. To collect data efficiently, we developed a distributed web crawler on the Apache Nutch framework. The designed distributed web crawler helps to collect text-based content (posts, comments, news, etc.) and images from

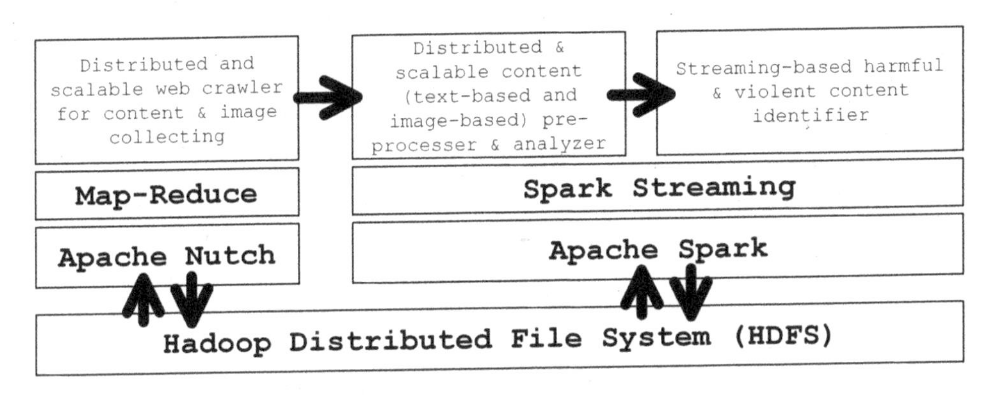

FIGURE 11.1 Illustrations of distributed, scalable, and streaming-based harmful and violent content recognition under different distributed processing frameworks.

the Internet. Then, the proposed harm-, abuse-, and violence-filtering mechanisms in this chapter are developed on Apache Spark to leverage the time-consuming performance and capability of handling the large-scale and high-velocity datasets.

11.3 METHODOLOGY

In this chapter, we present our approaches for constructing a distributed scalable high-speed harmful and violent content identification from multiple Internet resources. Our system is constructed on different distributed processing frameworks, which enable to handle large-scale and high-velocity datasets. The system is divided into three main parts (illustrated in Figure 11.1). The first one is the distributed web's crawler, which is responsible for collecting different types of contents, including text-based content (news, blogs, posts, comments, etc.) and multimedia content (images, videos, etc.) from different online resources (social networks, online news webpages, forums, etc.). The second part is a preprocessing model, which contains NLP-based (Spark NLP[2]) and image processing mechanism developed on the Apache Spark Streaming[3] framework. This part is in charge of handling preprocessing tasks for text (tokenizing, sentence splitting, NER, embedding, etc.) and multimedia contents (image analysis, processing, embedding, etc.). Finally, the third part is responsible for identifying the crawled and processed contents that might or might not contain harmful/violent contents. This part is also built on Apache Spark Streaming in order to enable the capability of large-scale high-velocity data handling.

11.3.1 Distributed Web Crawler System Based on Apache Nutch

For the first part of our contributions in this chapter, we develop a custom distributed-based web-crawler based on Apache Nutch framework. Apache Nutch is considered

[2] Spark NLP: https://nlp.johnsnowlabs.com/
[3] Apache Spark Streaming: https://spark.apache.org/streaming/

as the most well-known distributed framework for building large-scale data crawling systems. In order to collect data for training and testing process of our overall system, we need to collect abundant amount of data, such as news, blog's contents, comments, posts, image, etc. from multiple online resources such as social networks, online news website, blogs, etc. In the context of Big Data (Sivarajah et al., 2017; Bello-orgaz et al., 2016; Jin et al., 2015), the rapid growth of data from such online resources has raised many concerns, such as collecting, storing, and analyzing large amounts of unstructured data. Thanks to the tremendous developments of the Internet, nowadays, the amount of data that has been collected from the Internet is much larger than the traditional crawling and analyzing system can handle. In fact, the standalone-based infrastructures can't handle large amount of web's contents, images, etc., which rapidly increase in both size and velocity. For storage, this large-scale dataset also exceeds the abilities and performance of traditional databases, storage system as well as infrastructure's networks. Moreover, the need-to-crawl datasets not only is large in size but also arrives very fast such as posts/comments on social networks (> 510K comments/posts are posted on Facebook every minute, > 500 million tweets per minute in Twitter, etc.). Therefore, it requires the crawling and indexing systems that are capable to not only handle and store large amounts of various unstructured data but also process them very quickly. Just like other various unstructured contents, "harmful and violent contents" (Definition 11.1) are abundant on the Internet recently. In recent years, the rapid growths of digital contents (texts, images, videos, etc.) on the social networks (Chen et al., 2012; Ding et al., 2014; Eyben et al., 2013) lead to severe problems, which are related to negative influences on sensitive groups (children, teenager, etc.) as well as social interactions.

Definition 11.1: Online harmful and violent content. In short, online violent and harmful contents are defined as contents that are intensively uncontrolled spreading over the Internet and might cause distress or may become uneasiness for some or all kinds of people who have seen these contents, such as online abuse or bullying, self-harming or suicide, sexual harassment or threats, unwanted sexual or violent contents, etc.

In fact, harmful and violent content analysis and mining task are considered as complicated tasks due to the following several reasons:

- The harmful and violent contents might be composed of various forms such as texts (news, comments, posts, etc.), multimedia (images, voices, videos, etc.), etc. We must develop separated analysis mechanism for each type of data formatted form. Our works in this chapter focus on two types of data forms, which are:
  - **Text-based content:** For analyzing text-based contents, several techniques are needed to be applied for handling these data forms, like natural language processing (NLP) (word tokenizer, sentence splitting, name entity recognition, etc.) in combination with developing text-based harmful content recognition models.

- **Image-based content:** It is different from text-based contents. To recognize and identify images that contain violent and harmful contents, we need to apply image analysis and processing techniques in combination with neural-based learning models such as BPNN, CNN, etc. to classify these images.

- Large- and high-speed need-to-analyze data from the Internet such as social networks, online news, etc. lead to challenges related to the performance and scalability of constructing system.
- Up-to-date algorithms and approaches are not developed to work on distributed and scalable systems.

11.3.2 What Is Apache Nutch?

Apache Nutch is an intensive complete open-source application for web-based content crawler and search engine. It is developed and aimed to collect and index online resources such as web's pages, images, etc. from multiple online resources in World Wide Web effectively with scalable capability. Starting from a project of Doug Cutting and Mike Caffarella (creators of Hadoop and Lucene) (Iqbal & Soomro 2015) of developing a distributed scalable web crawling and indexing system, the Nutch project was developed based on the idea of distributed framework of Hadoop, which enables it to handle large-scale datasets. In fact, Nutch was built and run on the Map-Reduce mechanism (Definition 11.2) and integrated with distributed file system (DFS) such as GDFS (Google Distributed File System), HDFS (Hadoop Distributed File System), etc. In 2010, Apache Nutch was transferred and has been continuously developed by Apache Software Foundation (ASF) and now has become one of the top important projects of ASF's community.

Definition 11.2: Map-Reduce. It is considered as a processing mechanism that is introduced in Apache Hadoop framework. Map-Reduce is designed to run simultaneously on multiple nodes within Hadoop clusters. It facilitates distributed processing by dividing a large-scale dataset into smaller chunks which are assigned to different nodes and processed parallel (Map). Finally, the processed data from nodes are collected and aggregated to return the final output (Reduce).

11.3.3 Multityped Content Crawler on Apache Nutch

In our approach, we apply Apache Nutch to construct a distributed and scalable web's crawler to collect multityped content from large-scale online resources, such as: Facebook, Twitter, and Instagram. For text-based content such as news, posts, comments, etc., we can easily apply the default features of Apache Nutch to collect them from multiple online resources. However, for multimedia content such as images, we need to develop custom crawling feature in order to effective collect this kind of data from online resources. In recent years, images, graphical contents, and videos are being generated and posted to online resources such as social networks very quickly. In fact, the size of image-based content is much larger and difficult to process than text-based contents. In this part, we present our approach of building a custom image crawler, which is used to collect images as well as their metadata on the Apache Nutch framework. The constructed image-based crawler contains two main processes. In the first process, the crawler will perform the URL fetching and parsing the img tag to identify images in travelled web's pages. In this process, depending on the input URL seed list, the crawler also will try to parse all other related metadata (alt, name, id attributes) by investigating the img tag. Then, the images' links and their related metadata will be stored into ImageMetadataCrawlDB (in HDFS). Later on, the second crawler's processor will read these stored image's data and try to download them from given resources and stored to ImageCrawlDB (in HDFS). The purpose of dividing the image crawling process into two phases to prevent the overload of designed crawler when downloading both images and their metadata at the same time (illustrated in Figure 11.2).

In order to make easier for proficient storing and querying data from HDFS, we implement the indexing mechanism based on Apache Lucene (along with Apache Nutch) for both text-based content and image's metadata. The collected text-based contents (news, comments, posts, etc.) will be indexed and stored in different database, called TextCrawlDB (in HDFS). For image-based contents, only metadata (in ImageMetadataCrawlDB) is indexed via Lucene. The indexed metadata for image sets are used to check the duplications in the set of collected images.

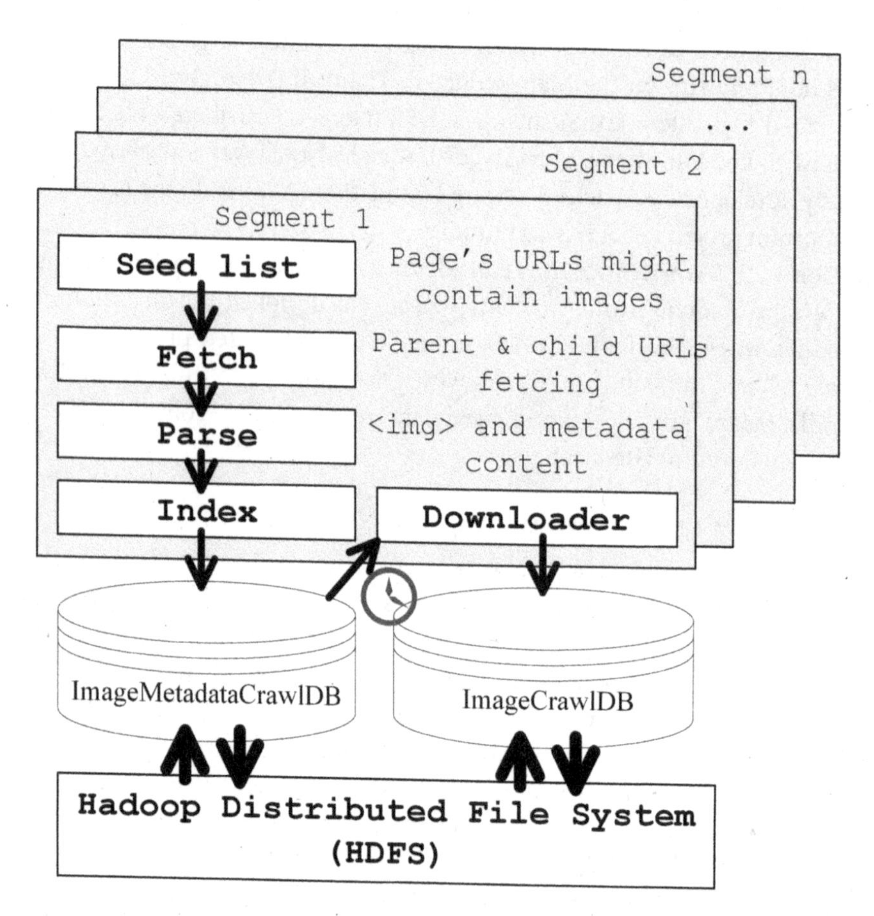

FIGURE 11.2 Image-based crawling system on Apache Nutch.

11.3.4 Data Pre-Processing on Apache Spark Streaming

11.3.4.1 Overview

In the era of Big data, a huge amount of data is generated continuously every minute. Data can be generated from many sources. In particular, IoT devices are the most important data sources. According to a new forecast from IDC, in 2025 there will be 41.6 billion connected IoT devices generating 79.4 zettabytes (ZB) of data (IDC.com, 2019). Almost IoT devices generate data continuously in streams. So, live stream data-processing techniques become a critical part of big data-processing techniques. Today, there are a lot of live stream processing frameworks like Apache Spark, Apache Storm, Apache Samza, Apache Flink, Amazon Kinesis Streams. Each one has different pros and cons, but Apache Spark is the most standout framework because it is a mature framework with a large community, proven in a lot of real projects, and readily supports SQL querying (Nasiri et al., 2019). In Apache Spark, there are two APIs support live stream data processing: Spark Streaming and Structured Streaming.

- Spark Streaming is an old API. It processes data streams using DStreams built on RDDs.
- Structured Streaming is a new API introduced from Apache Spark 2.x. This API processes structured data streams with relation queries on Datasets and DataFrames.

11.3.4.2 Spark Streaming API

Spark Streaming enables scalable, high-throughput, fault-tolerant stream processing of live data streams. Data can be received from many sources like Kafka, Flume, Kinesis, Twitter, TCP sockets, and can be processed using functions like map, reduce, join, and window. Finally, processed data can be pushed out to file systems, databases, and live dashboards. Figure 11.3 shows the Spark Streaming Model. In the real world, you can apply Spark's machine learning and graph-processing algorithms on data streams (Spark Streaming Programming Guide, 2019); (Kienzler, 2017).

Internally, Spark Streaming receives live input data streams and divides the data into batches, which are then processed by the Spark engine to generate the final stream of results in batches. Figure 11.4 shows the inner workings of Spark Streaming.

FIGURE 11.3 The Spark Streaming model. (Source: https://spark.apache.org).

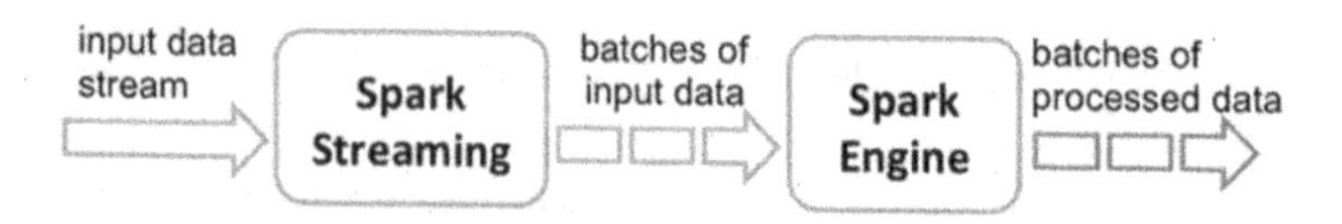

FIGURE 11.4 The inner workings of Spark Streaming. (Source: https://spark.apache.org)

Spark Streaming provides a data structure called discretized stream or DStream, which represents a continuous stream of data. DStreams can be created either from input data streams from sources such as Kafka, Flume, and Kinesis, or by applying high-level operations on other DStreams. In essence, a DStream is a sequence of RDDs. Each RDD in a DStream contains data from a certain interval, as shown in Figure 11.5. Operations applied on a DStream translate to operations on the underlying RDDs. Figure 11.5 shows the DStream structure.

11.3.4.3 Structured Streaming API

When processing the stream, we use Structured Streaming, which is a scalable and fault-tolerant stream processing engine built on the Spark SQL engine. We can use Dataset/DataFrame API of languages supported by Spark (such as Scala, Java, Python, etc.) to perform calculations on the stream (Kienzler, 2017); (Structured Streaming Programming Guide, 2019).

In essence, a structured Streaming query is processed by a micro-batch processing engine that processes a stream as a series of a small-batch job. This helps achieve low latency of about 100 milliseconds and ensures fault tolerance. Since Spark 2.3, a new low-latency processing mode named Continuous Processing is supported. It can achieve latencies of about 1 millisecond. We can choose the mode depending on the specific application without changing operations on Dataset/DataFrame in our queries.

Structured Streaming considers a stream as an unbounded table that is being continuously appended shown in Figure 11.6. Therefore, a streaming query is also similar to a query on a static table. Spark will process the streaming query as an incremental query on the unbounded input table.

The result of a streaming query is the “Result Table”. After each interval (e.g., every 1 second), new rows are appended to the Input Table, and new results rows will be appended to the Result Table. Figure 11.7 shows the Programming Model for Structured Streaming.

Structured Streaming is not the whole input table. It only reads the latest available data from the streaming data source, processes it incrementally to update the result, and then discards the source data.

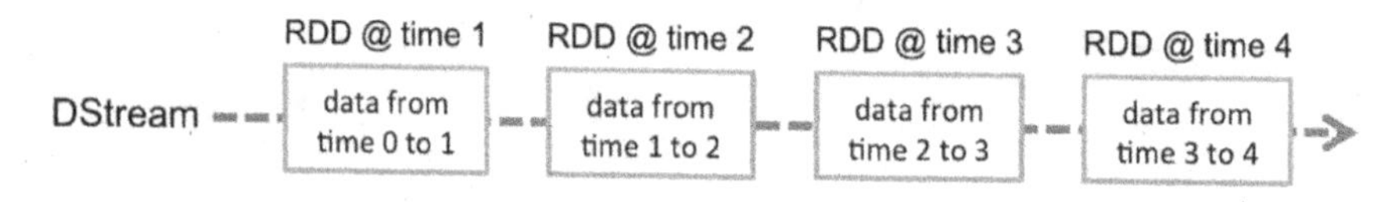

FIGURE 11.5 The DStream structure. (Source: https://spark.apache.org)

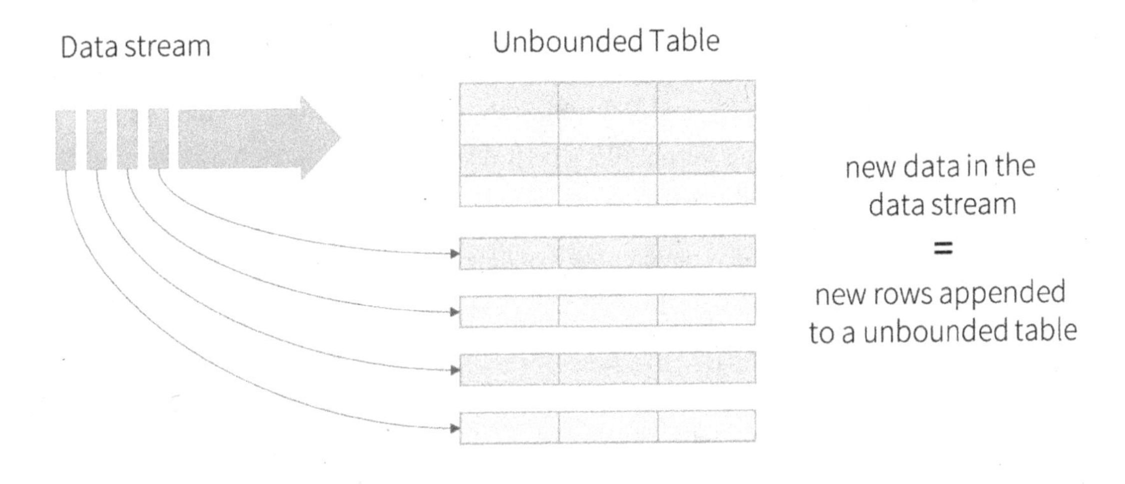

FIGURE 11.6 Data stream as an unbounded table. (Source: https://spark.apache.org)

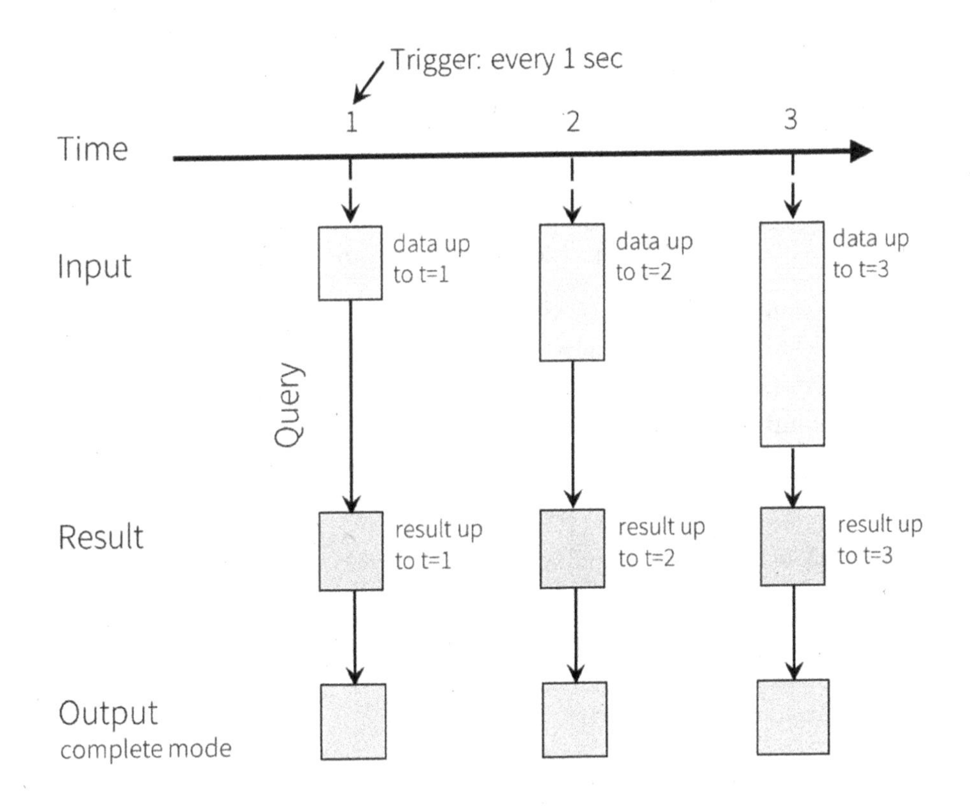

FIGURE 11.7 Programming model for structured streaming. (Source: https://spark.apache.org)

The following are the sources of Structured Streaming:

- **File source**: Reads files (text, csv, json, orc, parquet) in a directory as a data stream.
- **Kafka source**: Reads data from Kafka with broker versions from 0.10.0.
- **Socket source** (for testing): Reads UTF8 text data from a socket connection.
- **Rate source** (for testing): Generates rows of data based on time intervals.

11.3.4.4 Spark Streaming vs. Structured Streaming

11.3.4.4.1 Real Streaming Spark Streaming works on small batches. Spark polls the stream source after each interval of time and then a batch is created for receiving data. Each batch is an RDD of a DStream.

For Structured Streaming, there is no batch concept. The received data after each interval of time is appended to a DataFrame. New input data rows are processed and the result is updated into the result table. Therefore, Structured Streaming is superior to Spark Streaming.

11.3.4.4.2 RDD vs. DataFrames/DataSet Spark Streaming is processed by the DStream API, which is internally using RDDs. Structured Streaming is processed by the DataFrame/ Dataset APIs. DataFrames are built on RDDs, and have many improvements inside. Therefore, Structured Streaming is better.

11.3.4.4.3 Processing Event-Time and Late Data A challenge in stream handling is real-time data processing. It is inevitable that there may be latencies in data generation and data transmission. Spark Streaming only works with the timestamp when receiving data. This leads to the result that may not be less accurate. On the contrary, Structured Streaming uses the event-time when receiving data. So, latencies are not a problem, and data is processed more accurately.

11.3.4.4.4 End-to-End Guarantees Applications require fault tolerance and guarantee in data transfer. They must have a recovery mechanism in case of failure so as not to lose or duplicate data. For providing fault tolerance, Spark Streaming and Structured Streaming both use checkpoints. However, this method still has many drawbacks that can lead to data loss. On the other hand, Structured Streaming also applies additional conditions to be able to recover better when errors.

11.3.4.4.5 Example of Structured Streaming in Scala The following is a code snippet in Scala that counts words of text files in the *"/data"* directory in HDFS:

```
// Make the necessary imports
import org.apache.spark.sql.functions._
import org.apache.spark.sql.SparkSession

// Create a SparkSession object, the entry point for Spark
applications
val spark = SparkSession
  .builder
  .appName("WordCount")
  .getOrCreate()
```

```
// For implicit conversions
import spark.implicits._

// Read all the text files written automically in a directory
val df = spark.readStream
  .format("text")
  .option("maxFilesPerTrigger", 1)
  .load("/data")

// Split the lines into words
val words = df.as[String].flatMap(_.split(" "))

// Count words
val wordCounts = words.groupBy("value").count()

// Start running the query that prints the running counts to the
console
val query = wordCounts.writeStream
  .outputMode("complete")
  .format("console")
  .start()

query.awaitTermination()
```

The *df* DataFrame represents an unbounded table containing the streaming text data. This table contains one *String* column named "*value*", and each line in the streaming text data becomes a row in the table. Note that this is not currently receiving any data as we are just setting up the transformation, and have not yet started it. Next, we have converted the DataFrame to a Dataset of String using *.as[String]*, so that we can apply the *flatMap* operation to split each line into multiple words. The words Dataset contain all the words. Finally, we have defined the *wordCounts* DataFrame by grouping by the unique values in the Dataset and counting them. Note that this is a streaming DataFrame, which represents the running word counts of the stream.

We have now set up the query on the streaming data. All that is left is to actually start receiving data and computing the counts. To do this, we set it up to print the complete set of counts (specified by *outputMode("complete")*) to the console every time they are updated. And then start the streaming computation using *start()*.

After this code is executed, the streaming computation will be started in the background. The query object is a handle to that active streaming query, and we have decided to wait for the termination of the query using *awaitTermination()* to prevent the process from exiting while the query is active.

In order to execute this example, first we will create the "*/data*" directory in HDFS. Then, we will put text files in the "*/data*" directory for counting.

For both image-based and text-based contents, after crawling from online resources and indexing by Lucene, these data then will be stored to HDFS and we will wait for the next processing steps (Figure 11.8). In this step, the data that have been collected and stored in HDFS will be pushed to the second mechanism for analysis and pre-processing steps.

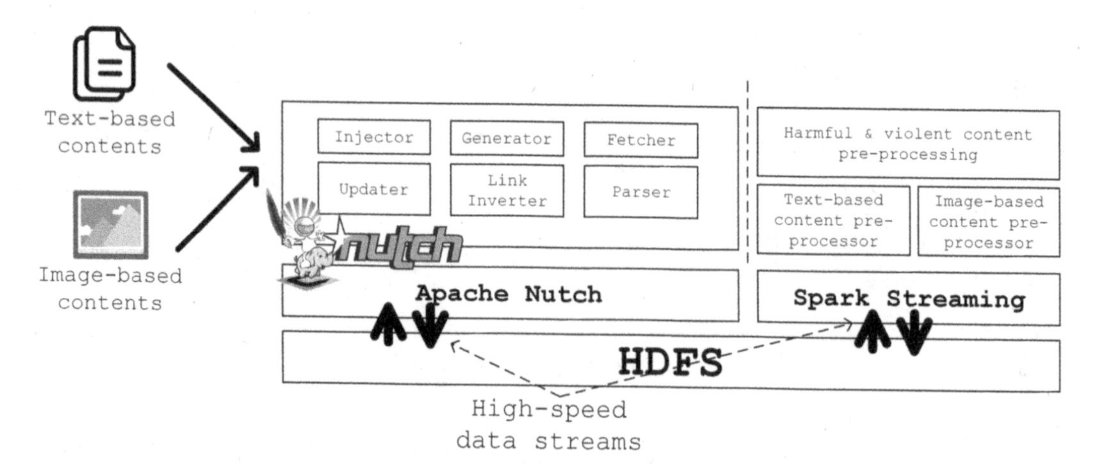

FIGURE 11.8 Illustration of our custom web crawler for collecting and analyzing harmful and violent contents from multiple Internet resources on the distributed scalable Apache Nutch framework.

11.4 APPLYING CNN IN HARMFUL AND VIOLENT CONTENT RECOGNITION

In fact, there are many research studies that have focused on attempting to identify the violent contents in both text-based and image-based that sensitive groups of people might encounter while surfing the Internet. In fact, harmful and violent content detection from online resources such as social network (Facebook, Twitter, etc.) is considered as the most laborious and time-consuming process because a large amount of content are needed to be reviewed every day. Large companies such as Facebook and Google spend a lot of money every year on third parties to review their customers' uploaded contents. Due to the large scale and high velocity of posted contents on these large social networks, human efforts seem hard to manage effectively and efficiently all contents (text and image) that might contain harmful or violent information. Recently, some automated harmful and violent content filtering mechanism have been introduced and proved to work well on recognizing harmful and violent contents from the Internet. However, these approaches merely mention about the performance on large-scale and high-speed data's arrival. To deal with these limitations, in this part, we present our approach of applying Convolutional Neural Network (CNN), which are constructed on Apache Spark for training and categorizing harmful/violent text-based and image-based contents in the context of large-scale and high-velocity (Zikopoulos et al., 2011; Ranjan, 2014) data.

11.4.1 What Is Convolutional Neural Network (CNN)?

Neural network (NN)-based learning model such as traditional full connected with back-propagation (BPNN) has been becoming more and more popular nowadays in most of researches. NN-based learning model has been applied in multiple disciplines to solve problems related to classification and pattern recognition from observational features of given datasets. Recently, many advanced studies in applying NN-based learning for solving

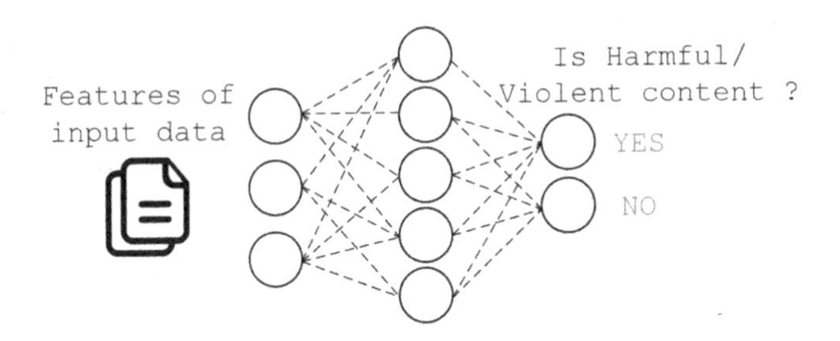

FIGURE 11.9 Illustration of neural network-based learning model.

more complex classification and recognition problems such as facial recognition, pattern/object detections in autopilot vehicles/robots, etc. have been proposed. These advanced studies apply superior architecture of NN with multiple layers to cover as much features of data as possible, also called deep learning approaches (Zhou et al., 2017; Sudhakaran & Lanz, 2017; Dong et al. 2016). Deep learning is considered as a powerful set of techniques for learning features of data by applying multilayer neural networks.

Definition 11.3: Artificial Neural Network. It, also called neural network or neural net (for short), is a computational strategy/method in machine learning, that inspires from the human biological neural network to enable the computer to learn from features of given datasets. A neural network is designed to understand and translate the input features into the desired outputs (see Figure 11.9).

In deep learning terminology, CNN is considered as an advanced and special type of multilayer NN approach. CNN has been designed to automatically extract features of given datasets. CNN is widely applied in multiple NN-based learning areas, especially in

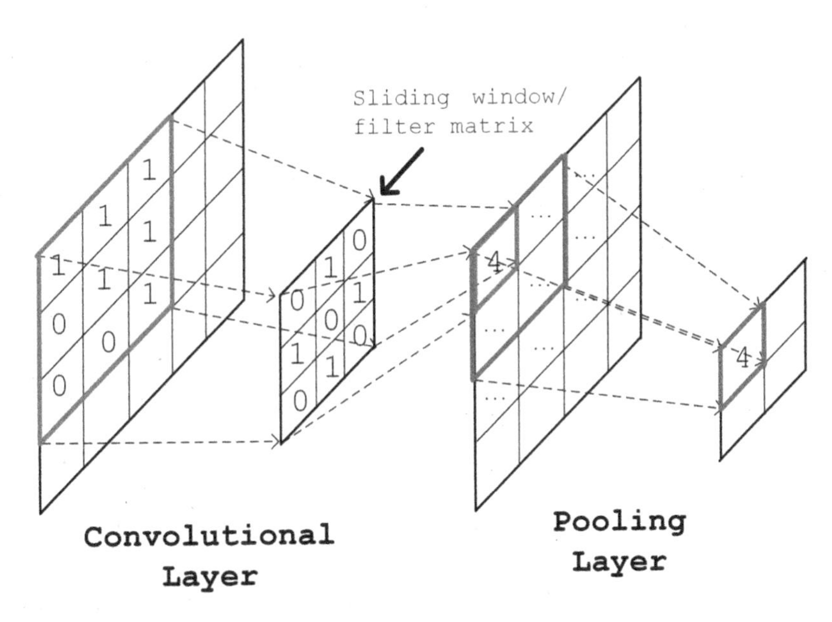

FIGURE 11.10 Illustration of the convolutional/pooling layer of CNN.

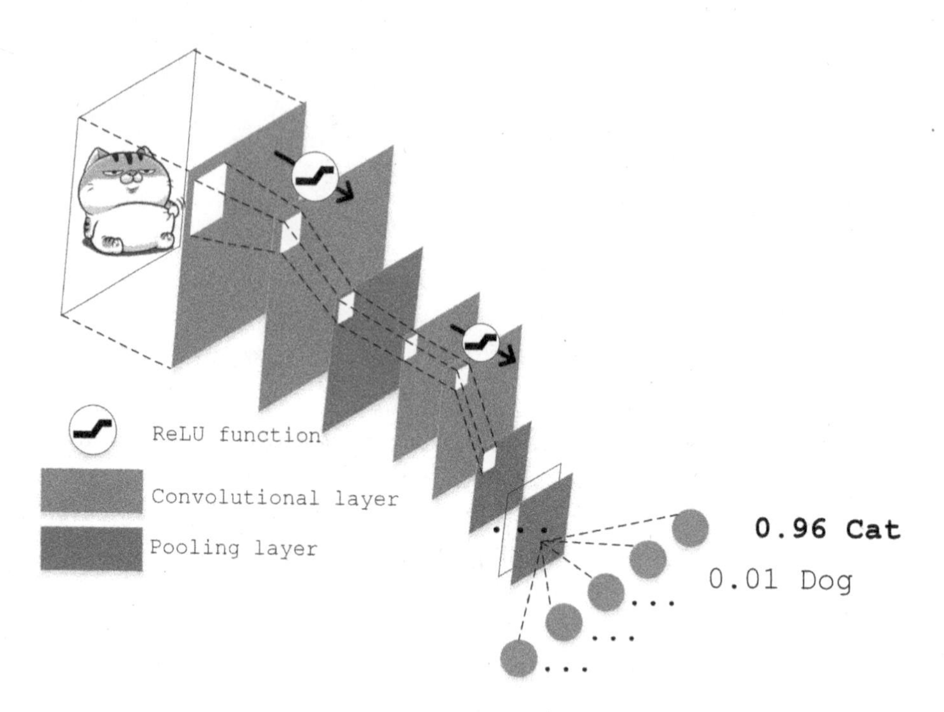

FIGURE 11.11 Illustration of overall CNN architecture.

image processing, object detection, image segmentation, fine-grained categorization, etc. The CNN contains two main layers: the convolution/pooling and full-connected back-propagation (BPNN) (see Figure 11.11). They are:

- **Convolutional/pooling layer:** It contains two processes that are convolution and pooling layers. The convolutional layer is used to extract variable features of given dataset via the sliding windows matrix (is also called filter). Normally, the ReLU (Rectify Linear Unit) function is applied as the activation function between convolutional layers. The pooling layer is used to reduce and extract most important feature of given areas in previous convolutional layer (Figure 11.10). The number of convolutional layers is depended on learning objectives.
- **BPNN layer:** This layer is located at the end of CNN and plays the role as the supervised learning mechanism to identify the desired outputs of CNN. Similar to BPNN or Multiple Layer Perceptron (MLP) approach, the backpropagation calculation strategy is applied in this layer to update the model learning weights.

Many researches have demonstrated that the CNN is a powerful deep learning architecture for many challenging problems in image (Milletari et al., 2016; Gatys et al., 2016) as well as text-based (Lai et al., 2015; Zhang et al., 2015b; Conneau et al., 2016) analysis/processing. Through comprehensive experiments, the CNN has been proved of having many critical advantages over traditional BPNN approaches. Back to our problem of recognizing harmful/violent content in text-based and image-based datasets, to leverage the accuracy performance of our system, we also apply the CNN to train our recognition model. The CNN is developed on the Apache Spark framework to enable handling training and predicting processes with large-scale datasets (training/testing).

11.4.2 Harmful/Violent Content Recognition Model via CNN

All text-based contents that we have collected and pre-processed in previous parts to transform them into feature vectors by applying Doc2Vec (Le & Mikolov, 2014) model. The Doc2Vec model supports to transform the documents into real-number vectors depending on their contextual features. There are two mechanisms for extracting document's features from documents that are continuous Distributed Memory version of Paragraph Vector (PV-DM) and Distributed Bag of Words version of Paragraph Vector (PV-DBOW) Le & Mikolov, 2014), which are inspired from the well-known skip-grams and Continuous Bag-of-words (CBOW) models in Word2Vec (Mikolov et al., 2013). These feature vectors will be predicted that might contain harmful/violent content or not by using the trained CNN classification model. For constructing the CNN classification model, we built our own training set with over 1 million text documents, comments, posts, etc., which have been labelled as [0] for not containing harmful/violent content and [1] for the opposite. Then these training data is also transformed into feature vector by applying Doc2Vec algorithm. Then the CNN architecture is applied as a supervised learning model to train the harmful/violent content classification model with two classes at the output [0] and [1].

Similar to text-based content, we also collected a numerous amount of images with its related metadata from multiple resources (Facebook, Twitter, etc.) and labelled them with two classes ([0] for images that don't contain harmful/violent content and [1] for the opposite). Then these images are applied with the image representation learning technique (Wang & Yeung, 2013) to embed the image data into feature vectors. Then these feature vectors are used to feed the CNN to conduct the harmful/violent image-based content classification model. Then the trained CNN is applied to predict the existence of harmful/violent content in new collected images (illustrated in Figure 11.12).

11.5 CONCLUSION AND FUTURE WORKS

Until this time, a very few research studies and their implementations have been proposed for analysis and filtering of large-scale and high-velocity harmful, abusive, and violent data. To solve this problem, in this chapter we propose a novel approach of using Apache-family platforms for building a distributed scalable harmful and violent content filtering system. Our contributions in this chapter can be categorized into three main parts. The first part is about constructing a distributed web crawler for collecting text-based (posts, comments, news, etc.) and image-based contents from multiple online resources, such as social networks, online news, forums, etc. Our distributed crawler is working on Apache Nutch platform, which enables to collect large-scale datasets and then store to distributed storage platform of HDFS (Hadoop Distributed File System). In the second part,

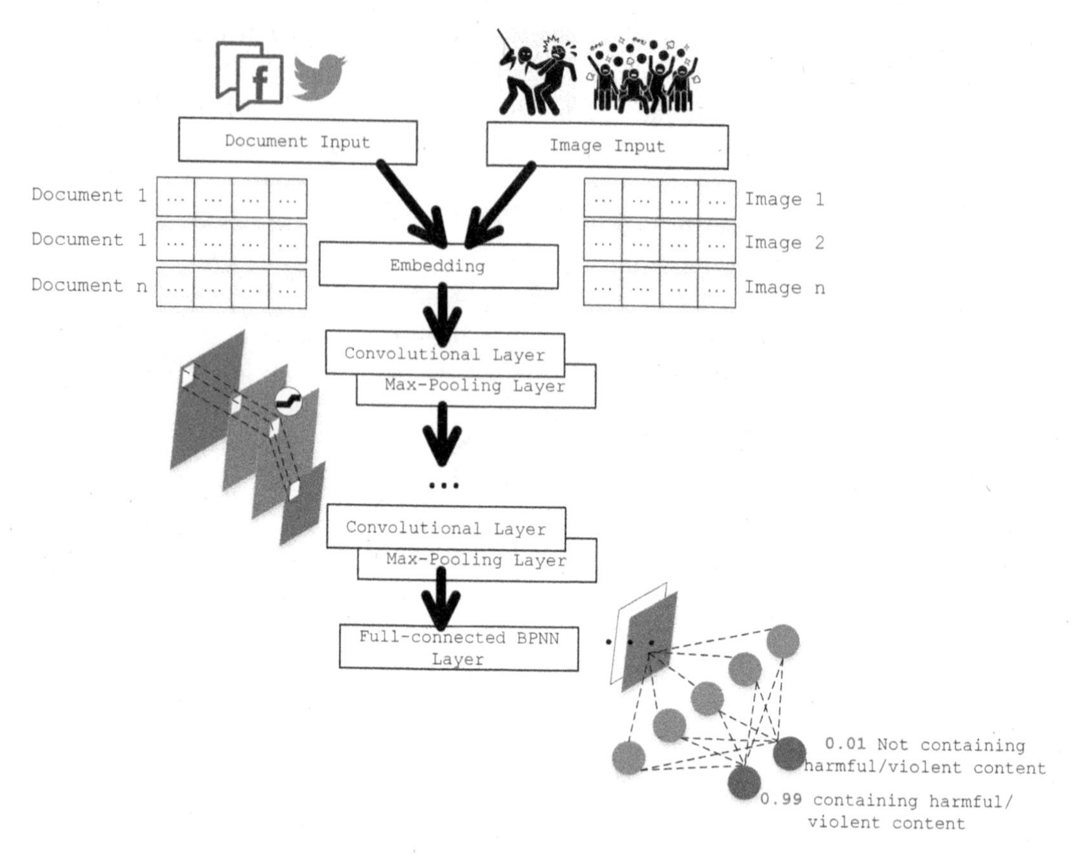

FIGURE 11.12 Illustration of our approach for harmful/violent text-based/image-based content recognition on Apache Spark.

we introduce an implementation of our content-processing mechanism on Apache Spark Streaming platform for pre-preprocessing both text-based (posts, comments, news, etc.) and image-based contents. The Spark Streaming-based processing mechanism enables to handle data as fast as possible in the context of high-velocity continuous incoming data from social networks. Finally, we build a novel analysis and detection mechanism on Apache Spark with CNN deep learning architecture for classifying the harmful, abusive, and violent contents that might be contained in the given collected datasets. All system parts are designed and developed as the pluggable modules, which consists of several child components. The purpose of splitting the system into multiple pluggable modules is for easier extension and replacement. Using the proposed system architecture, the harmful and violent contents on social networks will be conducted quickly and more accurately.

ACKNOWLEDGMENT

This research is funded by Vietnam National University HoChiMinh City (VNU-HCM) under grant number DS2020-26-01.

REFERENCES

Ali, A., & Senan, N. (2016). A review on violence video classification using convolutional neural networks. *International Conference on Soft Computing and Data Mining* (pp. 130–140). Springer, Cham.

Bányai, F., Zsila, Á., Király, O., Maraz, A., Elekes, Z., & Griffiths, M. D. (2017). Problematic social media use: Results from a large-scale nationally representative adolescent sample. *PLoS One, 12(1).*

Bello-orgaz, G., Jung, J. J., & Camacho, D. (2016). Social big data: Recent achievements and new challenges. *Information Fusion, 28*, 45–59.

Benevenuto, F., Rodrigues, T., Almeida, V., Almeida, J., & Gonçalves, M. (2009). Detecting spammers and content promoters in online video social networks. *Proceedings of the 32nd international ACM SIGIR Conference on Research and Development in Information Retrieval*, 620–627.

Bischke, B., Borth, D., & Dengel, A. (2019). Large-scale social multimedia analysis. *Big Data Analytics for Large-Scale Multimedia Search*, 157.

Chen, F., Tan, P. N., & Jain, A. K. (2009). A co-classification framework for detecting web spam and spammers in social media. *CIKM*, 1807–1810.

Chen, Y., Zhou, Y., Zhu, S., & Xu, H. (2012). Detecting offensive language in social media to protect adolescent online safety. *2012 International Conference on Privacy, Security, Risk and Trust and 2012 International Confernece on Social Computing*, 71–80.

Chowdury, R., Adnan, M. N. M., Mahmud, G. A. N., & Rahman, R. M. (2013). A data mining based spam detection system for YouTube. *Eighth International Conference on Digital Information Management (ICDIM 2013)*, 373–378.

Conneau, A., Schwenk, H., Barrault, L., & Lecun, Y. (2016). Very deep convolutional networks for text classification. *arXiv preprint arXiv:1606.01781.*

Dadvar, M., & De Jong, F. (2012). Cyberbullying detection: a step toward a safer internet yard. *Proceedings of the 21st International Conference on World Wide Web*, 121–126.

Das Sarma, A., Das Sarma, A., Gollapudi, S., & Panigrahy, R. (2010). Ranking mechanisms in twitter-like forums. *Proceedings of the Third ACM International Conference on Web Search and Data Mining*, 21–30.

Diakopoulos, N., & Naaman, M. (2011). Towards quality discourse in online news comments. *Proceedings of the ACM 2011 Conference on Computer Supported Cooperative Work*, 133–142.

Ding, C., Fan, S., Zhu, M., Feng, W., & Jia, B. (2014). Violence detection in video by using 3D convolutional neural networks. *International Symposium on Visual Computing.* Springer, Cham, 551–558.

Dong, Z., Qin, J., & Wang, Y. (2016). Multi-stream deep networks for person to person violence detection in videos. *Chinese Conference on Pattern Recognition* (pp. 517–531). Springer, Singapore.

Ediger, D., Jiang, K., Riedy, J., & Bader, D. A. (2010). Massive streaming data analytics: A case study with clustering coefficients. *2010 IEEE International Symposium on Parallel & Distributed Processing, Workshops and Phd Forum (IPDPSW)*, 1–8.

Eirinaki, M., Gao, J., Varlamis, I., & Tserpes, K. (2018). Recommender systems for large-scale social networks: A review of challenges and solutions.

Eyben, F., Weninger, F., Lehment, N., Schuller, B., & Rigoll, G. (2013). Affective video retrieval: Violence detection in Hollywood movies by large-scale segmental feature extraction. *PloS One, 8*(12).

Ferrara, E., Varol, O., Davis, C., Menczer, F., & Flammini, A. (2016). The rise of social bots. *Communications of the ACM 59*(7), 96–104.

Fire, M., Goldschmidt, R., & Elovici, Y. (2014). Online social networks: Threats and solutions. *IEEE Communications Surveys & Tutorials, 16*(4), 2019–2036.

Gatys, L. A., Ecker, A. S., & Bethge, M. (2016). Image style transfer using convolutional neural networks. *Proceedings of the IEEE Conference on Computer Vision and Pattern Recognition*, 2414–2423.

Golbeck, J. (2016). Combining provenance with trust in social networks for semantic web content filtering. *International Provenance and Annotation Workshop*, 101–108.

Heymann, P., Koutrika, G., & Garcia-Molina, H. (2007). Fighting spam on social web sites: A survey of approaches and future challenges. *IEEE Internet Computing 11*(6), 36–45.

IDC.com (2019). The Growth in Connected IoT Devices Is Expected to Generate 79.4ZB of Data in 2025, According to a New IDC Forecast. (2019, 01 18). Retrieved from IDC: https://www.idc.com/getdoc. jsp?containerId=prUS45213219

Iqbal, M. H., & Soomro, T. R. (2015). IG data analysis: Apache storm perspective. *International Journal of Computer Trends and Technology 19*(1), 9–14.

Ismailov, K. Y. (2017). Means to detect harmful for children content and measures to prevent its spread. *International Scientific and Practical Conference World Science 1*(4), 15–17.

Jin, X., Lin, C. X., Luo, J., & Han, J. (2011). Socialspamguard: A data mining-based spam detection system for social media networks. *Proceedings of the International Conference on Very Large Data Bases.*

Jin, X., Wah, B.W., Cheng, X., & Wang, Y. (2015). Significance and challenges of big data research. *Big Data Research*, *2*(2), 59–64.

Kienzler, R. (2017). *Mastering Apache Spark 2.x: Scalable analytics faster than ever.* Packt Publishing.

Kim, H. G., Lee, S., & Kyeong, S. (2013). Discovering hot topics using Twitter streaming data social topic detection and geographic clustering. *2013 IEEE/ACM International Conference on Advances in Social Networks Analysis and Mining (ASONAM 2013)*, 1215–1220.

Krause, B., Schmitz, C., Hotho, A., & Stumme, G. (2008). The antisocial tagger: Detecting spam in social bookmarking systems. *AIRWeb*, 61–68.

Lai, S., Xu, L., Liu, K., & Zhao, J. (2015). Recurrent convolutional neural networks for text classification. *Twenty-ninth AAAI Conference on Artificial Intelligence.*

Le, Q., & Mikolov, T. (2014). Distributed representations of sentences and documents. *International Conference on Machine Learning*, 1188–1196.

Li, M., & Tagami, A. (2014). A Study of contact network generation for cyber-bullying detection. *2014 28th International Conference on Advanced Information Networking and Applications Workshops*, 431–436.

Lumsdaine, A., Gregor, D., Hendrickson, B., & Berry, J. (2007). Challenges in parallel graph processing. *Parallel Processing Letters 17*(1), 5–20.

Markines, B. Cattuto, C., & Menczer, F. (2009). Social spam detection. *AIRWeb*, 41–48.

Mikolov, T., Chen, K., Corrado, G., & Dean, J. (2013). Efficient estimation of word representations in vector space. *arXiv preprint arXiv:1301.3781.*

Milletari, F., Navab, N., & Ahmadi, S.-A. (2016). V-net: Fully convolutional neural networks for volumetric medical image segmentation. *2016 Fourth International Conference on 3D Vision (3DV)*, 565–571.

Mu, G., Cao, H., & Jin, Q. (2016). Violent scene detection using convolutional neural networks and deep audio features. *Chinese Conference on Pattern Recognition* (pp. 451–463). Springer, Singapore.

Nair, L. R., & Shetty, D. S. D. (2015). Streaming twitter data analysis using spark for effective job search. *Journal of Theoretical & Applied Information Technology 80*(2).

Nasiri, H., Nasehi, S., & Goudarzi, M. (2019). Evaluation of distributed stream processing frameworks for IoT applications in Smart Cities. *Big Data.*

Noll, M. G., Au Yeung, C. M., Gibbins, N., Meinel, C., & Shadbolt, N. (2009). Telling experts from spammers: expertise ranking in folksonomies. *Proceedings of the 32nd international ACM SIGIR Conference on Research and Development in Information Retrieval*, 612–619.

Peel, L., & Clauset, A. (2015). Detecting change points in the large-scale structure of evolving networks. *Twenty-Ninth AAAI Conference on Artificial Intelligence.*

Ranjan, R. (2014). Streaming big data processing in datacenter clouds. *IEEE Cloud Computing 1*(1), 78–83.

Reynolds, K., Kontostathis, A., & Edwards, L. (2011). Using machine learning to detect cyberbullying. *2011 10th International Conference on Machine Learning and Applications and Workshops 2*, 241–244.

Riedy, J., & Bader, D. A. (2013). Multithreaded community monitoring for massive streaming graph data. *2013 IEEE International Symposium on Parallel & Distributed Processing, Workshops and Phd Forum*, 1646–1655.

Sivarajah, U., Kamal, M.M., Irani, Z., & Weerakkody, V. (2017). Critical analysis of Big Data challenges and analytical methods. *Journal of Business Research 70*, 263–286.

Song, W., Zhang, D., Zhao, X., Yu, J., Zheng, R., & Wang, A. (2019). A novel violent video detection scheme based on modified 3D convolutional neural networks. *IEEE Access 7*, 39172–39179.

Sosic, R., & Leskovec, J. (2015). Large scale network analytics with SNAP: Tutorial at the World Wide Web 2015 Conference. *Proceedings of the 24th International Conference on World Wide Web*, 1537–1538.

Spark Streaming Programming Guide. (2019). Retrieved from Apache Spark: https://spark.apache.org/docs/latest/streaming-programming-guide.html

Steinberg, A., Tonkelowitz, M., Deng, P., Mosseri, A., Hupp, A., Sittig, A., & Zuckerberg, M. (n.d.). Filtering content in a social networking service. *U.S. Patent No. 9,110,953*. Washington, DC: U.S. Patent and Trademark Office.

Structured Streaming Programming Guide. (2019). Retrieved from Apache Spark: https://spark.apache.org/docs/latest/structured-streaming-programming-guide.html

Sudhakaran, S., & Lanz, O. (2017). Learning to detect violent videos using convolutional long short-term memory. *2017 14th IEEE International Conference on Advanced Video and Signal Based Surveillance (AVSS). IEEE*, 1–6.

Vanetti, M., Binaghi, E., Carminati, B., Carullo, M., & Ferrari, E. (2010). Content-based filtering in on-line social networks. *International Workshop on Privacy and Security Issues in Data Mining and Machine Learning* (pp. 127–140). Springer, Berlin, Heidelberg.

Vanhove, T., Leroux, P., Wauters, T., & De Turck, F. (2013). Towards the design of a platform for abuse detection in OSNs using multimedial data analysis. *2013 IFIP/IEEE International Symposium on Integrated Network Management*, 1195–1198.

Wang, A. H. (2010). Don't follow me: Spam detection in twitter. *2010 International Conference on Security and Cryptography (SECRYPT). IEEE*, 1–10.

Wang, D., Irani, D., & Pu, C. (2011). A social-spam detection framework. *Proceedings of the 8th Annual Collaboration, Electronic messaging, Anti-Abuse and Spam Conference*, 46–54.

Wang, N., & Yeung, D.-Y. (2013). Learning a deep compact image representation for visual tracking. *Advances in Neural Information Processing Systems*, 809–817.

Wang, X., Zhang, Y., Leung, V. C., Guizani, N., & Jiang, T. (2018). D2D big data: Content deliveries over wireless device-to-device sharing in large-scale mobile networks. *IEEE Wireless Communications 25*(1), 32–38.

Xiang, G., Fan, B., Wang, L., Hong, J., & Rose, C. (2012). Detecting offensive tweets via topical feature discovery over a large scale twitter corpus. *Proceedings of the 21st ACM International Conference on Information and Knowledge Management*, 1980–1984.

Yamaguchi, Y., Takahashi, T., Amagasa, T., & Kitagawa, H. (2010). Turank: Twitter user ranking based on user-tweet graph analysis. *International Conference on Web Information Systems Engineering (pp.* 240–253). Springer, Berlin, Heidelberg.

Yue, D., Wu, X., Wang, Y., Li, Y., & Chu, C. H. (2007). A review of data mining-based financial fraud detection research. In *2007 International Conference on Wireless Communications, Networking and Mobile Computing*, 5519–5522.

Zhang, Q., Zhang, S., Dong, J., Xiong, J., & Cheng, X. (2015a). Automatic detection of rumor on social network. *Natural Language Processing and Chinese Computing*, 113–122.

Zhang, X., Zhao, J., & Lecun, Y. (2015b). Character-level convolutional networks for text classification. *Advances in Neural Information Processing Systems*, 649–657.

Zhou, P., Ding, Q., Luo, H., & Hou, X. (2017). Violent interaction detection in video based on deep learning. *Journal of Physics: Conference Series*. IOP Publishing, 012044.

Zikopoulos, P., et al. (2011). Understanding big data: Analytics for enterprise class hadoop and streaming data. McGraw-Hill Osborne Media.

Index

K

L

M

N

P